AIR POLLUTION

AIR POLLUTION

M N RAO
Professor-in-Charge
Technical Teachers' Training Institute
Extension Centre,
Bangalore
Formerly, Professor of Civil Engineering
Regional Engineering College
Durgapur

H V N RAO
Reader in Civil Engineering
R.V. College of Engineering
Bangalore

McGraw Hill Education (India) Private Limited
NEW DELHI

McGraw Hill Education Offices
New Delhi New York St Louis San Francisco Auckland Bogotá Caracas
Kuala Lumpur Lisbon London Madrid Mexico City Milan Montreal
San Juan Santiago Singapore Sydney Tokyo Toronto

McGraw Hill Education (India) Private Limited

50th reprint 2016
RZAACRLQRQAXC

ISBN (13 digit): 978-0-07-451871-7
ISBN (10 digit): 0-07-451871-2

Published by McGraw Hill Education (India) Private Limited, P-24, Green Park Extension, New Delhi 110 016 and printed at Sai Printo Pack Pvt. Ltd., New Delhi 110020

Visit us at: www.mheducation.co.in

Preface

This book presents the fundamentals of air pollution in a lucid manner understandable to those most concerned with such knowledge. Since air pollution is of topical interest, it is important to students and professionals in a variety of fields. Chief among these are environmental engineers, civil engineers, chemical engineers, meteorologists, medical practitioners, architects, and town and regional planners. Personnel in charge of pollution control in industries and government departments often have a vital interest in air pollution. Men indirectly concerned can be found in the fields of mining, petroleum engineering, forestry, public health and law among others. Although it is difficult to present a subject fitted to the requirements of such a diversity of interests, the common need of all is an understanding of the fundamental principles, methods, and problems encountered in the field as a whole. Thus, this book is an effort to make available a unified presentation of air pollution control.

The book is the outcome of the authors' experience gained while dealing with the problems of air pollution in institutes of higher learning and in industrial consultancy. The important principles and practices of air pollution sampling, analysis and control have been fully delineated so as to make it a simple textbook for students of the subject. The types, origin, sources, atmospheric movement and effects of air pollutants are also dealt with in great detail.

The authors are thankful to Dr C A Sastry, Professor of Chemical Engineering, IIT Madras and to Dr G Ramprasad, Professor of Civil Engineering, S V U College of Engineering, Tirupati, for their valuable suggestions. The authors referred to many works during the course of this work and the references given at the end would be very helpful to the readers desiring more information on the subject. The authors are grateful to all those who helped them directly and indirectly in bringing out this book.

The authors are also thankful to the Bureau of Indian Standards for allowing them to use their data. Thanks are due to M/s Voltas Limited, Bombay and M/s Thermax Private Limited, Pune for permission to use their material and photographs of equipments in the text.

Great appreciation should also be given to Mrs H V N Rao and Mrs M N Rao, for their encouragement and patience in bearing with the situation when the authors were fully occupied during the preparation of this book.

M N Rao
H V N Rao

Acknowledgements

A few paragraphs of this book have been reproduced from some Indian Standards on Air Pollution with the permission of Bureau of Indian Standards. These standards are available for sale from Bureau of Indian Standards, New Delhi and its regional, branch and inspection offices at Ahmedabad, Bangalore, Bhopal, Bhubaneshwar, Bombay, Calcutta, Chandigarh, Guwahati, Hyderabad, Jaipur, Kanpur, Madras, Patna, Pune and Trivandrum.

Contents

1

Introduction

The atmosphere, which makes up the largest fraction of the biosphere, is a dynamic system that continuously absorbs a wide range of solids, liquids, and gases from both natural and man-made sources. These substances travel through air, disperse, and react with one another and with other substances both physically and chemically. Most of these constituents, eventually find their way into a depository such as the ocean, or to a receptor such as man. Some substances such as helium, however, escape from the biosphere. Others such as carbon dioxide, may enter the atmosphere faster than they enter a reservoir and thus gradually accumulate in the air.

Clean, dry air contains 78.09% nitrogen by volume and 20.94% oxygen. The remaining 0.97% is composed of a gaseous mixture of carbon dioxide, helium, argon, krypton, nitrous oxide and xenon, as well as very small amounts of some other organic and inorganic gases whose amount in the atmosphere vary with time and place. Various amounts of contaminants continuously enter the atmosphere through both natural and man-made processes that exist upon the earth. That portion of these substances which interacts with the environment to cause toxicity, disease, aesthetic distress, physiological effects or environmental decay, has been labelled by man as a 'pollutant'.

In general, the actions of people are the primary cause of pollution and as the population increases, the attendant pollution problems also increase proportionately. The first significant change in man's effect on nature came with his discovery of fire. Prehistoric man built a fire in his cave for cooking, heating, and to provide light. The problem of air pollution came into existence at this time.

The British Parliament passed an Act in 1273, forbidding the burning of coal in London because it was beginning to choke the atmosphere. In 1300, King Edward I issued a royal proclamation, "Whosoever shall be found guilty of burning coal shall suffer the loss of his head." In 1306, a man was executed for violating this regulation. Later, the law fell into disuse as the industrial revolution took place in England.

Air pollution is basically the presence of foreign substances in air. Some specific definitions of air pollution are given below.

"Air pollution means the presence in the outdoor atmosphere of one or more contaminants, such as dust, fumes, gas, mist, odour, smoke, or vapour, in quantities, with characteristics, and of durations such as to be injurious to human, plant, or animal life or to property, or which unreasonably interfere with the comfortable enjoyment of life and property."

—Engineers Joint Council (U.S.A.)

"Air pollution is the presence in ambient atmosphere of substances, generally resulting from the activity of man, in sufficient concentration, present for a sufficient time and under circumstances which interfere significantly with the comfort, health or welfare of persons or with the full use or enjoyment of property."

—Indian Standards Institution IS-4167 (1966).

"Air pollution is the excessive concentration of foreign matter in the air which adversely affects the well being of the individual or causes damage to property."

—American Medical Association

Since air pollution has been mainly caused by rapid industrialisation in some western countries, some critics comment on air pollution as 'the price of industrialisation'. Air pollution caused by automobiles has been described as the 'disease of wealth'.

Air pollution means different things to different people. To the householder it may be an eye irritation and soiled clothing, to the farmer damaged vegetation, to the pilot dangerously reduced visibility and to industries problems of process control and public relations.

Further, the problem of air pollution varies from place to place. For example, air pollution in Tokyo is not the same as that in Bombay.

Today, the natural terrain that surrounds large cities is recognised as having a significant bearing on the problem of air pollution. However, this is not an altogether new concept either. Historians tell us that Los Angeles which in recent years has become a national symbol of comparison for excessive smog levels, was known as the 'valley of smokes' when the Spaniards first arrived there.

It has been found that a significantly increasing volume of particulate matter entering the atmosphere, scatters the incoming sunlight. This reduces the amount of heat that reaches the earth and tends to reduce its temperature. The decreasing mean global temperature of recent years has been attributed to the rising concentration of air-borne particles in the atmosphere. A counter acting phenomenon commonly referred to as the "greenhouse effect" is caused by the increasing amounts of carbon dioxide found in the atmosphere. It has been estimated that if the carbon dioxide content in the atmosphere generated in combustion processes continues to increase at the present rate, the mean global temperature could rise by 4°C in the next five decades. There has been conjecture that this might become a matter of great importance because small temperature increase could cause a partial melting of the ice caps of the earth causing continental flooding and devastating effects on man.

Air pollution can cause death, impair health, reduce visibility, bring about vast economic losses and contribute to the general deterioration of both our cities and country-side. It can also cause intangible losses to historical monuments such as the Taj Mahal which is believed to be badly affected by air pollution.

On account of large scale industrial activities in advanced countries notably the USA, UK, and other European countries, fall of acid rain has been reported in Scandinavia. This has reduced forest growth in Scandinavia. In Canada, thousands of lakes have been destroyed due to acid rain. Apart from the international issues involved, the basic ecology is affected. Large scale deforestation apart from creating an imbalance in the oxygen proportion of the atmosphere, affects weather and rain patterns as well. Industrial activity, particularly in thermal power stations, cement plants, oil refineries, chemical complexes, metallurgical industries, steel plants and fertiliser complexes, causes major problems of air pollution. The effects of air pollution are felt more by the elderly and chest and respiratory complaints are very common among them. Tragic instances of death have also been reported in air pollution episodes such as the London smog of 1952, and the Bhopal gas tragedy of 1984.

It is, therefore, a matter of great importance that engineers of all disciplines consciously incorporate in their designs sufficient constraints and safeguards to ensure that they do not contribute to atmospheric pollution. In addition, they must apply their ingenuity and problem-solving abilities to eliminate air pollution where it exists and restoring the natural environment.

There are three methods of identifying air pollution:

1. Sensory recognition
2. Physical measurement of pollution
3. Effects on plants, animals, and buildings.

These methods are widely used and they have their merits and demerits.

1. Sensory recognition—Usually the first awareness of an air pollution problem is through some effects on the individual. These are:

(a) Strong or unusual odours
(b) Reduction in visibility
(c) Eye irritation
(d) Acid taste in the mouth
(e) Feel of grit under foot

These are highly subjective phenomena and vary from individual to individual.

2. Physical measurement—While sensory perception may provide the first indication of the presence of most of the contaminants in the air, it is often not possible to detect trace quantities of many air-borne toxic substances or the presence of radioactive matter through the senses. Their identification requires physical measurement by standard methods of sampling and analysis.

3. Effects on plants, animals and buildings—Effects of air pollution can be observed on the growth of plants and health of animals. Similarly, its deleterious effect on buildings can also be observed. Thus plants, animals, and buildings act to some extent as indicators of certain atmospheric impurities.

These and other aspects of air pollution are dealt with in detail in the subsequent chapters.

2

Sources and Classification of Air Pollutants

2.1 Introduction

An inventory of air contaminants is a necessary first step towards control of air pollution. Air pollutants can be either natural or may be the result of various activities of man like industrial operations. The industrial contaminants can be either by-products of external combustion like smoke, dust, and sulphur oxides or by-products of internal combustion like the reactions in petrol and diesel engines. Further, the emissions can be either primary pollutants or secondary pollutants. The various sources of pollutants can also be broadly grouped under either stationary sources or mobile sources. All these are explained in detail in the following pages.

2.2 Classification of Pollutants

Air pollutants can be classified as follows:

1. Natural contaminants
   e.g., natural fog, pollen grains, bacteria, and products of volcanic eruption.
2. Aerosols (particulates)
   e.g., dust, smoke, mists, fog and fumes.
3. Gases and vapours

The various gases and vapours, which are important air contaminants are given in Table 2.1.

Table 2.1 *Air Contaminants*

No.	Group	Examples
1.	Sulphur compounds	SO_2, SO_3, H_2S, mercaptans
2.	Nitrogen compounds	NO, NO_2, NH_3
3.	Oxygen compounds	O_3, CO, CO_2
4.	Halogen compounds	HF, HCl
5.	Organic compounds	Aldehydes, hydrocarbons
6.	Radioactive compounds	Radioactive gases

In addition to the above, studies by various investigators have indicated, that some of these contaminants undergo chemical reactions when they enter the atmosphere. As a result, the end products formed are more harmful than the original contaminants. For example, it is reported that unsaturated hydrocarbons react with nitrogen dioxide in sunlight to form smog.

2.2.1 Natural Contaminants

Among natural contaminants pollen is important because of its peculiar properties irritating to some individuals. Pollen grains are the male gametophytes of gymnosperms and angiosperms and they are discharged into the atmosphere from weeds, grasses and trees. Because of wind pollination, thousands of pollen grains are liberated. While air transported pollen grains range chiefly between 10 and 50 μ (micron) in size, some have been found to be as small as 5 μ and as large as 100 μ in diameter.

From the point of view of pollution, air-borne pollutants are significant because of the allergic responses produced in sensitive individuals. Many people suffer from asthma or hay fever. While most victims have an uncomplicated type of hay fever in which the symptoms disappear at the end of the pollen season, some develop bronchitis, bronchial asthma, and dermatitis. (See Allergic Agents, Sec. 5.5.12.)

2.2.2 Aerosols

Aerosols refer to the dispersion of solid or liquid particles of microscopic size in gaseous media, such as dust, smoke, or mist.

An aerosol can also be defined as a colloidal system in which the dispersion medium is a gas and the dispersed phase is solid or liquid.

The term 'aerosol' is used during the time it is suspended in the air. After it has settled either by virtue of its weight, by agglomeration, or by impact on a solid or liquid surface, the term no longer applies. Thus, particulate matter is an air pollutant only when it is an aerosol. However, it is a nuisance both as an aerosol (visibility reduction) and as settled or deposited matter (soiling of surfaces, corrosions).

Aerosols differ widely in terms of particle size, particle density and their importance as pollutants. Their diameters generally range from 0.01 μ or less, up to about 100 μ. The following are the various aerosols.

Dust

Dust is made up of solid particles predominantly larger than those found in colloids and capable of temporary suspension in air or other gases. They do not tend to flocculate except under electrostatic forces; they also do not diffuse but settle under the influence of gravity.

Dust is produced by the crushing, grinding, etc., of organic and inorganic materials. Generally they are over 20 μ in diameter, although some are smaller. Fly ash from chimneys varies from 80–3 μ; cement, from 150–10 μ; foundry dust, from 200–1 μ. Most of the dust particles settle to the ground as dust fall, but particles 5 μ or smaller tend to form stable suspensions.

Smoke

Smoke consists of finely divided particles produced by incomplete combustion. It consists predominantly of carbon particles and other combustible materials. Generally the size of the particles is less than 1 μ. The size of coal smoke particles range from 0.2–0.01 μ and oil smoke particles from 1.0–0.03 μ.

Mists

This term refers to a low concentration dispersion of liquid particles of large size. In meteorology, it means a light dispersion of minute water droplets suspended in the atmosphere. Natural mist particles formed from water vapour in the atmosphere are rather large, ranging from 500–40 μ in size. The particles may coalesce.

Fog

Fog refers to visible aerosols in which the dispersed phase is liquid. Formation by condensation is usually implied. In meteorology, it refers to dispersion of water or ice in the atmosphere near the earth's surface reducing visibility to less than $\frac{1}{2}$ km. In natural fog the size of the particles ranges from 40–1.0 μ.

Fumes

These are solid particles generated by condensation from the gaseous state, generally after volatilisation from melted substances, and often accompanied by a chemical reaction such as oxidation. Fumes flocculate and sometimes coalesce.

Table 2.2 indicates the major sources of atmospheric dust.

2.2.3 Gases

Following are the gases in air pollutants:

Sulphur Dioxide

This is one of the principal constituents of air pollutants. The main source of sulphur dioxide is the combustion of fuels, especially coal. Therefore, its concentration in the atmosphere depends upon the sulphur content of the fuel used for heating and power generation. The sulphur content of fuels varies from less than 1% for good quality anthracite to over 4% for bituminous coal. In the USA in recent years, there has been a progressive decrease in the average atmospheric concentrations of sulphur dioxide, of several cities, owing to the increased use of coal with a low sulphur content. Air pollution control measures in St. Louis prohibit the use of high-sulphur coal.

Most crude petroleum products contain less than 1% sulphur; a few

Table 2.2 *Sources of Atmospheric Dust*

Sources	Examples
1. Combustion	Fuel burning (coal, wood, fuel oil) Incineration (house and municipal garbage) Others (open fires, forest fires, tobacco smoking)
2. Materials handling and processing	Loading and unloading (sand, gravel, coal, ores, lime, cement) Crushing and grinding (ores, stone, cement, rocks, chemicals) Mixing and packaging (chemicals, fertilisers) Food processing (flour, corn starch, grains) Cutting and forming (saw mills, wall board, plastics) Metallurgical (foundries, smelters) Industrial (paper, textiles manufacture)
3. Earth-moving operations	Construction (road, buildings, dams, site clearance) Mining (blasting) Agriculture (soil filling, land preparation) Winds
4. Miscellaneous	House cleaning Mud road cleaning Crop spraying Poultry feeding Engine exhaust

contain up to 5%. Refining processes tend to concentrate sulphur compounds in the heavier fractions. Fuel gases also contain sulphur, but, in small quantities.

About 80% of the sulphur in coal and nearly all that in liquid and gaseous fuels is found in flue gases in the form of sulphur dioxide. The remaining sulphur in coal is present as inorganic sulphur and thus remains in the ash. Generally, the concentration of sulphur dioxide in flue gases ranges from 0.05–0.25%, and occasionally is as high as 0.4%.

Another common source of sulphur dioxide in the atmosphere, is metallurgical operations. Many ores, like zinc, copper and lead, are primarily sulphides. During the smelting of these ores, sulphur dioxide is evolved in stack concentrations of 5–10% (SO_2). But, this can be recovered in the form of sulphuric acid.

Among the miscellaneous operations releasing sulphur dioxide into the atmosphere are sulphuric acid plants and paper manufacturing plants. The quantities are usually low and therefore easily amenable to control measures.

The open burning of refuse and municipal incinerators also contribute some amount of sulphur dioxide to the atmosphere.

Hydrogen Sulphide

Hydrogen sulphide is a foul smelling gas. The sources of its natural emission include anaerobic biological decay processes on land, in marshes and in the oceans. Volcanoes and natural water springs emit hydrogen sulphide to some extent.

One of the major sources of hydrogen sulphide is the Kraft pulp industry, which uses a sulphide process for manufacturing paper. The other industrial sources of hydrogen sulphide are petroleum refineries, coke-oven plants, viscose rayon plants, and some chemical operations.

Other sulphur compounds that are of interest in air pollution, principally because of their strong odours, are methyl mercaptan (CH_3 SH), dimethyl sulphide (CH_3 S CH_3), dimethyl disulphide (CH_3 SS CH_3) and their higher molecular homologs. The mercaptans are emitted in mixtures of pollutants from some pulp mills, petroleum refineries, and chemical manufacturing plants.

Hydrogen Fluoride

The major sources of fluorides are the manufacture of phosphate fertilisers, the aluminium industry, brick plants, pottery, and ferro-enamel works. Small amounts are also emitted from other metallurgical operations, such as zinc foundries and open-hearth steel furnaces. Small amounts are also liberated in the burning of coal, which normally contains about 0.01% fluorine.

Hydrogen fluoride is an important air contaminant even in extremely low concentrations of 0.001–0.10 ppm by volume. In these concentration levels, hydrogen fluoride is more important in terms of injury to vegetation and animals than in terms of injury to humans (see Sec. 7.5.2 and 6.2.1). The high degree of toxicity of fluorine compounds renders the control of such emissions imperative for industries manufacturing aluminium and phosphate fertilisers.

Chlorine and Hydrogen Chloride

Chlorine is found in polluted atmospheres as the element itself (chlorine), as hydrogen chloride, as chlorine-containing organic compounds such as perchloroethylene and as inorganic chlorides. The last mentioned compounds are solids and hence found in particulate form; the other materials mentioned are present as gases.

The most common sources of chlorine in the atmosphere are from operations in which it is manufactured or used to produce other chemicals. Also as chlorine is used in water purification plants, in sewage plants and in swimming pools, equipment failure sometimes leads to leakage of chlorine into the atmosphere.

Hydrogen chloride is evolved in numerous industrial chemical processes, but it is so easy to recover, that little reaches the atmosphere.

The main effects of chlorine and its compounds are respiratory irritation from chlorine, corrosion by hydrogen chloride, and damage to vegetation from chlorine and hydrogen chloride.

Oxides of Nitrogen

It is probable that oxides of nitrogen are the second most abundant atmospheric contaminants in many cities, ranking next to sulphur dioxide. Generally, the highest concentration of nitrogen oxides in gaseous emissions occurs in effluents from industries where nitric acid is produced or used in chemical reactions. The next highest concentration is in automobile exhausts. Then come effluents from large power plants, and then to a small extent those from low heat burners and furnaces.

Out of seven oxides of nitrogen (N_2O, NO, NO_2, NO_3, N_2O_3, N_2O_4, N_2O_5), only nitric oxide and nitrogen dioxide arise from many human activities and are classified as pollutants. In atmospheric analyses they are usually reported as 'total oxides of nitrogen' or "NO_x". It is a standard practice in the chemical industry to absorb and recover significant quantities of oxides of nitrogen.

Ammonia and ammonium salts are not important air contaminants. Ammonia is an important raw material in fertiliser and organic chemical industries and in the manufacture of nitric acid by the oxidation process. Its recovery is a matter of fundamental importance in the economical operation of such processes.

Carbon Monoxide

Carbon monoxide, an odourless and colourless gas has its major origin in the incomplete combustion of carbonaceous materials. It is a highly poisonous gas and is generally classified as an asphyxiant.

The chief source of carbon monoxide in the atmosphere is combustion, especially due to automobile exhausts. However, except for motor vehicles and other internal combustion engines, very little carbon monoxide is found in the gaseous emissions from properly adjusted, properly operated installations. Although certain industrial operations, such as electric and blast furnaces, some petroleum refining operations, gas manufacturing plants, and coal mines, are potential contributors of carbon monoxide to the atmosphere, automobile exhausts are by far the most important source.

Ozone

The origin of ozone that is found in the air has not been clarified, but it is likely that combustion and sunlight are involved in its production. (See Chapter 12— Photochemical air pollution). Ozone is poisonous and smelly. It exists in great abundance under natural conditions in the upper atmosphere.

Aldehydes

These are produced by the combustion of gasoline, diesel oil, fuel oil, and natural gas. Incomplete oxidation of motor fuel and lubricating oils leads to the formation of aldehydes. Lower aldehydes may be present in the atmosphere in concentrations about as high as sulphur dioxide. They may also be formed in the air because of photochemical reactions. Formaldehyde is irritating to the eyes.

Organic Vapours

These contaminants include a large number of chemical compounds, including paraffins, olefins, acetylenes, aromatic hydrocarbons, chlorinated hydrocarbons, etc. They are produced by combustion processes (particularly in automobiles), household incinerators, and petroleum processes. It is also probable that they undergo changes in the atmosphere which contribute to the formation of smog.

Radioactive Gases

A major source of radioactive gases and particulate is the nuclear power reactor and related fuel handling facilities. The other sources are experimental accelerators, testing of nuclear bombs in the atmosphere, agricultural, industrial and medical use of radioactive isotopes. Another source of radioactive particulates and gases that is increasing in importance, is the nuclear fuel reprocessing plant.

Table 2.3 indicates the various sources of air pollution

Table 2.3 *Sources of Air Pollution*

No.	Class	Aerosols	Gases and vapours
1.	Combustion processes (domestic burning, thermal power plants, cars, trucks, aeroplanes, and railways. Also refuse burning)	Dust, fume, smoke	SO_2, NO_2, CO, organic vapours, odours
2.	Chemical processes (paper mills, cement, fertilisers, etc.)	Dust, fume, mist	Process-dependent (SO_2, CO, NH_3, NO_2, organic vapours, odours)
3.	Petroleum operations	Dust, mist	SO_2, H_2S, NH_3, CO, hydro-carbons, mercaptans
4.	Metallurgical processes (aluminium refineries, steel plants)	Dust, fume	SO_2, CO, fluorides, organic vapours
5.	Mineral processing	Dust, fume	Process-dependent (SO_2, CO, fluorides, organic vapours)
6.	Food and feed operations	Dust, mist	Odourous materials
7.	Agricultural activities		
	(a) Crop spraying	Dust, mist	Organic phosphates, chlorinated hydro-carbons
	(b) Field burning	Smoke, fly ash	sulphur oxides, organic vapours
8.	Nuclear energy programmes		
	(a) Fuel fabrication		Fluorides
	(b) Ore preparation	Dust	Iodine-131 and Argon-41
	(c) Bomb explosion		Radioactive gases (Sr-90, Cs-137, C-14, etc.)

2.2.4 Primary and Secondary Air Pollutants

Air pollutants can also be broadly classified into two general groups—primary air pollutants and secondary air pollutants.

Primary air pollutants are those emitted directly from identifiable sources.

Examples of primary air pollutants:

1. Finer particles (less than 100 μ in diameter)
2. Coarse particles (greater than 100 μ in diameter)
3. Sulphur compounds
4. Oxides of nitrogen
5. Carbon monoxide
6. Halogen compounds
7. Organic compounds
8. Radioactive compounds

Finer aerosols include particles of metal, carbon, tar, resin, pollen, bacteria, etc.

Secondary air pollutants are those which are produced in the air by the interaction among two or more primary pollutants, or by reaction with normal atmospheric constituents, with or without photoactivation.

Examples of secondary air pollutants:

1. Ozone
2. Formaldehyde
3. PAN (peroxy acetyl nitrate)
4. Photochemical smog
5. Formation of acid mists (H_2SO_4) due to reaction of sulphur dioxide and dissolved oxygen, when water droplets are present in the atmosphere.

Smog

Smog is a synchronym of two words—smoke and fog. Smog can be of two types—photochemical or coal induced.

Photochemical smog is restricted to highly motorised areas in metropolitan cities, e.g., Los Angeles. It occurs under adverse meteorological conditions when the air movement is restricted. Smog is caused by the interaction of some hydrocarbons and oxidants under the influence of sunlight giving rise to dangerous peroxy acetyl nitrate (PAN). Its main constituents are nitrogen oxides, PAN, hydrocarbons, carbon monoxide and ozone. It reduces visibility, causes eye irritation, damage to vegetation and cracking of rubber.

The fog from burning coal covers urban areas at night or on cold days when the temperature is below 10°C and when calm meteorological conditions prevail, e.g., London (December 1952). This fog consists of smoke, sulphur compounds and fly ash. Prolonged exposure to smog may result in a high mortality rate especially among the elderly and those who have histories of chronic bronchitis, asthma, broncho-pneumonia and other lung or heart diseases.

2.2.5 Stationary and Mobile Sources

Emissions (air pollutants) may be classified by source, as stationary or mobile. Another method of classifying emission sources is by:

1. Point sources (large stationary sources)
2. Area sources (small stationary sources and mobile sources with indefinite routes)
3. Line sources (mobile sources with definite routes)

Table 2.4 gives a picture of emission inventory source classification.

Table 2.4 *Emission Inventory Source Classification*

Total Sources			
Stationary sources		*Mobile sources*	
Point sources	*Area sources*	*Line sources*	*Area sources*
1. Industrial processing	1. Residential heating	1. Highway vehicles	1. Motor vehicles
2. Power plants	coal	2. Railroad locomotives	light-duty
3. Fuel Combustion (industrial)	gas	3. Channel vessels	medium-duty
coal	oil		heavy-duty
oil	2. Institutional & commercial heating		2. Railyard locomotives
gas	coal		3. Port vessels
4. Solid waste disposal	oil		4. Aircraft (airports)
Municipal incinerators	gas		5. Miscellaneous
Open burning	3. On site incineration		
5. Miscellaneous	4. Open burning		
	5. Evaporative losses		
	6. Miscellaneous		

QUESTIONS

1. Classify air pollutants into different categories, indicating their sources.
2. What are the types and sources of particulate matter causing air pollution? Briefly explain them.
3. Identify the various pollution sources of the following air contaminants
(a) Sulphur dioxide
(b) Hydrogen sulphide
(c) Hydrogen fluoride
(d) Carbon monoxide
(e) Oxides of nitrogen
(f) Hydrocarbons
4. What is meant by 'Smog'? Discuss its cause and effects.

5. List the sources of atmospheric dust.
6. List out the industries that cause major air pollution problems.
7. Distinguish between
   (a) Primary and secondary air pollutants
   (b) Stationary and mobile sources of air pollutants
8. Explain the following terms with respect to air pollutants
   (a) Point sources
   (b) Area sources
   (c) Line sources
9. Write short notes on:
   (a) Aerosols
   (b) Smog
10. Describe the various air pollution sources in an industrialised city like Bombay.

3

Meteorology and Air Pollution

3.1 Meteorological Factors Influencing Air Pollution

The degree to which air pollutants discharged from various sources concentrate in a particular area depends largely on meteorological conditions. The application of dispersion theory and a knowledge of local weather conditions are necessary to determine the required stack height for an emission, and to evaluate the intensity of air pollution. Even though the total discharge of contaminants into the atmosphere in a given area remains constant from day to day; the degree of air pollution may vary widely because of differences in meteorological conditions. In a specified place the emission of pollutants may be the same but it is the weather that can trigger an air pollution episode.

The important meteorological parameters that influence air pollution can be classified into primary parameters and secondary parameters.

Primary parameters are:

1. Wind direction and speed
2. Temperature
3. Atmospheric stability
4. Mixing height

Secondary parameters are:

1. Precipitation
2. Humidity
3. Solar radiation
4. Visibility

The parameters vary widely as a function of latitude, season and topography.

Just as weather affects the severity of air pollution, air pollution, may, in turn, affect weather conditions. Air pollution may influence the weather in several ways. Visibility may be reduced, fog frequency and duration may be increased and the incoming solar radiation may be decreased, particularly in the ultra-violet end of the spectrum.

3.1.1 Wind Direction and Speed

The direction and speed of surface winds govern the drift and diffusion of air pollutants discharged near the ground level. The higher the wind speed at or near the point of discharge of pollution, the more rapidly are the pollutants carried away from the source. The pollutants so dispersed will not exist at the same concentration but will rapidly be diluted with greater and greater volumes of air. On the other hand, when wind speeds are low, pollutants tend to be concentrated near the area of discharge and the longer the periods of such light winds, the greater will be the concentration of pollutants. Further, gustiness, a very important characteristic of surface winds is directly proportional to its speed and determines the extent to which the pollutants are mixed and diluted with the surrounding air. Other things being equal, the concentration downwind from a source will be inversely proportional to wind speed. In rough terrain, it connot be assumed that the wind direction and speed near the source govern the subsequent motion of the contaminants. Hills may deflect the air flow either horizontally, vertically, or both, the amount of deflection depending on the vertical stability of the atmosphere. In valleys, the winds carrying a pollutant tend to flow either up or down the valley, following its meanderings. The deeper the valley, the more pronounced is this channelling effect.

3.1.2 Atmospheric Stability and Temperature Inversions

In well mixed air which is dry, for every 1000 ft (300 m) increase in altitude, the temperature decreases by about 3.3°F (about 1.8°C). This vertical temperature gradient is known as the 'lapse rate' and the value given is the normal lapse rate. When the reverse or negative lapse rate occurs, a dense cold stratum of air at ground level gets covered by lighter warmer air at higher level. This phenomenon is known as 'inversion'. During inversion, vertical air movement is stopped and pollution will be concentrated beneath the inversion layer, i.e., in the denser air at ground level. As a result, during temperature inversion, the atmosphere is stable and very littte turbulence or mixing takes place. Under such conditions, pollutants in the air do not disperse.

Inversion is a frequent occurrence in the autumn and winter months and the accumulation of smoke and other contaminants further aggravates pollution by preventing the sun's rays from warming the ground and the adjacent air. Fog is commonly associated with inversions, because the temperature of the air at ground level falls below the dew point of the water vapour in the air. Narrow valleys are favourable to inversions since horizontal air movement is restricted. At the time of inversions, visibility is greatly reduced and contaminants are at a maximum.

Types of Inversion

(a) Radiation inversion: It usually occurs at night, when the earth loses heat by radiation and cools the air in contact with it. If the air is moist and its temperature is below the dew point, fog will form. The cool air stratum is covered by warmer air, and the vertical movement is stopped until the sun warms the lower air, next morning. This type of inversion is more common in winter than in summer because of the longer nights. Valley areas, because of the restriction of horizontal air movement by surrounding high ground, may frequently have such inversions.

In India, because of intense solar heating of the ground, inversions are broken within a few hours after sunrise. However, simultaneous occurrence of fog or mist prolongs the duration of inversion by cutting out sunlight reaching the ground.

(b) Subsidence inversion: It occurs at modest altitudes and often remains for several days. It is caused by sinking or subsiding of air in anti-cyclones (high pressure areas surrounded by low pressure areas). The air circulating around the area descends slowly at the rate of about 1000 m per day. As the air sinks, it is compressed and gets heated to form a warm dense layer. This acts as a lid to prevent the upward movement of contaminants. The inversion height may vary from the ground surface to 1600 m. When it drops to less than 200 m, extreme pollution occurs.

Some times both radiation inversion and subsidence inversion may occur simultaneously. Such a phenomenon is known as 'double inversion'.

Adiabatic Lapse Rate

The change of air temperature with height has a profound influence on the upward lift of the air pollutants discharged into the atmosphere and hence on their utlimate dispersion and dilution. The lapse rate of a parcel of dry air as it moves upward in a hydrostatically stable environment and expands slowly to lower environmental pressure without exchange of heat, is known as the 'adiabatic lapse rate'. The dry adiabatic lapse rate is 0.98°C/100 m. Under conditions of adiabatic lapse rate, a smoke plume will rise directly into the atmosphere by virtue of low density because of higher temperature until it reaches air of similar density. In many situations, however, because of external heating or cooling effects, the lapse rate may be greater or less than the adiabatic rate. The two most important conditions from the point of air pollution are the super adiabatic lapse rate (rate more than adiabatic) and the negative lapse rate (inversion). On a clear summer day, rapid heating of the earth by the sun warms the air near the surface, to the point where the lapse rate is superadiabatic. Under this condition the atmosphere is said to be in unstable equilibrium, and marked vertical mixing of the air results. This is a condition when pollutants are dispersed rapidly.

3.1.3 Mixing Height

The fourth, primary meteorological parameter is the mixing height. It can be defined, as that height above the earth's surface to which related pollutants will extend, primarily through the action of atmospheric turbulence. It is usually related to one or more of the three factors: wind direction, wind speed, and wind turbulence. Under certain circumstances, it may be related to all three.

3.1.4 Precipitation

Rainfall or precipitation exerts a two-fold cleansing action on the pollutants discharged into the atmosphere. It accelerates the deposition of particulate matter on the ground and hence its removal from the atmosphere. It also helps to remove the concentration of gaseous pollutants which are soluble in water. The washout by precipitation of air-borne radioactive wastes is a matter of particular concern. Precipitation can be determined by using various types of rain gauges.

3.1.5 Humidity

The moisture content of the atmosphere influences the corrosive action of the air pollutants and indicates the potentiality for fog formation in relation to the degree of air pollution. Of the various means by which humidity may be expressed, the relative humidity is most frequently used in air pollution studies.

3.1.6 Solar Radiation

Depending on the location, solar radiation can have a pronounced effect on the type and rate of chemical reactions in the atmosphere. The photochemical smog formation at Los Angeles is a typical example of the effect of solar radiation on air pollution.

The application of these meteorological factors may be considered in the control of pollution from an industrial plant, in the selection of its location, in the design of equipment and in its day-to-day operation. In addition, these factors are to be taken into account in laying out zones for industrial use, in identifying causal factors in existing pollution problems and in establishing air quality criteria. The influence of both regional and local weather should be considered.

3.2 Methods for Measurement of Meteorological Variables

A meteorological instrument must possess an acceptable degree of accuracy and sensitivity. Also, it should be simple and durable. In our country the National Environmental Engineering Research Institute (NEERI) has developed some devices for measuring the meteorological variables.

3.2.1 Wind Direction Recorder

The patented instrument of the NEERI for measuring wind direction is a simple one, which employs the conventional wind vane to sense the

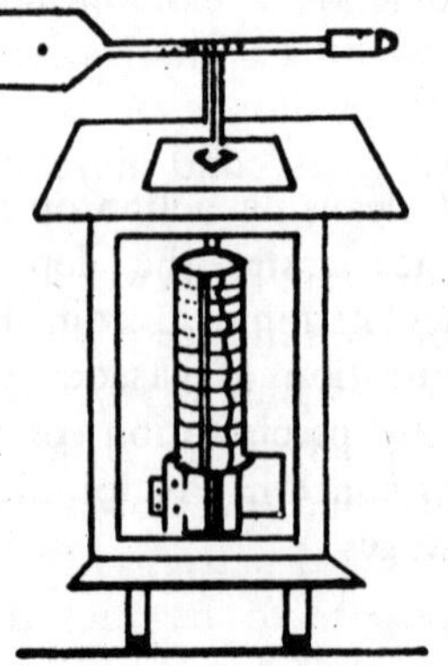

Fig. 3.1 *Wind direction recorder*

direction. It is automatic and operated mechanically without any power supply. The instrument continuously records on an attached chart the direction of the wind with time (Fig. 3.1).

Surface Wind Direction

There are four main types of wind vanes for measuring direction

1. Flat plate vane
2. Splayed vane
3. Aerofoil vane
4. Running average anemograph

Flat plate vane: In this type the sensing element which governs the azimuth angle of a vertical shaft specifying the wind direction, is a vertical plate which is mounted at one end of a horizontal rod. There is a counter weight at the other end of the horizontal rod. The rod is fastened to the vertical shaft. Wind pressure acting on the flat plate keeps the counter weight heading into the wind.

Splayed vane: In this instrument, two flat plates which are joined at a small angle at one end of a horizontal rod act as the wind direction sensor.

Aerofoil vane: In this instrument the vane has an aerofoil cross-section, with the span often being three or four times the chord.

Running average anemograph: It is often advantageous to be able to directly obtain a recording of the running average of the wind speed or direction or both. In averaging the wind direction, a difficulty arises from the discontinuity 360°—0° as the wind direction fluctuates around north. An anemograph which automatically produces the running averages of both wind speed and direction has been developed.

Wind Direction Aloft

The wind direction at the height of a plume from one or more stacks is important in some investigations. All the available following methods discussed involve considerable time and effort.

(a) Pilot balloons (Pibals): In this method a small balloon inflated with hydrogen or helium is left aloft and the direction of the wind aloft is determined by tracking the balloon by means of one or more theodolites on the ground. The average wind direction is obtained using triangulation techniques at successive height intervals. The pilot balloon technique of determining wind aloft is of limited value because the balloon may be lost with low cloud, fog or smoke.

(b) Tetroons: A tetroon is a constant volume Mylar balloon in the shape of a tetrahedron which is kept in a zero lift condition and carried by the wind in an almost horizontal direction. Zero lift balloons are produced by filling pilot balloons with a mixture of two gases, one more dense and the other less dense than air. As the mixture of gases leaks from the balloon, the loss of the heavier gas causes a decrease in its weight. Thus an approximate mixture of gases will keep the balloon in a zero lift or balanced condition for lengthy periods of time. Such a balloon, when accurately tracked, acts as a sensor for wind direction. The advantage of this type of balloon lies in the fact that it can indicate wind trajectories near cities, over shorelines and in valleys where complicated patterns of airflow are commonly observed. If substantial distances are to be tracked, a radar installation with a transponder suspended from the tetroon will be required which is an expensive method.

(c) Kite balloons: A kite balloon is an elongated captive balloon with fins at one end. It acts as an ordinary captive spherical balloon in light winds and as a kite in stronger winds, thus maintaining altitude under both conditions. The azimuth angle of the horizontal projection of the tethering cable is measured at the ground to determine the wind direction at the height of the balloon. The inflating gas lost by slow leakage is replaced by refilling it at intervals.

(d) Radio and radar: In the radio method, a small radio transmitter is carried aloft by a freely rising balloon and is tracked. With the radar system, pulses of electrical energy emitted by the radar are reflected back to it by a target carried by the free balloon. The distance to the target as well as azimuth and elevation angles are measured by the system. This gives more accurate results than the radio direction finder. However, operation and maintenance costs of radio and radar theodolites are prohibitively high.

(e) Smoke trails: Information on wind directions aloft may be obtained by finding at intervals the position in space of smoke trails released above the ground by a rising rocket or an aeroplane. However, observations in fog, smoke, or at night, are not possible. Also, this method requires more operating personnel than the pilot balloon method.

3.2.2 Wind Speed Recorder

Instruments for measuring wind speed are called anemometers. If they are recording instruments they are known as anemographs. The most common type is the cup anemometer. The rate of rotation of the shaft to which the cups are attached indicates the wind speed and this is transmitted to a recorder or an indicating panel by either mechanical, optical or electrical means.

The NEERI has developed and patented a wind speed recorder (Fig. 3.2). In the instrument a four cup rotor is employed to sense the wind. The motion of the cup is transferred after reducing its speed by a gear system,

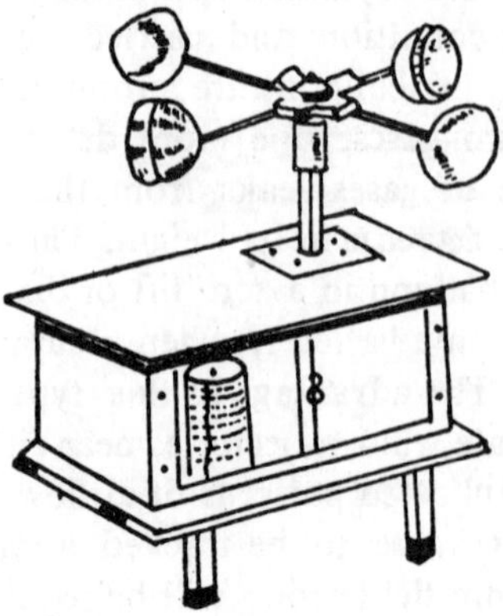

Fig. 3.2 *Wind speed recorder*

to the pen which makes a continuous rise and fall impression on chart paper. This rate of rise or fall is proportional to the wind speed.

The instrument gives a 24-hour record in one setting. Wind speed at a particular time and the average w i n speed can be found out from this record.

3.2.3 Humidity Measurement (Whirling Psychrometer)

One of the simplest and most reliable instruments is the whirling psychrometer. Two thermometers, one with a wet cloth surrounding the bulb, are whirled in the air. The temperatures of both dry bulb and wet bulb thermometers are noted. From the temperature difference between the dry bulb and wet bulb thermometers, the relative humidity of the air can be found by referring to psychrometric table.

Other instruments that can be used for measuring humidity are:

(1) Hair hygrometer
(2) Infra-red hygrometer.

3.2.4 Temperature Measurement

A common instrument for measurement of temperature is the thermometer. Mercury thermometers depend on thermal expansion while bimetallic thermometers are based on the differential expansion of two metals. The electrical resistance thermometer is based on the variation in electrical resistance of a metallic wire with change in temperature and a thermocouple is based on the electrical current which flows when two electrical conductors made of two different metals are joined together. All four types are used.

To obtain accurate readings, the thermometer must be shielded from radiant energy. Proper aspiration is also necessary to avoid the stagnation of the air inside the radiation shield. A motor aspirated radiation shielded temperature sensor for temperature inversion studies developed by NEERI is described.

In this temperature probe, a glass probe-bead type thermistor is used. The constructional details of the probe are shown in Fig. 3.3.

Two core shielded electrical contacts are taken from the thermistor leads. The thermistor is first wrapped in a shock absorbing material and is put into an exactly fitting stainless steel tube. A thin walled conical shaped metallic tip is fixed at one end of the stainless steel tube. The thermistor is pushed until its sensitive point makes a positive thermal contact with the

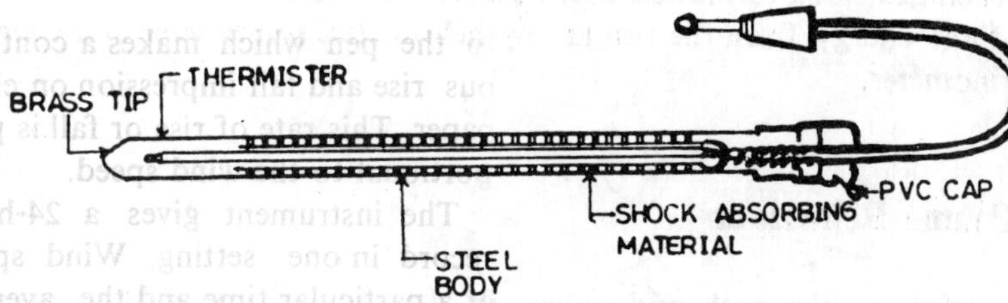

Fig. 3.3 *Probe details*

thin walled metallic tip. The open end of the stainless steel tube is closed with a PVC cap and the joint is made watertight with some adhesive.

Figure 3.4 gives the details of the radiation shield. It consists of two concentric PVC pipes *A* and *B* separated from each other by an air layer and supported at two positions by two polyvinyl chloride (PVC) rings R_1 and R_2. PVC is preferred as it is a good heat insulator and at the same time is also a light material with sufficient strength. The temperature probe (TP) is positioned on the axis of the inner pipe with the help of a PVC sheet (*S*) and clamps C_1 and C_2. The tip of the probe is kept 5–8 cm inside from the edges of the outer pipe and about 2.5 cm from the edges of the inner pipe. The outer surface of the outer pipe and the ends of the inner and outer surfaces of both the pipes are covered with aluminium foil having good emissivity.

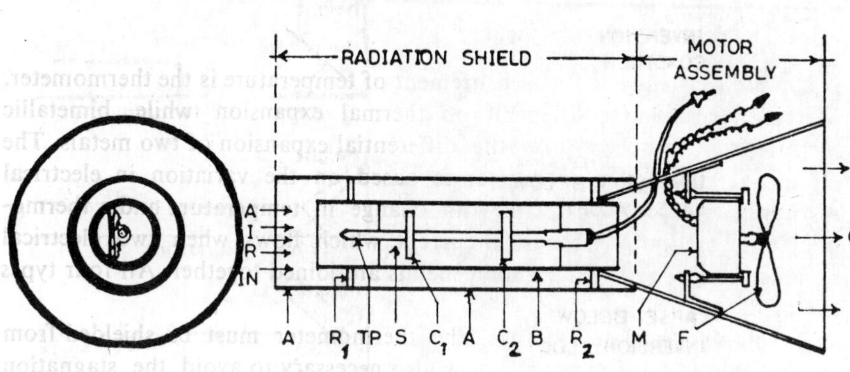

Fig. 3.4 *Motor aspirated radiation shielded temperature sensor*

A fan assembly is fixed at the other end of the pipes. A small fan (*F*) driven by a motor (*M*) working on 9–12 volts (dc) and mounted in a funnel shaped support, aspirates the air over the probe.

3.2.5 Solar Radiation Measurement

A knowledge of sunlight intensity is important specially in places where photochemical smog formation takes place. Instruments used to measure the radiant energy from the sun are pyrheliometer, solarimeter and chemical actinometer.

3.3 Plume Behaviour

Plume refers to the path and extent in the atmosphere of the gaseous effluents released from a source, usually a stack.

The behaviour of a plume emitted from any stack depends on localized air stability. Typical situations as shown in the Fig. 3.5, are generally encountered in the lower atmosphere (less than 300 m above ground). Effluents from tall stacks, are often injected to an effective height of several hundred metres above ground because of the cumulative effects of buoyancy and velocity on plume rise. Other factors influencing plume behaviour

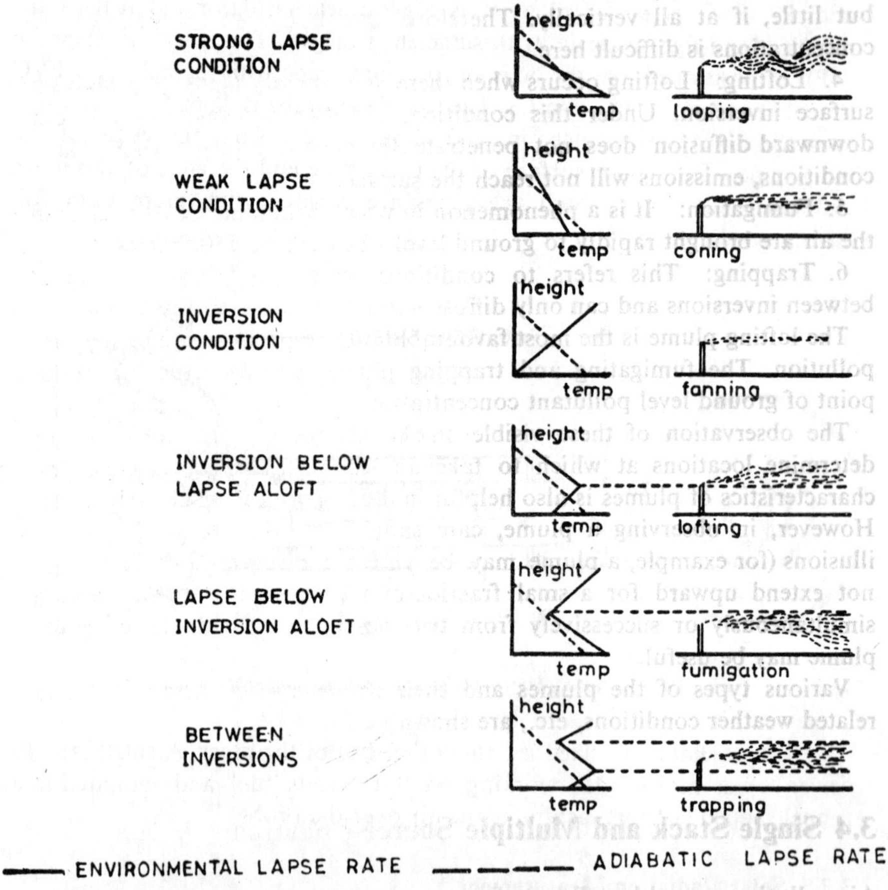

Fig. 3.5 *Types of plume behaviour*

are the diurnal variations in the atmospheric stability and the long term variations which occur with changing seasons.

Six types of plume behaviour are shown in Fig. 3.5. The spread of the plume is directly related to the vertical temperature gradient as shown on the left hand side of the figure.

1. Looping: It is a type of plume which has a wavy character. It occurs in a highly unstable atmosphere because of rapid mixing. The high degree of turbulence helps in dispersing the plume rapidly but high concentrations may occur close to the stack if the plume touches the ground.

2. Coning: It is a type of plume which is shaped like a cone. This takes place in a near neutral atmosphere (adiabatic condition) when the wind velocity is greater than 32 km/h. However, the plume reaches the ground at greater distances than with looping.

3. Fanning: It is a type of plume emitted under extreme inversion conditions. The plume, under these conditions will spread horizontally, but little, if at all vertically. Therefore, the prediction of ground level concentrations is difficult here.

4. Lofting: Lofting occurs when there is a strong lapse rate above a surface inversion. Under this condition, diffusion is rapid upward, but downward diffusion does not penetrate the inversion layer. Under these conditions, emissions will not reach the surface.

5. Fumigation: It is a phenomenon in which pollutants that are aloft in the air are brought rapidly to ground level when the air destabilizes.

6. Trapping: This refers to conditions where the plume is caught between inversions and can only diffuse within a limited vertical height.

The lofting plume is the most favourable with respect to minimising air pollution. The fumigating and trapping plumes are very critical from the point of ground level pollutant concentrations.

The observation of these visible smoke plumes is a useful practice to determine locations at which to take air samples. The knowledge of the characteristics of plumes is also helpful in dealing with invisible pollutants. However, in observing a plume, care should be taken to avoid optical illusions (for example, a plume may be visible for several kilometres yet not extend upward for a small fraction of a kilometre). For this, viewing simultaneously or successively from two points at right angles from the plume may be useful.

Various types of the plumes and their characteristics, occurrence and related weather conditions, etc., are shown in Table 3.1.

3.4 Single Stack and Multiple Source Pollution

Air pollution caused by effluents from a single or small group of stacks is a local problem. Effluent concentrations of concern generally occur at

Table 3.1 *Various Types of Plumes and their Characteristics*

Type of plume	Description of visible plume	Typical occurrence	Temperature profile and stability	Associated wind and turbulence	Dispersion and ground contact
Looping	Irregular loops dissipates in patches and relatively rapidly with distance.	During day time with clear or partly cloudy skies and intense solar heating.	Adiabatic or super adiabatic lapse rate. Unstable.	Light winds with intense thermal turbulence.	Disperses rapidly with distance, large probability of high concentrations sporadically at ground relatively close to the stack.
Coning	Cone shaped with horizontal axis, dissipates further downwind than looping plume.	During windy conditions, day or night. Layer type cloudiness favoured in day.	Lapse rate between dry adiabatic and isothermal. Neutral or stable.	Moderate to strong winds. Turbulence largely mechanical rather than thermal.	Disperses less rapidly with distance than looping plume, large probability of ground contact some distance downwind. Concentration less but persists longer than looping plumes.
Fanning	Narrow horizontal fan. No vertical spreading for kms downwind. If effluent is warm, plume rises slowly, then drifts horizontally.	At night and in early morning, any season, usually favoured by light winds.	Inverted and isothermal lapse rate. Very stable.	Light winds. Very little turbulence.	Disperses slowly, concentration aloft high at relatively great distance downwind, small probability of ground contact, though increase in turbulence can result in ground contact.
Lofting	Loops or cone with well defined bottom. Diffuses to top.	During change from lapse to inversion condition, usually near sunset on fair days.	Adiabatic lapse rate at stack top and above. Inverted below stack. Lower layer stable, upper layer neutral or unstable.	Moderate winds and considerable turbulence aloft, very light winds and little turbulence in layer below.	Probability of ground contact is small unless inversion layer is shallow, considered to be the best condition for dispersion since pollutants are dispersed in

					upper air with small probability of ground contact.
Fumigation	Fan or cone with well defined cone and dragged or diffused bottom.	During change from inversion to lapse condition, may occur with sea breeze in late morning or early afternoon.	Adiabatic or super adiabatic lapse rate at stack top and below. Isothermal or inverted lapse rate above. Lower layer unstable or neutral upper layer stable.	Winds light to moderate aloft and light below. Thermal turbulence in lower layer, little turbulence in upper layer.	Large probability of ground contact in relatively high concentration especially after plume has stagnated aloft.

distances ranging from the immediate vicinity of the stack to those of the order of several kilometres. At greater downwind distances, plumes formed by stack effluents become so diluted by diffusion in the ambient atmosphere, that concentrations may become negligibly small.

The combined effect of a large number of stack effluents spread over a large area (usually urban) produces a different type of air pollution problem which may have its principal area of concern at much greater distances. While the local pollution from any one stack may not be a problem, the combined effect of a large number of stacks produces serious urban pollution.

3.5 Wind Rose

A wind rose is defined as, "Any one of a class of diagrams designed to show the distribution of wind direction experienced at a given location, over a considerable period". In other words, the wind rose shows the prevailing direction of wind.

The most common form consists of a circle from which eight or sixteen lines emerge, one for each direction. The length of each line is proportional to the frequency of wind from that direction and the frequency of calm conditions is entered in the centre. There are many variations in the construction of wind roses. Some indicate the range of wind speeds from each direction, and some relate wind direction with other meteorological conditions.

Wind roses may be constructed from the data obtained over a given time period such as a particular month or season or a year. In constructing or interpreting wind roses, it is necessary to keep in mind the meteorological convention that wind direction refers to the direction from which the wind is blowing. A line or bar extending to the north on the wind rose indicates the frequency of winds blowing from the north.

The wind rose diagram is prepared using an appropriate scale to represent percentage frequencies of wind directions and appropriate index shades, lines etc., to represent various wind speeds. Observations corresponding to wind speed below 1 km/h are recorded as calm (Fig. 3.6).

Special wind roses are sometimes constructed like:

1. Precipitation wind rose
2. Smoke wind rose
3. Sulphur dioxide wind rose
4. Hydrocarbons wind rose

Instead of wind speed the parameters of precipitation, smoke, sulphur dioxide, hydro carbons etc. are attached to the wind direction. These are known as 'Pollution Roses'.

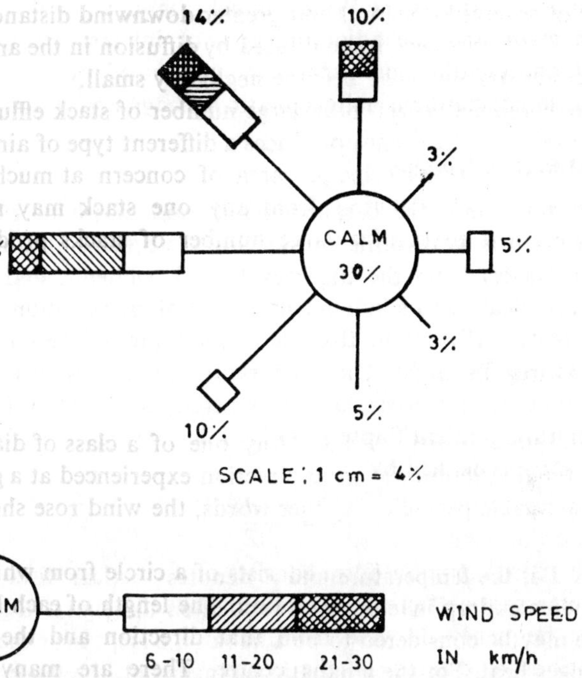

Fig. 3.6 *Wind rose construction*

3.6 Stack Effluent Dispersion Theories

The dilution of atmospheric wastes from elevated stacks is considered to occur in two stages, namely, plume rise and dispersion. In the first stage, the hot plume from the stacks goes up to a certain distance (defined as 'plume rise') into the atmosphere, due to buoyancy and momentum. Subsequently in the second stage, the plume spreads both vertically and horizontally by the dispersion process. The latter phase, i.e., the phase of pollutant dispersion can be described by the mathematics of diffusion, and has been the subject of considerable theoretical and experimental work by many workers in this field. On the other hand, all the plume rise equations are mostly empirical in nature, and none is uniformly accepted, However, the available theoretical and empirical equations can be effectively utilized to estimate the concentration of pollutants downwind from an emission source. These theories and equations, involving plume temperature, rate of emission, stack parameters such as height and diameter, meteorological factors like wind speed and direction, and the atmospheric stability, as well as the parameters representing, the topography of the region, have been employed successfully in many stack design and pollution control problems.

Another method of estimation of the effectiveness of the diffusion of the atmospheric waste, is the Wind Tunnel technique. In this method, a model of the stack and the adjoining area of interest is constructed in a wind tunnel, and the diffusion experiment conducted on the stack jets in that tunnel, satisfying the usual geometric and dynamic scaling parameters.

3.6.1 Effect of Dilution

Before going into the detailed description of plume rise and dispersion equations, it is worthwhile considering the process of dilution of the atmospheric wastes from the high level stack effluents, and also the principal meteorological factors influencing potential air pollution.

We have earlier seen, that the temperature of the atmosphere decreases with increase in height. The temperature gradient is closely interlinked with the stability for air masses and in the study of pollution in atmosphere, the temperature gradient (lapse rate) is of special significance. The temperature lapse rate γ is defined by

$$\gamma = \frac{-dT}{dZ} \tag{3.1}$$

where, T is the temperature and Z denotes vertical distance. (The negative sign indicates drop in temperature and often this sign is neglected.) The atmosphere may be considered to be adiabatic, except near the surface where air exchanges heat. For the adiabatic arrangement, the lapse rate $\gamma_d = 1°C$ per 100 metres.

These conditions correspond to the neutral equilibrium in which a quantity of air can be displaced without showing any tendency to return to its original height or to move further away from it. The value of $0 < \gamma < \gamma_d$ refers to a stable condition. By a stable arrangement we mean the condition in which a quantity of air when displaced from one height to another is forced back to its original position. When the displaced air tends to move farther away from its original position, the condition is known as unstable, i.e., $\gamma > \gamma_d$. The lapse rate can be negative so that temperature increases with height and this condition is exceedingly stable and the meteorologists call such a distribution an inversion, i.e. $\gamma_d < 0$. Humidity has an influence on the distribution of temperature in the atmospheric air which is in general mixed with water vapour. If the water is in a gaseous state in air, its presence in small quantities makes little difference in the behaviour of the air mass as regards its stability. If, on the other hand, water vapour condenses, it affects the behaviour of the air since there is a great amount of heat release due to condensation.

The lapse rate, which may vary widely even in normal day conditions, plays an important role in the case of dilution of stack effluents in the atmosphere. The negative lapse rate or temperature inversion (the inversion layer may begin at ground level and extend up to a few hundred metres) may occur on clear nights with light winds when the air next to the surface is cooled by the cold ground surface. A low-level inversion always acts as a barrier to vertical dilution of the pollutants even in an unstable surface layer and thus leads to an increase in the intensity of air pollution near the ground.

Gustiness of wind plays an important role in the dilution of atmospheric wastes. Gusts are a manifestation of atmospheric turbulence and a quantitative measure of gustiness gives the magnitude of turbulence. In other words, gustiness, g', is the measure of the fluctuation of the mean velocity of a turbulent wind in a particular direction. If $\bar{u}$ is the mean wind velocity in the x-direction, and v' is the component fluctuation of the mean wind in y-direction, then, the gustiness in the y-direction, g'_y is given by

$$g'_y = \frac{1}{\bar{u}}\sqrt{\overline{v'^2}} \tag{3.2}$$

Another important factor, wind-speed profile in the vertical plane, is a good measure of the degree of turbulence and atmospheric stability. In an unstable atmosphere, the variation in wind-speed profile with height is much less than under stable condition. If $\bar{u}_1$ is the mean wind at a particular elevation z_1, and $\bar{u}$ be the mean wind at any elevation z, $\bar{u}$ can be expressed as follows as given by Sutton:

$$\bar{u} = \bar{u}_1\left(\frac{z}{z_1}\right)^{n/(2-n)} \tag{3.3}$$

in which n is the turbulence parameter, $0 < n < 1$. The approximate value of n for a large lapse rate is 0.20 and for marked inversion is 0.50. Other meteorological factors which affect the turbulence or mixing of pollutants are humidity, vertical gradient in humidity, etc.

It is apparent that the form of stack emissions depend on the wind profile, turbulence of the wind, temperature distribution, etc.

3.6.2 Estimation of Plume Rise

The height of the stack is an important factor in determining the level of pollution at a given location. The 'effective height' of a stack is the sum total of height of stack and the rise of the plume beyond the stack exit. The plume rise depends on many factors like exit velocity, wind speed, diameter of the stack, temperature of the plume, lapse rate, etc. Several formulas are available to predict plume rise (Δh) from stacks. Most of the available plume rise formulas are empirical in nature, as the theory has not been developed adequately, and can be given by

$$\Delta h = K\frac{Q^{\alpha}}{\bar{u}^{\beta}} \tag{3.4}$$

in which α, β and K (dimensional) are constants, Q is the heat emission rate from stack. In the CCRL (Canadian Combustion Research Laboratory) equation, the values of $\alpha = 1/4$, $\beta = 1$, $K = 66.4$ where Q is expressed in kcal/s and $\bar{u}$ in m/s. An extensive series of observations at the Tilbury Power Station show $\alpha = 1/4$, $\beta = 1$, $K = 450$–500 and Q and $\bar{u}$ are expressed in MW and m/s respectively. It is also observed that K is a function of height of the source of emission.

Alternatively, the plume rise may be considered due to buoyancy and momentum. For instance,

$$\Delta h = \Delta h \text{ buoyancy} + \Delta h \text{ momentum} \tag{3.5}$$

Following is the empirical formula of Moses and Carson.

$$\Delta h = C_1 \frac{V_s d}{\bar{u}} + C_2 \frac{Q_h^{1/2}}{\bar{u}} \tag{3.6}$$

where

Δh = Plume rise (m)
V_s = Stack exit velocity (m/s)
$\bar{u}$ = Wind speed (m/s)
d = Stack diameter (m)
Q_h = Heat emission rate (kcal/s)
C_1, C_2 = Plume rise regression coefficients, which depends on atmospheric stability.

It is seen from the formula, that the first term evaluates the vertical momentum of the gases leaving the stack, and the second term evaluates the buoyancy force of the plume which is a function of heat content of the plume. Therefore, the magnitude of the plume rise is inversely proportional to wind speed, directly proportional to the sum of mass ejection rate, and the square root of heat content of the gases.

The Bureau of Indian Standards has suggested that the following Brigg's formula be used to compute plume rise for practical use (IS: 8829).

(a) For hot effluents with heat release of the order of 10^6 cal/s or more

$$\Delta h = 0.84\,(12.4 + 0.09h)\, \frac{Q_H^{1/4}}{\bar{u}} \tag{3.7}$$

where

Q_H = heat release in calories per second,
h = height of the stack in metres,
$\bar{u}$ = wind speed in m/s

(b) For not very hot releases and which can be counted as momentum sources above.

$$\Delta h = \frac{3 W_0 D}{\bar{u}} \tag{3.8}$$

where W_0 is the efflux velocity of the same units as $\bar{u}$ and D is the stack exit diameter.

It may be noted that although the available plume rise formulae predict well in a particular place for some particular conditions, none can be universally accepted due to empirical character. The lack of a complete theory concerning the rise of plumes in the atmosphere is probably a handicap to environmental engineers in the prediction of downwind pollution levels.

3.6.3 Dispersion Equations

The dispersion of pollutants is due to the turbulent flow of wind, and is dependent on the atmospheric stability, gustiness, etc. The turbulent diffusion models for atmospheric pollutants are based on Fick's law of molecular diffusion. Assuming a continuity in the atmosphere and incorporating the mass balance equation in the diffusion equation we get,

$$\frac{\partial c}{\partial t} + u\frac{\partial c}{\partial x} + v\frac{\partial c}{\partial y} + w\frac{\partial c}{\partial z}$$

$$= \frac{\partial}{\partial x}\left(K_x\frac{\partial c}{\partial x}\right) + \frac{\partial}{\partial y}\left(K_y\frac{\partial c}{\partial y}\right) + \frac{\partial}{\partial z}\left(K_z\frac{\partial c}{\partial z}\right) \tag{3.9}$$

in which K_x, K_y and K_z are diffusion coefficients in the x, y and z directions respectively, c is the concentration of the pollutants, u, v, and w denote the component velocities of wind in x, y, and z directions respectively and t refers to time.

For simultaneous deposition of particulate matters from a plume, the above equation is modified slightly by the addition of the vertical advective term $\bar{w}\ \partial c/\partial z$, on the right hand where $\bar{w}$ is the velocity of the falling particles.

For a functional form for the distribution of pollutants, in addition to Fick's law, the knowledge of the shape of the distribution of concentration is required.

For continuous source the following type of Gaussian Distribution is assumed normally.

$$C(x, y, z) = \frac{Q}{\pi\sigma_y\sigma_z\bar{u}} \exp - \{(h^2/2\sigma_z^2) + (y^2/2\sigma_y^2)\} \tag{3.10}$$

Where

C = Concentration (g/m³)
Q = Pollutant release rate (g/s)
σ_y, σ_z = Crosswind and vertical plume standard deviations (m)
$\bar{u}$ = Mean wind speed (m/s) at h_s, the height of the stack
h = Effective stack height (m)
x, y = Downwind and crosswind distances.

From the above equation it is seen that C, the ground level ($Z = 0$) concentration of a pollutant at a point (x, y) downwind from a source is proportional to the rate of emission (Q). Further, it is inversely proportional to the wind speed ($\bar{u}$), and to the parameters σ_y and σ_z. Also, it can be shown that the maximum ground level concentration of a gaseous pollutant is inversely proportional to the square of the effective stack height.

Many solutions of the above equation are now available which predict the ground level concentration of pollutants, both for gaseous and particulate pollutants.

But most of such formulas are derived only for simpler cases of constant velocity, steady state, and constant diffusion coefficients, though the variation of wind and diffusion coefficient with height is a recognised fact. It may be noted, that all point source models fail if there is a weak and variable wind. The models are also inadequate for inversion layers.

On the other hand, the concentration of pollutants may be determined solving Fick's Law numerically considering the wind profile, lapse rate, gustiness, topography, variation of diffusion coefficient or other relevant factors. Of late, the numerical technique has been employed in some investigations. However, it is important to note that success of the method depends on the appropriate consideration of influencing factors while solving the Fickian equation. To this end, the work of Hino deserves a special mention. The author has proposed a method for estimating smoke diffusion over a complicated topography.

It is important that mean wind speed at the centreline of the plume be used in dispersion equations. Where data at least at stack height, is not available, upper wind speeds may be calculated by the following equation

$$\bar{u}_h = \bar{u}_1\left(\frac{h}{z_1}\right)^n \tag{3.11}$$

where subscripts represent two different heights. Smith recommends $n = 0.25$ for unstable and $n = 0.50$ for stable conditions.

Nearby buildings may affect plume behaviour by causing mechanical turbulence which may bring portions of the plume to ground level near the stack. The effect is most noticeable when wind speeds are high and the stack is downwind of the building. The effect may be overcome by using the conservative thumb rule which states that stacks 2 to 2½ times the height of nearby buildings will clear the building wake.

Irregular terrain may also affect ground level concentrations. The dispersion formulas usually used are meant for flat, level areas. Generally, rough terrain promotes dispersion.

3.6.4 Dispersion Models

Mathematical approaches have been made to the dispersion of pollutants in the atmosphere and atmospheric dispersion models have been developed based on Fick's molecular diffusion equation and on statistical theory.

For instantaneous point source, dispersion models have been developed based on Fick's molecular diffusion equation and on Sutton's statistical concepts.

For continuous point source, the dispersion models are based on plume behaviour, eddy diffusion, wind speed, and amount of dilution etc., because the continuous emission of pollutants results in a plume which is carried by the wind speed and spreads by the intensity of turbulence eddy.

Based on stability classification in the field of atmospheric diffusion the three well-known models in use are:

1. Pasquill model
2. ASME model
   (American Society of Mechanical Engineers' model)
3. McElroy model

Details of these models are given in IS: 8829.

3.6.5 Wind Tunnel Method

The available formulas for estimating or predicting the pollutants are usually obtained for uniform topographic and meteorological conditions. These difficulties led to the adoption of model study of what is popularly known as the wind tunnel method. This technique employs the method of similitude between the actual flow condition and that of the model in the wind tunnel.

In the case of a strong wind when turbulent mixing keeps the lapse rate almost adiabatic, it may not be difficult to obtain the usual geometric and dynamic similitude. But the simulation of scales and intensities of turbulence, and of variation of mean wind velocity with height, is extremely difficult. Conventional tunnels fail in correct prediction when the atmospheric lapse rate controls the plume behaviour, as the temperature within the model is essentially isothermal. Many works are in progress to develop suitable simulation methods which will in turn promise the applicability of the wind tunnel technique in solving the atmospheric diffusion problem.

3.6.6 Stack Height

The height of the stack and the height of rise of the plume above the stack play a major role in the ground level concentration expected on the down wind side.

The actual stack height (H) is easy to determine while effective stack height (H_e) is quite difficult to estimate with any degree of accuracy. The increased stack height is considered to diffuse the pollutants better. Immediately above the stack, the rise of pollutant is proportional to the emission velocity of gases and to the temperature differences between the gases and the surrounding atmosphere.

For a given stack height, the concentration at ground level decreases with increase in wind velocity. However, there is a 'critical velocity' when the ground concentration attains its maximum value which is given by

$$C_m = \frac{M \cdot K}{H^2} \sqrt[3]{\frac{1}{V \cdot A_T}} \tag{3.12}$$

where,

C_m = Maximum concentration of pollutants on ground level

H = Height of the stack

M = Mass of the pollutant gases discharged per unit time

V = Volume of the pollutant gases discharged per unit time

A_T = Difference in temperature of stack gases and surrounding air

K = Constant which accounts for factors such as

1. Horizontal and vertical mixing of pollutants in the air;
2. settling velocity of polluting substances;
3. discharge characteristics of the gases from stack.

It has been reported that in the direction of wind, the maximum concentration is attained at a distance X_m ranging from 10–40 times the stack height.

The empirical formula for the stack height emitting maximum concentration of pollutants is given as follows

$$C_{max} = \frac{AFMm}{H^2}\left(\frac{n}{V \cdot t}\right)^{1/3} \tag{3.13}$$

where,

C_{max} = Maximum concentration of pollutants
A = Coefficient, depends upon atmospheric condition
= 200 for Indian conditions
F = Coefficient, depends upon dust precipitation efficiency, ranges from 0.8 to 0.95
M = Quantity of flue gases in g/s
m = Coefficient depends upon velocity of flue gases
H = Height of stack or chimney in m
n = Number of emitting sources
V = Volume of gases leaving the chimney in cu. m/s.
t = Difference in temperature between flue gases and atmospheric temperature at the top of the stack.

This formula is in use in the USSR and coefficients are being derived for Indian conditions.

Example

A factory uses 2,00,000 litres of furnace oil (specific density 0.97) per month. If for one million litres of oil used per year, the particulate matter emitted is 3.0 tonnes per year, SO_2 emitted is 59.7 tonnes per year, NO_x emitted is 7.5 tonnes per year, hydrocarbons emitted are 0.37 tonnes per year, and carbon monoxide emitted is 0.52 tonnes per year, calculate the height of the chimney required to be provided for safe dispersion of the pollutants.

As per emission regulations (July 1984) part I published by the Central Board for Prevention and Control of Water Pollution, New Delhi, the chimney height is to be calculated according to the formula:

$$H = 74(Q)^{0.27} \tag{3.14}$$

where

Q = particulate matter emission in tonnes per hour
H = height of chimney in metres

The particulate emission is equal to 3.0 tonnes per million litres of oil per year. Consumption of oil is equal to

$$2{,}00{,}000 \times 12 = 24{,}00{,}000 \text{ l/year}$$
$$= 2.4 \text{ million l/year}$$

Therefore,

Total particulate emission = 2.4 × 3.0 = 7.2 tonnes per year.

$$= \frac{7.2}{300 \times 24} \text{ tonnes per hour (assuming 300 working days and 24 hours per day).}$$

Now,

$$H = 74\left(\frac{7.2}{300 \times 24}\right)^{0.27}$$

$$= 11.47 \text{ m}$$

The height of the chimney for effective dispersion of SO_2 is to be calculated as per the formula $H = 14(Q)^{0.3}$ (as given in Sec. 19.5 under guidelines for minimum stack height)
where,

Q = SO_2 emission in kg/h
H = Height of chimney in metres

Q in the example = 59.7 × 2.4 = 144 tonnes per year
= 20 kg/h

Therefore $H = 14(20)^{0.3} = 34.4$ m

So, adopt a height of 34.4 metres.
(Since the emission of SO_2 is much more than that of NO_x, CO, and Hydrocarbons, the calculation of stack height is done based on SO_2 emission data only).

3.6.7 Scope for Future Work

Many experimental and theoretical works are in progress to investigate the complicated atmospheric structure, both near the ground and well above the ground. Many interesting meteorological features, from the point of atmospheric pollution, are now recognised. In view of the importance of many meteorological factors and topographical features besides the stack geometry, and also in view of the increased available memory capacities of digital computers, there is no reason why we should not take up the numerical method of solution of the problem of atmospheric pollution thereby avoiding any unnecessary simplification of the problem.

QUESTIONS

1. List out the meteorological factors influencing air pollution.
2. Explain the role of meteorological elements in the dispersion of air pollutants in the atmosphere.
3. Explain the cause and effects of 'inversion of atmosphere'.
4. Explain stable and unstable atmosphere and inversion of the atmosphere.
5. Explain the following atmospheric conditions:
   (a) Super-adiabatic
   (b) Sub-adiabatic

(c) Neutral
(d) Inversion
How do they influence the dispersion of pollutants in the atmosphere?

6. With the help of neat sketches, explain the working of
   (a) Wind speed recorder
   (b) wind direction recorder.
7. Define a wind rose. Explain the importance of wind roses in air pollution studies.
8. Explain the terms
   (a) Environmental lapse rate
   (b) Adiabatic lapse rate
   (c) Wind rose
9. Describe with neat sketches, how different atmospheric conditions give rise to different kinds of plumes.
10. Write explanatory notes on
   (a) Dispersion models
   (b) Estimation of plume rise
   (c) Wind tunnel techniques for the study of atmospheric diffusion
   (d) Calculation of stack height
11. Write short notes on:
   (a) Inversions
   (b) Wind rose
   (c) Lapse rate
   (d) Atmospheric dispersion
   (e) Plume behaviour

4

Industrial Plant Location and City Planning

4.1 Introduction

Till now the factors considered for locating a new industrial plant were the availability of raw materials, power, water supply, transportation facilities, labour and the market. Now one more critical factor has to be considered by the plant management, i.e., the factor of air pollution control. In fact, the neglect of this aspect of air pollution control in the past by many large industries has proved to be a costly error, for today they are facing heavy damage claims, litigation and the necessity of taking control measures after several years of operation.

4.2 Factors to be Considered for Industrial-plant Location

While selecting a site from the point of air pollution control, the following factors should be taken into consideration to avoid costly control measures, improve public relations, and prevent litigation.

1. Existing levels of air contaminants
2. Potential effects on the surrounding area
3. Meteorological factors and climate
4. Topographical features
5. Clean air available
6. Planning and zoning

4.2.1 Existing Levels of Air Contaminants

If the new plant is to be located in an area which is already industrialised, it is a good practice to undertake a pre-operational survey to know the existing levels of contaminants, under prevailing meteorological conditions. This type of survey gives an idea regarding the nature of pollution due to existing industries, i.e., whether the existing level of pollution is high, medium or low. The results of such a survey, with respect to known operational data on the magnitude of contemplated emissions from the new sources, would provide information on the extent to which waste products could be safely discharged into the atmosphere without resulting in too much contamination.

A preliminary survey is also useful if the site to be chosen is located in a rural or suburban area. This is because, the area under consideration may be exposed to exotic pollution, i.e., contamination from distant sources. It is important to determine what those concentration levels are in relation to the proposed scale of operations. Suspended particulate matter, sulphur dioxide, fluorides, etc., may be transported over great distances from large industrial complexes to predominantly rural areas. The rural community may tolerate the existing conditions until a new plant commences activities in the immediate neighbourhood.

4.2.2 Potential Effects on the Surrounding Area

Another important factor from the point of site selection is to have a knowledge of the specific effects of the major pollutants likely to be discharged into the atmosphere in relation to the population and land use of the area surrounding the site. For example, whether the pollutants will have any effect on the health of the people, whether it causes damage to the vegetation, and whether it affects the farm animals in that area, is to be considered. A rural and predominantly agricultural area is more affected by fluorides and sulphur dioxide than an urban population. This is because certain pollutants are more toxic and harmful to vegetation and animals than to people. Hydrogen sulphide has little effect on vegetation but is obnoxious and even dangerous to human life in comparatively low concentrations.

Thus it is seen that the effects of contaminants likely to be discharged from the proposed industrial plant is important, in particular, from the point of its effects on human health, animals and damage to crops. Incidentally, it is also important from the point of keeping good public relations, as any polluting industry leads to number of public complaints and litigations.

4.2.3 Meteorological Factors and Climate

The prime factors which have to be considered in order to minimise air pollution problems by site selection are the climate and meteorology of the location under consideration. It is important to know the prevailing wind direction, wind speed, and factors favourable for stable atmosphere, i.e., inversion conditions. The dispersive ability of the air at each possible site has to be determined. This can be done on the basis of the average values for wind movement and inversion conditions. Wind roses for each possible site have to be constructed and studied. Meteorological factors should be favourable for the air to dilute the pollutional load down to acceptable levels of contamination. These factors have been discussed in Chapter 3 in greater detail.

The ideal site for location of an industry is a level terrain in a region where the average wind speed is of the order of 16 km/h or more and where temperature inversions rarely occur. Also, it is always better, if a populated town or agricultural land is not on the downwind of the proposed site. Further, if possible, the plant site area should be large enough so that maximum ground level concentration of pollutants downwind, occur well within factory premises rather than on surrounding private property.

Areas where prolonged subsidence inversions occur frequently should be avoided for location of industries, if alternate sites are available. Individual valleys where prolonged conditions of stable atmosphere are frequent should be critically examined. Industrialisation of these areas must be minimised.

4.2.4 Topographical Features

Air movement is greatly influenced by the topography in the neighbourhood of the site under consideration, like valleys, mountains, etc. In fact, we have to give more attention to air pollution control in valley sites than in level terrain, especially when the average wind velocity is less than 16 kilometres per hour.

The location of industries in valleys, mountainous areas, and undulating terrain present difficult problems from the point of air pollution control. The air pollution disasters in Meuse Valley, Belgium, and at Donora, Pennsylvania, (Chapter 5) are good examples. These episodes occurred in valley areas, due to prolonged inversion conditions for four or five days. They resulted in many cases of illness and loss of life. If an industrial plant is located at the bottom of a narrow valley with mountains rising fairly steeply on either side, the situation becomes very critical. Probably, it will be one of the worst site conditions one can think of.

For efficient dilution and dispersion of air pollutants in the atmosphere, it is important to know the prevailing wind direction, wind velocity, degree of turbulence, etc. The probable influence of other industrial plants and tall buildings in the vicinity should also be considered. The eddies produced by nearby buildings may cause erratic dispersion of the gas. This must be taken into consideration in the design of stacks. Stacks must be tall enough so that the effective height of the plume will permit the air pollutants to be dispersed safely in the atmosphere. The necessary data may be obtained by micro-meteorological investigations, and conducting air pollution surveys.

In general, the adverse influence of topography and weather factors in relation to pollution control should be carefully considered, and critically examined, before selecting a site for a new industrial plant.

4.2.5 Clean Air Available

The requirement of many industrial processes for supplies of clean air introduces another important aspect of air pollution into the problem of site selection. For example, industries requiring clean air for manufacture are factories dealing with the manufacture of transistors, electronic components, antibiotics, and vaccines. Also, clean air is required for cooling the reactors of atomic energy plants, since, if polluted air were used, the impurities present would become radioactive and their escape into the atmosphere would create a hazard. In these cases, location of industries in areas of heavy air pollution will add materially to the cost of cleansing the air.

4.2.6 Planning and Zoning

Proper planning and zoning of industrial areas and residential areas can play an important role in the control of air pollution. Residential areas and certain heavy industries should not be located too close to each other. It is always better to have a green belt between industrial areas and residential areas. The concerned municipal authorities should encourage the creation of green belt. If there are any municipal laws and regulations regarding this aspect, they should be strictly enforced.

Recently, scientists have identified a dozen species of trees which have a capacity to absorb industrial pollutants from the air. Many of these trees grow in our country. For example, it has been reported that tamarind and margosa trees are capable of absorbing dust and gas from the atmosphere polluted by cement factories, chemical industries and quarries. This fact has been established by laboratory studies carried out jointly by the Toxicological Research Centre and the National Institute of Botanical Research, Lucknow. Therefore, growing suitable trees is an important way of solving the problem of air pollution.

In zoning of land for industrial use, direction, speed, and frequency of prevailing winds should be given due consideration. The topography of the area should also be considered.

In general, while considering each possible site for the location of a plant, the site advantages vis-a-vis the cost of control of air pollutants should be carefully investigated.

4.3 City Planning

Today, we are seeing many cities developing in a very haphazard manner. They are now paying the price for the failure to plan for future systematic and homogeneous development. Smoke, dust, fumes, odours and poor zoning practices have spoiled fine residential areas and created ugly sights in central parts of some cities. In many cases, good residential areas in the heart of the cities have degenerated into slums.

One of the basic principles of country planning is, it must be flexible, continuous and adopted to the local requirements. While planning, the growth trend in population and industries that may come up in the near future should be taken into consideration. Residential, commercial and industrial areas must be properly planned. Prime importance must be given for locating sufficient number of public parks and gardens, for they act as lung spaces, especially in the industrial cities. Also, there must be sufficient provision for traffic lanes and parking facilities.

For the proper growth and development of a city, a zoning law is required. The residential, commercial and industrial zones in and around the city must be clearly defined by the competent authority. And of course, the regulations must be strictly implemented. It is a sad commentary that many cities with good zoning restrictions have nullified them by granting exemptions whenever sufficient pressure has been brought to permit land use not in conformity with the regulations.

Every city or town requires its own zoning ordinance according to the local conditions. As land usage changes, the zoning regulations must be modified to meet new conditions. Satellite towns bordering a city must also be subjected to zoning to provide for orderly growth.

However, there can be no successful solution to the problem of eliminating ugly and unattractive sights in cities like black dense smoke given out by buses and lorries, until air pollution is properly controlled. As such, in many cases zoning and pollution control measures must go hand in hand, and must be carried out on a regional rather than a local basis. This requires active cooperation between adjoining cities, as well as between states.

In fact, is some western countries, the effects of air pollution on future zoning are being taken into considerations by city planners.

For effective control, larger air pollution zoning districts, or even interstate contracts or international agreements, may be required depending on meteorological conditions. For example, to prevent or reduce the effects of acid rain, this may be necessary.

Following are some of the measures that can be taken for air pollution control by planning and zoning:

1. Decentralisation of industry
2. Creation of a green belt between industry and receptor areas
3. Regulations over automobile exhausts
4. Traffic control
5. Creation of smokeless zones in selected areas by limiting industries and residences in those zones to the use of certain specific smokeless fuels
6. Prohibiting use of volatile fuels.

QUESTIONS

1. What steps can be taken in the planning stages of a city or an industry to avoid the problem of air pollution to the maximum possible extent?
2. Explain how topography and climate affect the planning of a town with reference to location of industries from the point of atmospheric pollution.
3. Explain the importance of proper planning and zoning of industrial and residential areas from the point of air pollution control.

4. List of factors that should be taken into consideration while selecting a site for an industry from the point of minimising air pollution.
5. Write short notes on
   (a) Exotic pollution
   (b) Green belt
   (c) Zoning law
   (d) City planning

5

Effects of Air Pollution on Human Health

5.1 Introduction

Air pollution is one of the greatest environmental evils. The air we breathe has not only life-supporting properties but also life-damaging properties. Under ideal conditions the air we inhale has a qualitative and quantitative balance that maintains the well-being of man. But when the balance among the air components is disturbed, or in other words, if it is polluted, it may affect human health.

An average man breathes 22,000 times a day and takes in 16 kg of air each day. It far exceeds the consumption of food and water. It has been estimated, that a man can live for five weeks without food and five days without water, but only for five minutes without air.

All the impurities in the inhaled air do not necessarily cause harm. Depending upon the chemical nature of the pollutants, some may be harmful when present in the air in small concentrations and others only if they are present in high concentrations. The duration of exposure of the body to polluted air is also an important factor. Therefore, the prime factors affecting human health are:

1. Nature of the pollutants
2. Concentration of the pollutants
3. Duration of exposure
4. State of health of the receptor
5. Age group of the receptor

Generally speaking, susceptibility to the effects of air pollution is great among infants, the elderly, and the infirm. Those with chronic diseases of the lungs or heart are thought to be at great risk. Pre-school and school children appear to be both sensitive and specifically reactive to air pollution health effects. Another point to be noted is that the effect of air pollution on human health is worst during the winter season, when pollution levels reach a climax.

An objectionable odour, visibility reduction, eye irritation or vegetation damage are useful guides to the likelihood or severity of health effects. A grey pall over a city or an industrial area can have a depressing effect and impair the enjoyment of life. There is no doubt, howe' 'r, that the urgency with which steps are taken to improve air quality will depend very much on how serious the risk of ill health from air pollution is thought to be.

Table 5.1 gives some of the disastrous air pollution events in the world.

Table 5.1 *Air Pollution Episodes*

S. No.	Month and year	Place	Mortality
1.	December, 1930	Meuse Valley (Belgium)	63
2.	October, 1948	Donora (Pennsylvania)	20
3.	November, 1950	Poza Rica (Mexico)	22
4.	December 1952	London	4000
5.	November, 1953	New York	220
6.	January, 1956	London	1000
7.	December, 1957	London	750
8.	December, 1962	London	700
9.	January, 1963	New York	300
10.	November, 1966	New York	168
11.	December, 1984	Bhopal (India)	2000

5.2 Mechanism of Action of Air Pollutants

The effects of air pollution on human health generally occur as a result of contact between the pollutants and the body. Normally, bodily contact occurs at the surfaces of the skin and exposed membranes. Contact with exposed membranous surfaces is of utmost importance because of their high absorptive capacity compared to that of the skin. Air-borne gases, vapours, fumes, mist, and dust may cause irritation of the membranes of the eyes, nose, throat, larynx, tracheo-bronchial tree and lungs (Fig. 5.1).

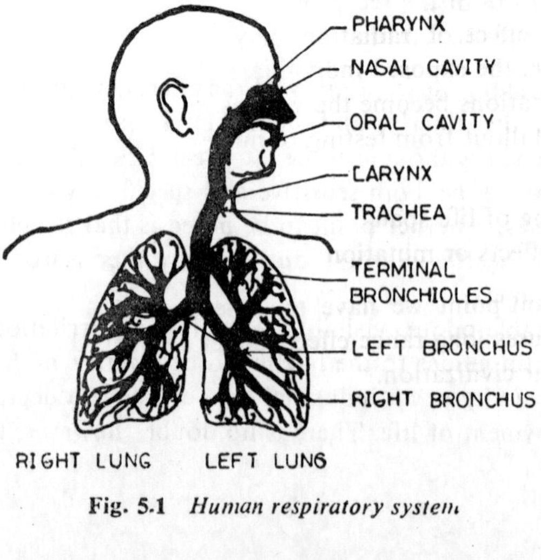

Fig. 5.1 *Human respiratory system*

Some irritants even reach the mucosa of the digestive tract. The details of the mechanism involved is beyond the scope of this book, as greater attention has been given to the engineering aspects.

5.3 Health Effects

1. Eye Irritation.
2. Nose and throat irritation.
3. Irritation of the respiratory tract.
4. Gases like hydrogen sulphide, ammonia and mercaptans cause odour nuisance even at low concentrations.
5. Increase in mortality rate and morbidity rate.
6. A variety of particulates particularly pollens, initiate asthmatic attacks.
7. Chronic pulmonary diseases like bronchitis and asthma, are aggravated by a high concentration of SO_2, NO_2, particulate matter and photochemical smog.
8. Carbon monoxide combines with the haemoglobin in the blood and consequently increases stress on those suffering from cardiovascular and pulmonary diseases.
9. Hydrogen fluoride causes diseases of the bone (fluorosis), and mottling of teeth.
10. Carcinogenic agents cause cancer.
11. Dust particles cause respiratory diseases. Diseases like silicosis, asbestosis, etc., result from specific dusts.
12. Certain heavy metals like lead may enter the body through the lungs and cause poisoning.

5.3.1 Effect of Radioactive Fallout

The biological effect of radiation may be somatic or genetic damage. In somatic damage, the exposed individual is affected, while in genetic damage the future generations become the victims.

Radioactive fallout from testing of nuclear weapons causes:

(a) Cancer
(b) Shortening of life span
(c) Genetic effects or mutation

One significant point we have to note about the effect of radioactive fallout is, it causes long range effects affecting the future of man and hence the future of our civilization.

5.4 Investigation of Health Effects of Air Pollutants

Three methods are available for determining the effect of various pollutants on people. They are:

1. Experimental exposures of men and animals
2. Clinical studies
3. Epidemiology

The experimental exposure of different types of animals under controlled conditions to various concentrations and dosages of air pollutants, can give valuable information regarding the mode of action of a pollutant and its effects. Of course, results must be subjected to rigorous statistical analysis. But the main problem here is, the extrapolation of results to human population.

Experimental exposures of men should be limited to concentrations and dosages of pollutants that will not result in serious illness. Types of effects noticed and measured include detection of odour, irritation of eyes, nose and throat, change in pulse rate, breathing frequency, reduction in physical activity and many other physiological responses.

Clinical studies mainly involve observations made on subjects who are or were exposed to atmospheric pollutants under uncontrolled conditions. Studies of truck drivers, traffic policemen, factory workers and other occupational groups may give valuable information. Also, patients visiting physicians to complain of symptoms supposedly caused by living in polluted areas may upon observation yield clinical data of great importance. Further, information has been collected by the follow up of the acute air pollution episodes. The data collected has indicated a relationship between air pollution and disease, particularly pulmonary disease.

In epidemiological studies, the relationships between the distribution of specific diseases in a human population and the factors that determine the distribution are found out. For example, one may compare morbidity records, mortality records, hospital admissions, absenteeism, and other health related data from various geographical areas with levels of air pollutants in the same areas to determine a correlation, if any. To avoid misinterpretation, the population under study must be carefully observed for smoking habits, occupational exposures, and any other factor that might prejudice the results of the study.

5.5 Specific Pollutants

5.5.1 Sulphur Dioxide

Sulphur dioxide is an irritant gas which affects the mucous membranes when inhaled. Under certain conditions, some of the air-borne sulphur dioxide gas is oxidised to sulphur trioxide. Each of these two gases, in the presence of water vapour or water, forms sulphurous and sulphuric acid respectively. Sulphur trioxide is a very strong irritant, much stronger than sulphur dioxide, causing severe bronchospasms at relatively low levels of concentration.

5.5.2 Carbon Monoxide

Carbon monoxide has a strong affinity for combining with the haemoglobin of the blood to form carboxyhaemoglobin, COHb. This reduces the ability of the haemoglobin to carry oxygen to the body tissues. CO has about two hundred times the affinity of oxygen for attaching itself to the haemoglobin, so that low levels of CO can still result in high levels of COHb. Carbon-monoxide also affects the central nervous system. It is also responsible for heart attacks and a high mortality rate.

5.5.3 Oxides of Nitrogen

Of the seven oxides of nitrogen known to exist in the ambient air, only two are thought to affect human health. These are nitric oxide (NO) and nitrogen dioxide (NO_2). While some questions remain about haemoglobin reactions with oxides of nitrogen, there is no positive evidence that nitric oxide exposure is a health hazard associated with community air pollution.

Nitrogen dioxide is known to cause occupational disease. Among occupations with NO_2 hazards are the manufacture of nitric acid, exposures of farmers to silage that has had high nitrate fertilisation, electric arc welding, and mining utilising nitrogen compounds as explosives. It is estimated that eye and nasal irritation will be observed after exposure to about 15 ppm of nitrogen dioxide and pulmonary discomfort after brief exposures to 25 ppm of nitrogen dioxide.

5.5.4 Hydrogen Sulphide and Mercaptans

Hydrogen sulphide is a foul smelling gas. It is well known for its rotten egg like odour. Exposures to hydrogen sulphide for short periods can result in fatigue of the sense of smell.

Other sulphur compounds that are of interest in air pollution mainly because of their strong odours are methyl mercaptan (CH_3SH) and ethyl mercaptan (C_2H_5SH). But it has been reported that at the concentrations at which they are odour nuisances, they have no other effect on human health. In fact, mercaptans are often added to natural or manufactured gas supplies so that leakage of gas will be noticed.

5.5.5 Ozone

Ozone is a gas that has an irritant action in the respiratory tract, reaching much deeper into the lungs than the oxides of sulphur.

5.5.6 Fluorides

Fluorides present in air, range from those which are extremely irritant and corrosive like hydrogen fluoride to relatively non-reactive compounds. But, fluorine is a cumulative poison even under condition of prolonged exposure and in sub-acute concentrations.

5.5.7 Lead

The main source of lead in urban atmospheres is the automobile. It creates urban concentration of inorganic lead of about 1–3 $\mu g/m^3$, with high values in areas of heavy traffic. Inorganic lead acts as an agent which causes a variety of human health disorders. The effects include gastro-intestinal damage, liver and kidney damage, abnormalities in fertility and pregnancy, and mental development of children gets affected.

5.5.8 Hydrocarbon Vapours

Some of the hydrocarbon vapours in the atmosphere have health implications. The effect of formaldehyde is primarily irritating. It is a major contributor to eye and respiratory irritation caused by photochemical smog.

5.5.9 Carcinogenic Agents

Carcinogenic agents are responsible for cancer. For example, the poly-cyclic organic compound, 3, 4-benzpyrene. The origin of these compounds is in the incomplete combustion of hydrocarbons and other carbonaceous materials. They are also reported to be present in exhaust discharges from I.C. engines. In addition to poly-cyclic organic compounds, it has been found that some aliphatic hydrocarbons are also carcinogenic.

5.5.10 Insecticides

Insecticides are not only harmful for insects but also poisonous for man, e.g., DDT (Dichloro diphenyl trichloroethane). They can affect the central nervous system and may attack other vital organs. In fact, DDT has been found in mother's milk in western countries and even in our own country. Hence, the use of DDT has been banned in the USA. The United Nations Environment Programme (UNEP) has warned against the undesirable effects of indoor spraying of DDT. It has also been reported that indoor spraying affects domestic livestock.

According to a study conducted at the Industrial Toxicology Research Centre, Lucknow (India), the accumulation of pesticides in the environment due to their growing use for agricultural purposes can also cause premature labour and abortion, due to high concentration of pesticides in the body of expectant mothers.

5.5.11 Radioactive Isotopes

The important radioactive isotopes that may reach ambient air are Iodine 131, Phosphorous 32, Cobalt 60, Strontium 90, Radium 226, Carbon 14,

Sulphur 35, Calcium 45 and Uranium. The major sources of radioactive air pollutants are:

(a) Nuclear reactors
(b) Experimental accelerators
(c) Scientific and medical use of radioactive isotopes
(d) Agricultural and industrial use of radioactive isotopes as tracers, and
(e) Testing of nuclear bombs in the atmosphere.

The serious health effects are anaemia, leukaemia and cancer. Radioactive isotopes also cause genetic defects and sterility, as well as embryo defects and congenital malformations. It also shortens the life span of an individual.

5.5.12 Allergic Agents

It is generally recognised by medical personnel that the air we breathe is the natural carrier of many microscopic organic materials which may act as allergens. Our body reactions to such allergens occur mainly in the skin and the respiratory tract. Organic allergens have their origins in living things like plants (pollen grains), yeasts or moulds, or in animal hair, fur, or feathers.

An interesting case in this respect, is the city of Bangalore. The garden city of Bangalore, known for its fine climate, is also known as the place of asthma for many people. The incidence of bronchial asthma is particularly high in this city. Instances have been cited where people who have never exhibited tendencies earlier, have, on coming to Bangalore, had attacks of asthma. Further, those who have had only mild attacks elsewhere experienced severe ones here. Also, it has been recorded that the percentage of asthmatics who reported relief when they left Bangalore is as high as 37.6%.

A systematic investigation of the pollen and fungal spore content of the atmosphere in Bangalore was undertaken for the first time from July 1976 to July 1978, by the Asthma Research Society. The study resulted in identification of 75 types of air-borne pollen and 120 types of spore. The pollen calendar for 1976-78 from 10 different sampling stations around Bangalore, showed that pollen grains were present all the year around in the city atmosphere. All the identified pollen types were attributed to local plants. According to the study, the pollen of parthenium was the highest in quantity (41%), followed by grass pollen (28.8%). The Cassia species (11.8%) take the next place.

The survey revealed that housewives (37.01%) and office workers (40.31%) are more susceptible to asthma attacks than industrial workers (7.11%) and retired persons (11.29%). This suggests that industrial pollutants can be almost ruled out as a causal factor for asthma and that attention should be focussed on pollen, spores and dust particles.

Although asthma is not a contagious disease, 32.27 per cent of hereditary cases came to light during the survey. But according to the Secretary of the Asthma Research Society, even though two factors (pollen and fungal spores) have been analysed for two years, final conclusions have not been drawn because clinico-immunological studies have not been done on asthmatics. Hence, further investigations and clinical studies are required to find the real cause for the high incidence of asthma among the people of Bangalore city.

Also, some industrial operations may add to the allergens. Industrial operations like cotton milling, fur processing, hair processing, feather processing, flour milling, jute processing, leather processing, wood working and tobacco processing, may be included under this category.

Although industrial operations contribute to the problem of air-borne allergens, they are primarily due to a natural origin. Some persons have been found to be allergic to sulphur dioxide. Finely divided cobalt and beryllium in the form of its compounds are also found to be allergic. Both cause lung disease. The commercial fur dye, paraphenylenediamine is an allergic agent capable of causing dermatitis in addition to bronchial asthma.

Factories processing the castor bean for the extraction of oil, discharge a powdery material into the air that is a strong allergen. This material has been the cause of bronchial asthma in people living near those factories.

5.6 Major Disasters

Several disastrous episodes have focussed attention upon air pollution as a health hazard. They have attracted attention because of illness and loss of life. The episodes have always occurred under extraordinary meteorological conditions and most of them were due to 'inversion phenomena'.

5.6.1 Meuse Valley (Belgium)

The first episode of 'modern' times was the disaster in Belgium's Meuse Valley. In December, 1930, an anticyclone (high pressure area) blanketed Belgium, resulting in fog and temperature inversion in Meuse Valley. In a 24-kilometre length, with hills 80–120 m high on either side, there are many industrial plants. Important among them are steel works, power plants, coke ovens, sulphuric acid plants, glass factories, zinc smelters and a fertiliser plant. Because of a thermal inversion, various pollutants got trapped in the valley. After three days of abnormal weather many people became ill, and about 60 died. The number of deaths was considered to be about 10.5 times the expected number for an equivalent period of time and season under ordinary circumstances. The symptoms were cough, throat irritation, shortness of breath (respiratory tract irritation), nausea, and vomiting. The illness affected people of all ages and both sexes. But older people with chronic heart or respiratory disease suffered the highest death rate. Even cattle were killed. After careful consideration of all the gases and aerosols that were discharged into the air from this industrial valley, it was believed that only an irritant air pollutant could have caused the illnesses and deaths, and it was probably a mixture of sulphur dioxide gas and sulphurtrioxide aerosol. Other air pollutants, viz., fluorides (from glass factories), zinc particulates, etc., were considered to have played only a minor role in the disaster.

5.6.2 Donora (USA)

Donora, Pennsylvania, lies about 45 km south of Pittsburgh in a horseshoe-shaped valley on the Monongahela River, with steeply rising hills on each side of the river. Anti-cyclonic weather conditions characterized by little or no air movement, temperature inversion and fog occurred for a period of four days in October, 1948. The industries present in 1948 were a steel mill, zinc plant, and sulphuric acid plant. Out of a population of 14,100 people, 43% became ill and 10% were severely affected. Twenty persons died. The fatal cases were confined to elderly persons, 13 of whom had histories of heart or respiratory disease. Symptoms included irritation of eyes, nose, and throat, cough, headache, vomiting, and respiratory irritation. According to an observer, during those four days, "everything was black with gas and soot; one could even taste it". It was concluded that the illness was probably due to the combined action of two or more of the air contaminants namely sulphur dioxide, together with oxidation products and particulate matter.

5.6.3 London

London, situated in Thames Valley, experienced its worst air pollution episode due to fog from December 5 to December 9, 1952. The 'killer-smog' began on December 4 as a high pressure area created a subsidence inversion over southern England. A white fog formed in the London area. Because of the extensive use of coal (sulphur content 1.5%) as fuel for space heating and electricity production, the particulate and sulphur dioxide levels in the atmosphere increased. The white fog became a black fog. Smoke concentrations during the fog were found to be five times greater than those found at other times. The average concentration of sulphur dioxide in the atmosphere during the episode was about six times the usual level.

During the period, the high pressure area become stationary. The build up of pollutants combined with fog, resulted in practically zero visibility. By December 6, pollutants were sufficiently concentrated to cause deaths. According to one observer, "one could not see one's hand in front of one's face—a white shirt collar became almost black within 20 minutes." The symptoms were cough, nasal discharge, sore throat, irritation of eyes and bronchi, and sudden attacks of vomiting.

Most of the people who died were old and those who had histories of chronic bronchitis, asthma, broncho-pneumonia and other lung or heart disease. Younger people exposed to the outside atmosphere were also affected. The fog lifted on December 9.

This London disaster of 1952 caused 4000 deaths. The excessive mortality and morbidity were caused by the irritation of the respiratory tract by contaminants in the fog and these contaminants were probably derived from coal and its products.

Similar air pollution episodes, but of a mild nature had taken place in London before the 1952 crisis, e.g., December 1873, January 1880, February 1882, December 1891, December 1892, November 1948, and in December 1950. In 1950, there were 1000 excess deaths.

Atmospheric conditions of a similar nature have occurred in London in the following years: January 1955, January 1956, December 1956, December 1957, January 1959, and December 1962.

The December 1962 air pollution episode was acute, with high levels of sulphur dioxide as in December 1952. However, the smoke levels were comparatively lower because of the clean up effected by the British Clean Air Act, 1956. As a result, deaths in 1962 were much lower than in 1952.

5.6.4 Los Angeles

Subsidence inversion, is responsible for the air pollution problem of Los Angeles. It results from a more or less permanent high pressure area over the Pacific Ocean which brings warm air over the land at low elevations. Los Angeles has mountains to the north and east and these act as a trap for obnoxious gases from thousands of automobiles as horizontal air movement is restricted by the mountains.

In Los Angeles, ozone is formed by the photo-chemical reactions of oxides of nitrogen and hydrocarbons (oxides of nitrogen and hydrocarbons are waste products from automobiles). Nitrogen dioxide absorbs energy from solar radiation and breaks into an oxygen atom and nitric oxide. The oxygen atom combines with oxygen to form ozone.

The temperature inversion increases the concentration of exhaust gases and photo-chemical smog is formed. The physical difficulties caused by smog are irritation of the eyes, nose, and throat. According to some investigators, eye irritation is due to peroxy acetyl nitrate (PAN), which is formed when oxides and hydrocarbons undergo photo-chemical reactions.

As far as Los Angeles is concerned, motor vehicles are responsible for 75 to 80% of the existing air pollution problems. One interesting fact reported about the Los Angeles air pollution problem is that school children become more unruly during days of high air pollution!

5.6.5 New York

New York has also had several episodes of air pollution which have been responsible for excess death. For example, in November 1962 and 1963. The 1966 Thanksgiving week-end episode which resulted in 168 deaths is possibly the best known. As usual, the death rate was higher among the elderly.

5.6.6 Poza Rica (Mexico)

Poza Rica is an oil refining and natural gas processing town in Mexico. On November 24, 1950, it was seriously affected by an accident spillage of hydrogen sulphide gas. The incident caused the illness of 320 persons and

22 deaths. The situation was complicated by a weather inversion associated with little lateral air movement.

Within 10 to 20 minutes after the spillage began, the gas reached the people resulting in severe illness. The source of gas was stopped 20 to 25 minutes after it first appeared. It was during this very short period that sickness began and deaths occurred. The effects of the gas were those characteristic of hydrogen sulphide gas inhalation. The symptoms were loss of sense of smell, cough, irritation, nausea, and severe headache. People of all ages and both sexes were affected.

5.6.7 Tokyo

Since 1967, increased level of oxidants in the atmosphere of Tokyo have been reported by the monitoring station of the Tokyo Metropolitan Public Health Research Laboratory. On June 18, 1970, a thick fog was present in the early morning. As time passed the fog dispersed but visibility was less than two kilometres. Between 11 a.m. and 1 p.m. complaints of eye irritation started, and in one school 45 children complained of smarting eyes, sore throat, and difficulty in breathing. Several had to be taken to hospitals where they were treated for 'smog poisoning'. The concentration of pollutants rose quite abruptly between 8 a.m. and noon. In the evening authorities concerned were sure that the symptoms and reduction in visibility were caused by a photochemical pollution. During subsequent days, many more people were affected. The number of persons reporting smarting eyes and sore throat exceeded 6000. Sulphur dioxide reached a maximum of 0.39 ppm. It was inferred from this that the photochemical oxidant reacted with sulphur dioxide to produce sulphuric acid mist which caused eye irritation.

The usefulness of studies of air pollution disasters is, that they indicate the levels of atmospheric pollution which cause immediate health effects even though the concentrations exist for only short periods of time, i.e., a few days. Hence, air pollution ambient standards should be placed well below these levels.

5.7 Relation of Air Pollution Effects to other Exposures

5.7.1 Cigarette Smoking

The dose of pollutants from cigarette smoking is high and intermittant, relative to that from community air pollution. For example, 400 ppm (400 mg/m^3) of carbon monoxide for five minutes with frequent repetition may be a representative cigarette smoking dose, while from 10–30 ppm (11–33 mg/m^3) of carbon monoxide for 4-8 hours may represent a frequent community air pollution exposure. Each of these alone may result in inactivation of up to 5% of the haemoglobin. Together, of course, the effect will be more. Similarly, the exposure to oxides of nitrogen present in cigarette smoke is about 100 ppm, whereas the average exposure to heavy air pollution over periods of about an hour will not exceed about 1.5 ppm (2.8 mg/m^3). But, as pointed out earlier, cigarette smoking produces intermittent exposures, whereas city air pollution is more constant. The other substances in cigarettes to which the body is exposed include carcinogens, aldehydes, hydrogen cyanide, lead, etc.

The effects of cigarette smoking are more important than effects of air pollution, as far as causing lung cancer or chronic pulmonary disease in the whole population is concerned. But, when both factors are present, they naturally have more than additive effect. This means that cigarette smokers are at unusual risk, if they live in areas with substantial air pollution and that the effects of air pollution on chronic pulmonary diseases are more likely to occur in cigarette smokers.

5.7.2 Domestic Pollution

Cooking and home heating appliances can generate a group of air pollutants (carbon monoxide, sulphur oxides, oxides of nitrogen, soot, and oily aerosols) whose health effects have commonly been overlooked. During periods of still weather and low winds, the dispersal of such domestic pollutants will be impaired. Generally, oxidants are about half as concentrated within buildings in polluted areas as they are outside, but carbon monoxide and nitric oxide are likely to occur at similar concentrations indoors and outdoors. Hence, if smoking or cooking occurs indoors, exposures to carbon monoxide and nitrogen oxides may be quite high.

5.7.3 Occupation

Many people are exposed to common air pollutants in their occupations. For example smoke, dust, carbon monoxide, sulphur dioxide and lead. The traffic policeman, the automobile mechanic, and the truck driver in a big city may all have substantial exposures to carbon monoxide and lead in association with their occupations. Therefore, such individuals have an unusually high risk from exposure to community air pollution.

QUESTIONS

1. Explain with a neat sketch the mechanism of action of air pollutants on human beings.
2. What are the harmful effects of polluted air on human beings?
3. Explain briefly the various methods adopted for determining the health effects of air pollutants on people.
4. What are the harmful effects of the following on human beings?

   (a) Sulphur dioxide
   (b) Carbon monoxide
   (c) Hydrocarbons
   (d) Insecticides

5. Briefly explain the causes and effects of 'episodes' of notable air pollution disasters in the world.
6. What are air-borne allergens? With examples, explain their origin.
7. Explain with an illustration the probable relation between 'asthma' and 'air pollution'.
8. Write short notes on
   (a) Radiation hazards
   (b) Air pollution episodes
9. Explain the effects of air pollutants with respect to:
   (a) Cigarette smoking
   (b) Domestic pollution

6

Effects of Air Pollution on Animals

6.1 Introduction

Interest in the effects of air pollution on animals has generally developed as a corollary to the concern about its influence on human health. Most of the information concerning the natural exposure of animals to air pollution is contained in the reports of some major air pollution disasters e.g., Donora, London and Poza Rica. Recently, considerable information has been reported from medical research laboratories which describes the results of experimental exposure of small animals to various air pollutants. Animals used for laboratory research work were mice, rabbits, rats, guinea pigs and monkeys.

6.2 Air Pollution Effects on Farm Animals

The process by which farm animals get poisoned is entirely different from that by which human beings exposed to polluted atmospheres are poisoned. In case of farm animals it is a two-step process:

1. Accumulation of the air-borne contaminant in the vegetation and forage
2. Subsequent poisoning of the animals when they eat the contaminated vegetation

In case of human beings working in polluted atmospheres in factories, the concern is for the harmful substances that are directly inhaled, whereas in the case of the farm animals, the danger obviously is not in inhaling the polluted air, but rather the ingestion of forage which has been contaminated with pollutants like fluorine from the air.

The three pollutants responsible for most livestock damage are fluorine, arsenic and lead. These pollutants originate from industrial sources or from dusting and spraying.

6.2.1 Fluorine

Of all farm animals, cattle and sheep are the most susceptible to fluorine toxicosis. Horses appear to be quite resistant to fluorine poisoning. Poultry are probably the most resistant to fluorine, of all farm animals.

Symptoms of Acute Fluorine Poisoning

Lack of appetite, rapid loss in weight, decline in health and vigour, lameness, periodic diarrhoea, muscular weakness and death, characterize the acute form of fluorine poisoning. It may also result in considerable increase of bone fluorine. But acute poisoning due to fluorides is unlikely in a majority of the cases.

Symptoms of Chronic Fluorine Poisoning

Fluorine is a cumulative poison under conditions of continuous exposure to sub-acute doses. Fluorine is also a protoplasmic poison. It has a marked affinity for calcium and interferes with normal calcification. Animals have been reported to be more resistant than humans to dental mottling. Cattle and sheep are the most frequently affected animals. Teeth in the process of formation are easily affected. Hence tooth symptoms are a sensitive and unique criterion of chronic fluorosis. Excessive wearing of incisor and molar teeth may occur at high levels of fluorine intake.

Bone lesions may develop at any age. A bony overgrowth may be observed on the leg bones, jaw bones and ribs. Their appearance indicates that there has been a high fluoride intake over a long period. Lameness may occur as a result of this overgrowth on the leg bones. This may be followed by stiffness of joints. Effects of fluorides on skeletal structures are usually of a permanent nature.

Symptoms of advanced fluorosis include lack of appetite, general ill-health due to malnutrition, lowered fertility, reduced milk production and growth retardation. The symptoms developed in other species such as rabbits, horses and poultry are similar to those in cattle and sheep, i.e., mottling, staining and wearing of the teeth, bony overgrowths on the skeleton, stiffness and lethargy and general ill-health from malnutrition and starvation.

While diagnosis of chronic fluorine poisoning is done, one has to be careful. This is because drinking water supplies may contain unduly high levels of fluorides. Also, mineral supplements may contribute appreciable amounts of fluorine to the diet. Hence, a check of all possible sources must be made when investigating fluorosis.

6.2.2 Arsenic

Arsenic occurs as an impurity in many ores and in coal. It has been reported to cause poisoning of livestock near various industrial processes and smelters. Just like most industrial air contaminants, arsenic may spread over a considerable area from a stack source. Arsenic in dusts or sprays on plants can lead to poisoning of cattle.

Symptoms of Acute Arsenic Poisoning

In acute cases the symptoms are severe salivation, thirst, vomiting, uneasiness, feeble and irregular pulse and respiration. The animal stamps, lies down and gets up. There is diarrhoea, and the faeces have a garlic odour and are sometimes bloody. The ears become cold, the body trembles and develops abnormal temperature and convulsion. Death may occur in few hours or days.

Symptoms of Chronic Arsenic Poisoning

Arsenic appears to have a depressing effect upon the central nervous system. The animal becomes dull, and exhibits a lack of appetite, with a resulting weight loss. Also, there may be chronic cough and diarrhoea may occur continuously. There may be thickening of the skin, anaemia and abortion or sterlity. Chronic poisoning can result in eventual paralysis and death.

Tolerance Limits

Arsenic as well as its soluble compounds are extremely poisonous. It has been reported that sheep have been poisoned by as little as 0.25–0.50 g of arsenic daily, whereas cattle and horses may tolerate 1.3–1.9 g daily. It has also been reported from laboratory experiments, that 10 mg of arsenic per kilogram of body weight per day has produced chronic arsenic poisoning in guinea pigs. A detailed investigation of the effects of arsenic poisoning on farm animals is necessary to establish standard tolerance limits.

6.2.3 Lead

Lead contamination of the atmosphere takes place on account of various industrial sources such as smelters, coke ovens and other coal combustion processes. Lead is also used in dust and sprays containing lead arsenate.

Symptoms of Acute Lead Poisoning

In case of acute lead poisoning the onset is sudden and the course relatively short. Prostration, staggering and inability to rise are prominent symptoms. The pulse is always fast but weak. Some animals may fall suddenly, stiffen the legs and have convulsions. There is complete loss of appetite, paralysis of the digestive tract and diarrhoea. Other nervous symptoms in cattle are grinding of the teeth and rapid chewing of the cud.

In case of horses, it may result in complete loss of appetite, nervous depression, lethargy and death.

Symptoms of Chronic Lead Poisoning

Chronic lead poisoning has been observed frequently in horses that have been grazing on forage near smelters, lead mines, and in orchards that have been sprayed. Paralysis of the muscles of the larynx and difficulty in breathing are the main symptoms. Convulsions may occur from the paralysis of the throat, and the difficulty in breathing may be unusually severe and persistent, during and after exercise.

Tolerance Limits

Lead is a cumulative poison. Therefore, the continuous ingestion of even very small daily doses will be finally as effective as one toxic dose. Hence in case of slight contamination death can occur after many months and in case of severe contamination death can occur within 24 hours.

It has also been reported that inadequate calcium in diet increases the retention of lead in the animal body. Low-calcium diets may result in lead storage of as much as five times as compared to that found in animals that receive sufficient amounts of calcium. However, extra quantities of calcium above the amount considered sufficient, will not offer additional protection against poisoning.

6.3 Pets and Pollution

It is reported that pollution is now beginning to affect pets and domestic animals in Tokyo and other smog affected cities. Canine patients are found to be suffering from bronchitis, asthma and lack of appetite. Many dogs also suffer from coughs, nose and throat diseases due to increasing air pollution. It is interesting to note that the average life of pets in Japan is seven to eight years, whereas in developing countries, they enjoy a life span of 12 to 13 years. It is also reported that some chemical laboratories in universities in Japan and some industrial firms keep canaries to warn the staff of poisonous gas in the air. Once the canaries cease to sing, the people know that something is wrong!

6.4 Exposure to Ionising Radiation

Radioactive fallout from nuclear bomb testing in the atmosphere results in ionising radiation, which has biological effects. The effects of radiation on animals are qualitatively similar to those in human beings. The effects may be either acute radiation effect or delayed long-term effect.

Symptoms of acute radiation injury develop within a period of hours to weeks, following exposure. Only fallout occurring close to the nuclear bombs test site can produce acute radiation effects, because of the high degree of radiation received by the animal body.

The long term effects are:

1. Cancer (including leukaemia)
2. Shortening of life span
3. Genetic or mutation effect

For manifestation of these effects, it generally takes several years, even generations in case of genetic changes.

QUESTIONS

1. What are the effects of air pollutants on animals?
2. Explain the mechanism by which farm animals get poisoned due to air-borne contaminants.
3. What are the symptoms of acute and chronic poisoning by the following pollutants on farm animals?
   (a) Fluorine
   (b) Arsenic
   (b) Lead
4. Write a note on the effects of exposure of animals to ionising radiation.

7

Effects of Air Pollution on Plants

7.1 Introduction

Air pollution has long been known to have an adverse effect on plants. At first, it was only sulphur dioxide that was considered a dangerous pollutant. Now, with the advent of various pesticides and new industrial processes, the range of harmful pollutants has multiplied tremendously. Sometimes, vegetation over 150 km away from the source of the pollutant has been found to be affected.

Industrial pollution, particularly from smelters, has caused complete destruction of vegetation in some cases e.g., at Ducktown, Tennessee. The international dispute between Canada and the United States over the damage caused by the Trail, British Columbia Copper Smelter is well known. Los Angeles smog has caused wide spread damage to some crops and forests in Southern California. In fact, many books and reports have been published in USA on air pollution injury to vegetation. In some cases, economics of the damage caused to crops and plants has been estimated. Compensation amounting to thousands of dollars has been paid for damage done to the crops. In our own country, there are many reports of the effect of pollutants like cement dust on plants.

7.2 Structure of a Normal Leaf

A knowledge of the structure of a leaf will help us understand why damage occurs to plants on account of air pollutants. A generalized diagram of the structure of a normal leaf is given in Fig. 7.1. If we examine a leaf, one will normally see a network of denser structures, the veins, all interconnected to the base or stem of the leaf. The leaf veins act as the transport system for water and food, just like blood vessels in animals. The leaf tissue is in layers with a skin or epidermis layers on top and bottom and the photosynthetic cells in between. The stomata are the entrances in the leaf bottom (and in some leaves in the top) through which CO_2 enters to play its role in photosynthesis. These openings are protected by pairs of specialized guard cells which open and close to allow gases to enter or leave the leaf. Such gases of course, include pollutants like sulphur dioxide.

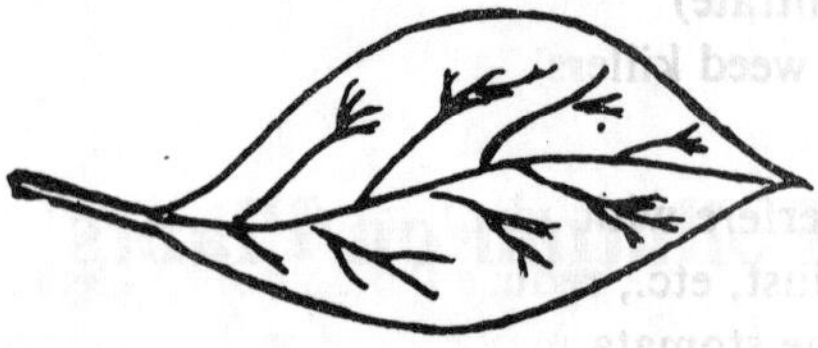

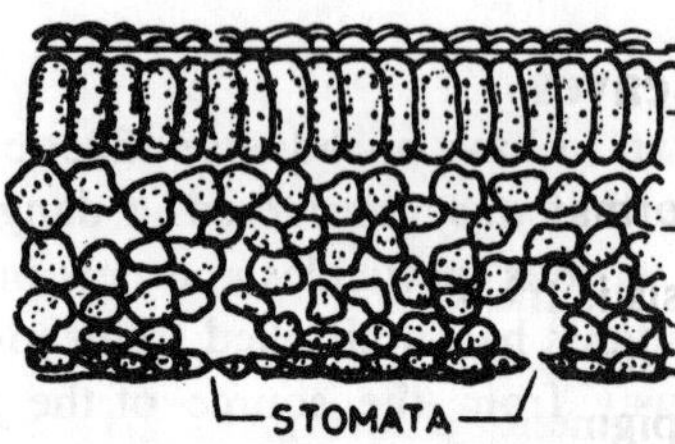

Fig. 7.1 *Cross-section of a leaf*

7.2.1 Effect of Environment on Plants

The primary factor which controls gas absorption by the leaves is the degree of opening of the stomata. When the stomata are wide open, absorption is maximum and vice versa. Consequently, the same conditions that enhance the absorption of the gas (CO_2 for photosynthesis), predispose the plant to injury (by absorbing a pollutant gas like SO_2). The conditions that cause the stomata to open are high light intensity (especially in the morning hours), high relative humidity, adequate moisture supply to the roots of the plant and moderate temperatures.

Most plants close their stomata at night and are therefore much more resistant at night than in the day time. But some plants like the potato, which do not close their stomata at night are as sensitive in the dark as in the light.

7.3 Air Pollutants Affecting Plants

1. Sulphur dioxide
2. Fluoride compounds (like hydrogen fluoride)
3. Ozone
4. Chlorine
5. Hydrogen chloride
6. Nitrogen oxides (NO, NO_2, etc.)
7. Ammonia
8. Hydrogen sulphide
9. Hydrogen cyanide
10. Mercury
11. Ethylene

12. PAN (peroxy acetyl nitrate)
13. Herbicides (sprays of weed killers)
14. Smog

The above pollutants interfere with plant growth and the phenomenon of photosynthesis. Smog, dust, etc., reduce the amount of light reaching the leaf and also by clogging the stomata may reduce carbon dioxide intake to some extent and thus interfere with photosynthesis.

7.4 Forms of Damage to Leaves

Damage to leaves takes several forms

1. Necrosis : Necrosis is the killing or collapse of tissue.
2. Chlorosis : Chlorosis is the loss or reduction of the green plant pigment, chlorophyll. The loss of chlorophyll usually results in a pale green or yellow pattern. Chlorosis generally indicates a deficiency of some nutrient required by the plant. In many respects, it is analogous to anaemia in animals.
3. Abscission : Leaf abscission is dropping of leaves.
4. Epinasty : Leaf epinasty is a downward curvature of the leaf due to higher rate of growth on the upper surface.

Notes: 1. Tissue severely injured by air pollutants often has a characteristic colour. Bleaching is associated with sulphur dioxide, yellowing with ammonia, browning with fluoride. A silvering or bronzing of the under surface of some leaves is associated with PAN injury.

2. Phytotoxicant is the name given to plant damaging substances.

7.4.1 Kinds of Injury to Plants

Acute Injury

It results from short-time exposure to relatively high concentrations, such as might occur under fumigation conditions. The effects are noted within a few hours to a few days and may result in visible markings on the leaves due to a collapse and death of cells. This leads to necrotic patterns, i.e., areas of dead tissue.

Chronic Injury

It results from long-term low level exposure and usually causes chlorosis or leaf abscission.

Growth or Yield Retardation

Here the injury is in the form of an effect on growth without visible markings (invisible injury). Usually a suppression of growth or yield occurs.

While deciding about the type of injury the plant has suffered, one must carefully distinguish between adverse effects caused by air pollution and adverse effects resulting from other factors. Plant diseases may show symptoms which are very similar to those caused by air pollution. High temperatures, poor plant care, shortage of nutrients and water may also cause an appearance which is similar to that of a plant damaged by air pollution. It may also be due to insect damage. Therefore, while diagnosing the effects of air pollution one must consider factors such as plant disease, nutrition history, weather damage, insect damage, and the nature of pollutants in the area.

7.5 Effects of Air Pollutants on Plants

The effects of various pollutants like sulphur dioxide, ozone, fluorides, etc., on plants is shown in Table 7.1.

Table 7.1 *Effects of Pollutants on Plants*

Pollutant	Dose	Effect
1. Sulphur dioxide	Mild	Interveinal chlorotic bleaching of leaves
	Severe	Necrosis in interveinal areas and skeletonized leaves
2. Ozone	Mild	Flecks on upper surfaces, premature aging and suppressed growth
	Severe	Collapse of leaf, necrosis and bleaching
3. Fluorides	Cumulative effect	Necrosis at leaf tip
4. Nitrogen dioxide	Mild	Suppressed growth, leaf bleaching
5. Ethylene	Mild	Epinasty, leaf abscission
6. PAN	Mild	Bronzing of lower leaf surface (upper surface normal), suppressed growth. Young leaves more susceptible

7.5.1 Effect of Sulphur Dioxide

Sulphur dioxide produces two types of injury on the leaves of plants—acute and chronic, depending on the concentration and period of exposure. The acute injury is characterized by the killing of marginal or interveinal areas of the leaf. Immediately after fumigation these areas will get a dull, water-soaked appearance. Subsequently they dry up and usually bleach to an ivory colour, though some species finally assume a brown or reddish-brown colour. Chronic injury is caused by the slow, long-continued absorption of sub-lethal amounts of gas or by absorption of an amount of gas somewhat less than that necessary to cause acute injury.

Sulphur dioxide is phytotoxic in concentrations above 0.1–0.2 ppm. Below about 0.4 ppm, it tends to be oxidised in the cells as rapidly as it is absorbed, but interferences with functions such as photosynthesis is slight. Chronic injury, if any, is generally exhibited with these small concentrations.

Above about 0.4 ppm, acute injury occurs more frequently. Sometimes temporary interference with photosynthesis or 'invisible injury' can occur. However, the mechanism by which sulphur dioxide affects the plant has not yet been well understood, though various explanations have been put forth by many investigators.

7.5.2 Effect of Hydrogen Fluoride

Hydrogen fluoride behaves somewhat similar to sulphur dioxide except that with a few species of plants it is effective in causing lesions and interfering with photosynthesis in concentrations two or three orders less than in the case of sulphur dioxide. With most species it is up to 10 times as effective as sulphur dioxide. However, recovery of plants from the fluoride effect is much slower than from sulphur dioxide. This difference in rate of recovery is probably explained by the fact that sulphite in the leaves is rapidly oxidised to the relatively non-toxic sulphate, whereas fluorides can be removed only by the slower process of volatization or by some obscure chemical reaction. Forage may be rendered unsafe for animal feeding if more than 50 ppm of fluorine is absorbed.

7.5.3 Effect of Ozone

Ozone is phytotoxic in exposures of a few hours at about 0.2 ppm. Injury due to ozone is quite different from typical smog injury. The lesions are generally confined to the upper surface. The effects of ozone on plants in mild and severe doses is indicated in Table 7.1.

7.5.4 Effects of various Fumigants on Plants

Chlorine

Apart from the fluorides, halogens and their compounds are relatively unimportant air pollutants which might cause injury to vegetation. However, chlorine is more toxic to vegetation than sulphur dioxide by a factor of two or three. Lesions are generally marginal and interveinal. Damage to vegetation caused by chlorine is rare, and most of the reported cases are due to accidents or excessive use of gas for sterilising.

Hydrogen Chloride

Hydrogen chloride is considerably less toxic to vegetation than sulphur dioxide. Hydrogen chloride causes first a chlorotic margin in the leaf, which may become necrotic. At higher concentrations, lesions are produced. The threshold concentration is about 10 ppm for a few hours exposure.

Nitric Oxides

Injury to plants due to nitric acid vapours has been observed near factories handling large amounts of this acid. The effects include brown margins and brownish-black spots on the leaves. Concentrations of about 25 ppm will cause these effects. Nitrogen oxides are important in photochemical reactions which cause smog.

Ammonia

Ammonia is a gas of intermediate toxicity: It is interesting to note that ammonia and hydrogen chloride almost have about the same toxicity.

Hydrogen Cyanide

Hydrogen cyanide is used to fumigate green houses and trees in orchards for pest control. Sometimes this fumigation injures the vegetation.

Ethylene

Ethylene causes injury to leaves of sensitive plants. The effects are epinasty, curling, chlorosis, leaf abscission, and growth retardation.

Herbicides

Injury to sensitive vegetation may be caused in the field through careless or uncontrolled use of herbicides and pesticides. It has been reported that weedy grain fields have sometimes been sprayed when there was wind, and the spray has been carried for several kilometres in sufficient concentration to injure crops like cotton and tomatoes. Many other plants which have exhibited symptoms of injury under similar conditions are roses, cabbage, pepper, grapes and tobacco.

7.5.5 Smog

London type smog is thought to be essentially a sulphur dioxide problem. But the gaseous constituents as well as the aerosols need further evaluation.

The Los Angeles type smog is now fairly well understood, but the actual compounds that cause these effects are still unknown. Two types of smog injury to vegetation have been recognized in Los Angeles, one due to gases (smog gas) and the other due to deposition on the leaves of fog droplets (smog fog). The smog causes characteristic leaf lesions which are quite different from those produced by other pollutants, including ozone, which may be a constituent of the smog. It also causes some 'invisible' injury.

7.6 Sensitivity of Plants to Air Pollutants

The sensitivity of plants to air pollutants is conditioned by many factors.

7.6.1 Genetic Factors

Plant response to pollutants varies between species of a given genus and between varieties within a given species. Such variation is simply a function of genetic variability as it affects the plants' morphological, physiological, and biochemical characteristics. Plants do not necessarily show similar susceptibility to different pollutants. For example, some plants are sensitive to fluoride but resistant to sulphur dioxide.

7.6.2 Climatic Factors

The important climatic factors affecting the response of vegetation to air pollutants are

1. Duration of light
2. Light quality (wave length)
3. Light intensity
4. Temperature
5. Humidity

7.6.3 Miscellaneous Factors

Soil, water, and fertility affect the sensitivity of plants to air pollutants. However, these factors have to be studied in detail before drawing any conclusions.

7.7 Time-Concentration Equations

The effect of concentration versus time is an important variable in assessing damage. A high dose for a short time may cause an acute injury, whereas the same total dose over a longer time may cause no visible effects at all.

The concentration-time equation may be expressed as,

$$t\,(C - C_0) = K \tag{7.1}$$

where,

t = time in hours to produce a certain effect on a certain species
C = concentration of a specific gas in ppm
C_0 = threshold concentration of the gas (ppm) to cause injury
K = experimentally determined constant

The equation can be rewritten as

$$C = C_0 + \frac{K}{t} \tag{7.2}$$

Then plotting C versus $\frac{1}{t}$, we get a straight line whose intercept C_0 is the threshold for injury. With the exception of some results for sulphur dioxide and ozone, sufficient data is not available at present to determine these parameters, assuming that a linear model is accurate. The equation is based on the exposure of a plant for a single pollutant. However, in the field, exposures are usually not due to individual gases but to a mixture of pollutants whose relative concentrations change as functions of time. Thus, there are almost infinite number of possible combinations of pollutants which could be tested.

Thomas, Hendricks, and Hill (1935) generalized this equation for the alfalfa plant, for any degree of leaf destruction and any degree of susceptibility. With maximum susceptibility, three equations were given as follows:

$$t(C - 0.24) = 0.94 \text{ traces of leaf destruction} \quad (7.3)$$

$$t(C - 1.4) = 2.1 \quad 50\% \text{ leaf destruction} \quad (7.4)$$

$$t(C - 2.6) = 3.2 \quad 100\% \text{ leaf destruction} \quad (7.5)$$

7.8 Leaf Destruction in Relation to the Yield

Many industries that emit phytotoxic gases are located near valuable agricultural lands. If damage is done to nearby crops, the farmer is legally entitled to recover his losses (especially in countries where air pollution control legislation is in force). Therefore, an objective method is required for determining the extent of losses. Many studies have shown that for sulphur dioxide, the reduction in the yield of the crop is proportional to the percentage of leaf area destroyed. The equation is of the form

$$y = a - bx \quad (7.6)$$

where,

y = yield expressed as the percentage of full yield
x = percentage of leaf area destroyed
a = constant (about 100%)
b = Slope of yield-leaf destruction curve.

In general, the leaf destruction yield functions are straight lines, starting with 100% yield y, at 0% leaf destruction x, and ending with a definite decrease in yield at 100% leaf destruction. But from the practical point of view, the reduction in the yield is difficult to measure experimentally if leaf destruction is less than about 5%.

7 9 Air Pollution Injury Report

A typical air pollution injury report with respect to crops can be as follows:

Grower's name :
Location of plants/crops :
Suspected pollutant SO_2. O_3,
HF, others (specify) :

Suspected source/sources :
Crop and variety :
Area affected : Hectares
Number of plants :
Date/month of incidence :
Average % of plants affected : 0–10, 11–30, 31–60, 61–80, 81–90, 91–99, 100
Average % of each plant affected : 0–10, 11–30, 31–60, 61–80, 81–90, 91–99, 100
Loss in quality : Loss in Quantity
Estimated crop loss : % OR Rs.
Remarks :

Reporter's name : Date

QUESTIONS

1. List the air pollutants affecting plants.
2. Explain the following terms with respect to leaf damage due to air pollution:
   (a) Necrosis
   (b) Chlorosis
   (c) Abscission
   (d) Epinasty
3. Explain the various kinds of injury to plants due to air pollutants.
4. Explain the effects of various air pollutants on plants.
5. Explain the effects of the following air pollutants on plants:
   (a) Sulphur dioxide
   (b) Ozone
   (c) Peroxy acetyl nitrate (PAN)
   (d) Hydrogen fluoride
   (e) Ethylene
6. Explain the effects of photochemical smog on plants.
7. Explain the importance of genetic and climatic factors in relation to the effects of air pollution on plants.
8. Write a note on: Time-concentration equations.
9. Explain how leaf destruction in relation to the yield can be expressed mathematically.
10. Give the specimen of a typical air pollution injury report.

8

Economic Effects of Air Pollution

8.1 Introduction

Air pollution damage to property is a very important economic aspect of pollution. In the United States of America, Great Britain, France and many other European countries, this problem has been investigated in detail and successful attempts have been made to translate observable air pollution damage into terms of economic impact. Air pollution damage to property covers a wide range—corrosion of metals, soiling and eroding of building surfaces, fading of dyed materials, rubber cracking, spoiling or destruction of vegetation, effects on animals, as well as interference with production and services. Another important economic effect of air pollution is deterioration of works of art. In our country today there is an urgent need to investigate and study this problem in detail and express the damage to property in economic terms, as very little work has been done in this direction so far.

8.2 Mechanism of Deterioration in Polluted Atmospheres

Air pollutants cause damage to materials by five mechanisms:

1. Abrasion: Solid particles of sufficient size and travelling at high velocities can cause abrasive action. Also, large sharp edged particles imbedded in fabrics can accelerate wear.

2. Deposition and Removal: Solid and liquid particles deposited on a surface may not damage the material itself but it may spoil its appearance. However, the removal of these particles may cause some deterioration. Although a single washing or cleaning may not cause noticeable deterioration, frequent cleaning ultimately does.

3. Direct chemical attack: Some air pollutants react directly and irreversibly with materials to cause deterioration. For example, the bleaching of marble by sulphur dioxide, the tarnishing of silver by hydrogen sulphide and the etching of a metallic surface by an acid mist.

4. Indirect chemical attack: Certain materials absorb some pollutants and get damaged when the pollutants undergo chemical changes. For

example, sulphur dioxide absorbed by leather is converted to sulphuric acid, which deteriorates the leather.

5. Corrosion: The atmospheric deterioration of ferrous metals is by an electrochemical process, i.e, corrosion. This is due to the action of air pollutants facilitated by the presence of moisture.

8.3 Factors Influencing Atmospheric Deterioration

There are four prime factors which influence the rate of attack of damaging pollutants. They are moisture, temperature, sunlight, and air movement.

1. Moisture: The presence of moisture in the atmosphere greatly helps the process of corrosion. Without moisture in the atmosphere, there would be little atmospheric corrosion, if any, even in the most severely polluted environments. In case of sulphur dioxide and various combinations of particulates, investigations have revealed that the rate of corrosion of metals will increase as relative humidity in the air increases.

2. Temperature: Temperature affects the rate of the chemical reaction and consequently it affects the rate of deterioration.

3. Sunlight: In addition to producing damaging agents such as ozone through a series of complex photochemical reactions in the atmosphere, sunlight can cause direct deterioration of certain materials. But in cases of fading of certain dyes or rubber cracking, damage due to direct sunlight cannot always be distinguished from that caused by ozone.

4. Air movement: Wind direction is an important factor to be considered in places where deterioration is caused by pollutants released from nearby factories. For example, damage to agricultural crops in the vicinity of factories. Similarly, wind speed is also an important factor in determining the impact of air pollutants on the receiving surfaces. Pollutants present in wind blowing at high speeds will have more abrasive effects and they may also be carried over long distances. Where leather bound books are stored, air movement is a critical factor when the air is contaminated by sulphur dioxide.

8.4 Economic Losses (Material Damage)

Various economic losses occur because of material damage due to air pollutants. Air pollution damages materials chiefly by corrosion of metals. The prime air pollutant responsible for metallic corrosion is sulphur dioxide. In the presence of oxygen, sulphur dioxide is slowly converted to sulphur trioxide, which in turn, may react with the moisture in the air to form sulphuric acid. Deposition of this acid on the metal parts of building roofs, eaves, downspouts, and other metal equipment results in a considerable loss from atmospheric corrosion in most urban communities.

Losses through atmospheric corrosion in New York City were estimated in 1955 as 6 million dollars per year. According to a survey conducted by the Smoke Nuisance Department of West Bengal, the financial loss due to corrosion in West Bengal is Rs. 5 crore annually. This figure has been arrived at by a survey made in Greater Calcutta and the districts of West Bengal for over five years.

Carbon soot appears to accelerate the process of corrosion, probably because of the absorption or adsorption of the gaseous pollutants on the particulate matter. Other important variables are temperature and relative humidity.

Metals adversely affected include steel, aluminium, copper, zinc and iron. The economic loss due to corrosion caused by polluted air must also take into consideration deterioration of objects like steel rails, steel tramway tracks, overhead wires, metal on bridges, electrical contacts, etc. Such costs can only be estimated and no definite figures are available. Also, additional costs of painting, replacement, and substitution would have to be considered in estimating the total economic loss due to corrosion because of direct and indirect effects. Various economic losses due to air pollution are as follows.

8.4.1 Building Materials

Building materials are corroded and disfigured by air pollutants in a number of ways in addition to the usual weathering process. Smoke and aerosols adhere to stone, brick, and other building surfaces to produce unsightly coatings.

Sulphur dioxide and sulphur trioxide in the presence of moisture can react with limestone ($CaCO_3$) to form calcium sulphate ($CaSO_4$) and gypsum ($CaSO_4$, $2H_2O$), both of which are soluble in water. This type of damage is notable especially in works of art. For example, the world famous Taj Mahal of Agra, which is constructed out of marble.

Similarly carbon dioxide in the presence of moisture, produces carbonic acid. This acid converts limestone into a water-soluble bicarbonate, which is then leached away.

Staining of building stone by sulphur compounds and particulates requires additional cleaning expense. From the point of economic repercussions of pollution, losses due to maintenance costs inside and outside houses and other buildings as well as due to depreciation of objects should be considered and properly evaluated.

8.4.2 Paints

Paint is also affected by air pollutants. Areas of high pollution require more frequent painting. Paints contain both pigments and a vehicle. Pigments provide colour, hiding power, and durability. The vehicle holds the pigment to the surface. Together, they enhance the attractiveness of the surface and protect the underlying material from corrosion or weathering. Air pollutants may limit both of these functions by damaging the protective coating and by exposing the underlying surface to attack. Some of the common pollutants that can cause this damage are sulphur dioxide, ozone, hydrogen sulphide and aerosols. In fact, hydrogen sulphide is a common cause of the darkening of surfaces covered with paint containing white lead.

8.4.3 Textiles

Sulphur oxides cause deterioration of natural and some synthetic textile fibres. For example, cotton, which is a cellulose fibre is weakened by sulphur dioxide. The most common form of permanent damage to textiles has been the deterioration of nylon hose. This has been usually traced to sulphuric acid mist.

Curtains suffer badly in polluted areas because they hang at open windows and to some extent serve as filters for dust, soot, and acid droplets.

The fading of textile dyes by air pollutants, primarily oxides of nitrogen, and ozone, has been a problem in the textile industry. It is reported, that the combination of ozone and high humidity causes pronounced fading of certain dyes on cotton and rayon fabrics.

8.4.4 Rubber

Another form of property damage which is becoming increasingly common in the USA and other countries is rubber cracking. The main affected areas are the side walls of tyres and various forms of electrical insulation. This type of damage is caused by ozone. The mechanism appears to be an attack at the double bonds in the hydrocarbon polymer compounds used in the rubber. The problem has become so acute in areas like Los Angeles that tyre manufacturers add a special anti-ozone compound to all tyres sold in the area. More rapid failure of rubber insulation in atmospheres of high ozone content has also been noticed in power-transmission sub-stations and in telephone exchanges.

8.4.5 Leather

Sulphur dioxide causes leather to lose much of its strength and ultimately disintegrate. The storage of leather bound books in libraries can pose a serious problem. The bindings of books stored in the open in rooms with polluted air were found to deteriorate much more rapidly than those stored in confined spaces or inside glass cases. Initially, chemical decay of book bindings can be observed by the cracking that takes place on the top inside hinges of the book. The cracks generally spread apart and as further material is exposed, the leather loses its resiliency and disintegrates to a reddish brown powder. After sometime, the entire book may get detached.

8.4.6 Paper

Paper made prior to about 1750 is not seriously damaged by sulphur dioxide. This is because, it is about this date onwards in history that chemical methods were introduced for manufacturing paper. Apparently, the small amounts of metallic impurities in 'modern' paper accelerate the conversion of absorbed sulphur dioxide to sulphuric acid, in the presence of moisture. The sulphuric acid content of some papers has been found to be as high as one per cent, which makes the paper extremely brittle. It has been found that exposure of books and writing paper to sulphur dioxide in concentrations 2–9 ppm for 10 days made them brittle and decreased their folding resistance.

8.4.7 Glass and Ceramics

Although glasses and enamels are especially resistant to the chemical action of air pollutants, it has been observed that during a long exposure for three years, porcelain enamels showed a change in their surface appearance. Moisture and atmospheric pollution by acidic substances seems to have played a major role in surface degradation.

Fluorides, especially hydrogen fluoride, are capable of attacking a wide range of ceramic materials and glass through their ability to react with silicon compounds. In the past, window glasses in areas near synthetic fertiliser and enamel plants have been rendered opaque through the action of fluorides. However, the concentrations required to produce this type of damage are far in excess of those necessary to kill sensitive vegetation. Restrictions imposed on emission limits of fluorides by legislation in many countries has actually eliminated this type of effect except perhaps in occasional localized situations. Hence, fluoride damage to materials appears to be an insignificant economic effect.

8.4.8 Electronics Industry

Particulate matter and gaseous pollutants have been constant sources of trouble to the electronics industry. Many kinds of electronic components and equipment have been damaged. Lower-power (millivolts, microamperes) electrical contacts, which are used in numerous electrical devices, are particularly sensitive components. Air pollutants cause thin insulating film to develop on contacts resulting in open circuits and malfunctioning of the equipment. Sulphur dioxide and hydrogen sulphide tarnish copper and silver contacts by producing sulphide films. These pollutants even cause gold–silver bonded contacts to fail. Atmosphere particles can settle on contacts and prevent intimate surface contact when closed. If the particles contain corrosive components, direct chemical action can occur.

Contact problems have made engineers make a special effort to specify appropriate contact materials for various applications. For example, one can think of a computer that controls the operations in a chemical plant. The computer can cease to function because of an inexpensive tarnished electrical contact. The resulting time loss could run into thousands of rupees. Therefore, because of the critical nature of contact materials, probably the only solution is to specify precious metals such as gold, platinum and pallac.um which of course, is an expensive alternative.

Similarly, power companies have a unique problem associated with air pollution. Air-borne particulate matter is deposited on high-voltage transmission line insulators. During conditions of high humidity, fog, or rain, the deposited material serves as a conductor and results in insulator flash over. This is a serious problem in areas downwind fro m industries emitting large amounts of particulate matter.

8.4.9 Laundering and Dry Cleaning

Temporary property damage results from the soiling of surfaces by pollutants. Here the economic impact is the cost of cleaning. Losses most commonly encountered are the additional cost of laundering and dry cleaning clothing and other fabrics. Most of these losses are caused by smoke, soot, and dust fall.

8.4.10 Vegetation

Damage to vegetation as a result of air pollution is a very important aspect of economic loss due to air pollution. This has already been explained in Chapter 7 under effects of air pollution on plants. According to a survey made by the Smoke Nuisance Department of West Bengal, the loss to agriculture because of air pollution in West Bengal, is about Rs. 11 crore annually. The survey was made in Greater Calcutta and the districts for over five years.

8.4.11 Intangible Loss

Another very important economic repercussion of air pollution is the intangible loss due to great damage to art objects throughout the world including our country. For example, effects on the Taj Mahal (Agra), Sri Channakeshava Temple (Belur), and the Parthenon (Athens). This has been explained in detail in Section 8.6.

8.5 Other Economic Losses

There are many other economic losses due to air pollution, which should also be considered. They are given below.

1. Loss due to effects of pollution on human health
2. Loss due to effects of pollution on farm animals
3. Increased transportation costs in periods of smog. Wastage of electricity due to premature twilight caused by smoke
4. Loss due to reduction in tourist traffic, because of the air pollutants effects on art treasures and tourist centres

5. Expenditure incurred by the adoption of technical measures for the reduction of smoke or emissions from factories
6. Increase in cost of electrical power due to dust removal and other treatment of smoke from power or nuclear stations
7. Losses due to unburnt residue in chimneys, motors and various engines which contribute to general pollution
8. Expenditure in connection with the administrative organisation of pollution control
9. Costs of medical, agricultural, chemical and physio-chemical research aimed at working out techniques for the measurement and control of air pollution.

Unfortunately, it is not possible immediately to translate all air pollution effects into economic terms. Odours are an example. Odours constitute subjective discomfort for persons, but it is difficult to evaluate such discomfort statistically. Similarly, harmful smoke affecting general health may be represented as a 'loss' on the assumption that it increases costs for care and treatment. But it is not easy to express the physical and physiological effects in terms of money.

Table 8.1 presents a generalized picture of air pollution damage to various materials.

Table 8.1 *Air Pollution Damage to Various Materials*

Materials	Principal air pollutants	Effects
Metals	SO_2, acid gases	Corrosion, spoilage of surface, loss of metal, tarnishing
Building materials	SO_2, acid gases, particulates	Discolouration, leaching
Paint	SO_2, H_2S, particulates	Discolouration
Textiles and Textile dyes	SO_2, acid gases and NO_2, Ozone	Deterioration, reduced tensile strength, and fading
Rubber	Oxidants, ozone	Cracking, weakening
Leather	SO_2, acid gases	Disintegration, powdered surface
Paper	SO_2, acid gases	Embrittlement
Ceramics	Acid gases	Change in surface appearance

8.6 Effects on Art Treasures

Air pollutants have caused great damage to art objects throughout the industrialized world. Some excellent monuments of antiquity have been silently eaten away by pollutants. This can be clearly seen in many works of art like sculptures and carvings on historic buildings and cathedrals. A few examples of such works of art which have been affected by air pollution both in India and abroad have been described below.

8.6.1 Effect on Art Treasures in India

(a) *Taj Mahal* The Taj Mahal a miracle in marble and a historical monument is facing grave danger from pollution created by existing foundries, power houses, railway yards and other industrial units. According to some environmentalists, the problem now seems to have been aggravated because of the commissioning of the Mathura Oil Refinery, about 40 km from the priceless monument. Whether the monument is going to be affected by the pollutants emitted from the refinery or not has created a controversy which is now being debated. Many reports on this have appeared in noted newspapers and journals. However, it is to be hoped that certain anti-pollution safeguards built into the refinery will minimise the effect of pollutants on the monument.

According to some reports, pollutants from existing foundries, power houses and railway yards are already affecting the Taj Mahal. Black and brown patches are disfiguring the shiny white surface of the Taj. Reports in newspapers also speak of the formation of gypsum, scaling and flaking effects.

Sulphur dioxide from the various sources mentioned, mixes with the atmospheric moisture and gets converted into sulphuric acid which settles down on the exterior of the Taj Mahal. The acid reacts with the marble (calcium carbonate) and corrosion and discolouration of the monument is the ultimate result.

It looks as though more money will have to be spent on protecting the Taj Mahal from the effects of pollution than it took Shah Jehan to construct it. Perhaps the immediate solution to save the Taj Mahal lies in shifting the foundries and power houses from Agra, and in effectively supervising anti-pollution measures adopted by the refinery. Further, the coal locomotives being used by the railways in that region can be replaced by diesel engines. In addition to these steps, a dense green belt of suitable tree species should be created around the Taj Mahal to absorb atmospheric pollutants as trees can act as effective air scrubbers.

(b) *Belur Temple* The renowned temple of Sri Channakeshava at Belur (Hassan district, Karnataka state), is threatened with a similar hazard. A plywood factory located close to the temple emits soot-laden fumes, which get deposited on the sculptures in the temple and discolour the surface, inside and outside This pollution effect is under investigation.

8.6.2 Effect on Art Treasures in Other Countries

The Parthenon in Athens, has been affected in recent years because of air pollution. Cleopatra's needle, the large stone obelisk moved from Alexandria, Egypt, to London, has suffered more deterioration in the damp, smoky, acid atmosphere of London in the last 90 years than in the earlier 3000 or more years of its history. Another ancient Egyptian obelisk in New York City, shows similar decay.

Similar cases have been reported from other parts of the world. The

Colosseum in Rome, and the San Marco Basilica in Venice, Italy, show signs of decay, possibly due to air pollution. In France, it is reported that environmentalists have removed statues from the exterior of cathedrals and replaced them with copies. Similarly, ancient temples and buildings in industrial areas of Japan are facing the hazards of atmospheric pollution. Paintings, books, fabrics, antique costumes and other art objects have suffered damage.

Of course, it is impossible to make adequate substitution or compensation for the losses suffered by works of art. They are irreversible losses. It is impossible to estimate the losses in economic terms for these intangible assets. In some cases it is possible to preserve them, by applying special preservatives. But, they are not always successful. Sensitive art objects displayed inside buildings can be placed in hermetically sealed containers. Air conditioning can also be used as a protective measure.

8.7 Cost Estimates

As already pointed out, it is difficult to estimate the costs of air pollution with regard to its damaging effects. Certain costs such as loss to agriculture have been estimated, but many of the economic effects of pollution are indirect or hidden and thus difficult to determine. For example, additional cleaning costs on account of soiling of clothes, medical expenses of the affected people, and damage to works of art. In spite of these difficulties some estimates have been made, especially in countries like the U.S.A., Great Britain, and France. It is estimated that damage amounting to $ 3 million has been done to crops in the Los Angeles region every year since 1953. Unfortunately in our country, so far this type of estimate has not been made for various cities and states. It has to be done immediately, so that people can be alerted about the invisible economic impact of air pollution. One available example of this type of statistics is given in Section 8.7.2.

8.7.1 Evaluation of Damage—Basis for Calculation

The damage to materials due to air pollution can be calculated according to:

(1) Type of object damaged (plants, metal-work, etc.)

(2) Extent or degree of damage (partial or total)

The type of pollutant causing damage should also be considered. An example of evaluation of damage due to air pollution is as follows:

8.7.2 Financial Loss Due to Air Pollution in West Bengal

Air pollution is causing a financial loss of Rs. 38 crore to West Bengal annually, according to a survey conducted by the Smoke Nuisance Department of the state government. The survey was made in Greater Calcutta and the districts for over five years.

A major part of the damage is accounted for by agriculture in terms of reduced yield per acre. According to the study, the loss to agriculture is Rs. 11 crore annually. Corrosion of metals costs Rs. 5 crore, medical expenses for the affected people account for Rs. 6.75 crore, damage to and repair of buildings Rs. 2 crore, damage to textiles Rs. 5 crore, and laundry charges Rs. 3 crore.

Agriculture was affected because rain water in many districts turned acidic following the high concentration of sulphur and other compounds. The survey also revealed that steel tramway tracks lasted only three years in the city before they got corroded, because of the acidity in the soil after the rain. The level of pollution reduces after the monsoon with the flushing of impurities. But in summer and winter, the problem became acute again.

The study explains how the increased laundry charges, because of pollution, were worked out. Account of sales of detergents and soaps were taken from shops spread all over the state. The population increase was also taken into account. If the increase in sales was far more than that warranted by the population rise, then clearly other reasons like pollution were at work, it was concluded.

Similarly, it was found that in hospitals, pollution victims accounted for 10% the total number of those admitted for treatment. The diseases were mainly cardiovascular problems, asthma, straining of eyes and reduced vision, lung afflictions, headache, and respiratory problems.

In Hooghly district, the survey also revealed that the atmospheric pollution had not only reduced the agricultural yield per acre, but also affected the quality of the sub-soil ground water sources. The productivity of livestock and cattle also suffered. There were special diseases from which the people of the district seemed to suffer, more than the others.

QUESTIONS

1. Explain how air pollutants cause damage to materials.
2. Explain the factors influencing atmospheric deterioration.
3. Discuss the effects of air pollution on property.
4. Explain with examples how air pollution affects building materials.
5. Explain how air pollution affects the following materials
   (a) Paints
   (b) Rubber
   (c) Paper
   (d) Textiles
   (e) Leather
   (f) Glass and ceramics
7. Explain how air pollution affects the electronics industry.
8. "Air pollution damages materials chiefly by corrosion of metals." Substantiate the statement.
9. With illustrations, explain the effects of air pollution on art treasures in India and other countries.

10. **List the various economic losses due to air pollution.**
11. **Explain how the various economic losses due to air pollution are estimated.**
12. **"An important economic repercussion of air pollution is the intangible loss due to damage to art objects throughout the world." Substantiate the statement with examples.**

9

Sampling Procedures

9.1 Introduction

This chapter discusses difficulties encountered in collecting representative samples and the principles governing the avoidance of sources of error. The methods used for measuring gaseous emissions from a stack or a vent depend on the nature of the compound and the purpose for making the measurement. In addition, the composition and the temperature of the carrier gas stream affect the selection of a sampling technique, analytical method and the sampling plan.

9.2 Classification of Sampling Methods

The sampling methods used for the study of air pollution can be classified under three different headings.

1. Sampling of impurities of every nature (ranging from particulate matter to gases)
2. Sampling under various environmental conditions (ranging from samples taken from chimneys to samples taken in the open air)
3. Sampling methods varying according to the time factor (ranging from intermittent to continuous sampling).

9.3 Difficulties Encountered in Sampling

1. Collecting samples of true representative character.
2. Errors arising from methods used for the collection and separation of the various components of pollution.
3. Difficulty in preventing any change in the concentration of particulate matter in suspension, as a result of sampling operations.

9.3.1 Preliminary Considerations and Stages of Sampling

Following principles should be followed to ensure correct sampling.

1. Statistical studies

They are of utmost importance to establish the basic data, e.g., size and frequency of sampling. Statistical studies use the basic principle of probability, i.e., they do not give absolute results but only indicate the percentage probability that a given series of data will be accurate within certain limits.

2. Size of samples

The samples should be large enough to make analysis possible.

3. Changes in the sample during and after sampling.

This is one of the fundamental causes of error. These possibilities should always be borne in mind by the investigator so that he can try to avoid such serious sources of error.

4. Continuous and intermittent sampling

An automatic, continuous recording apparatus is becoming more and more popular. Intermittent sampling can also be made, but continuous recording is always preferable. Many of the continuous recording instruments combine the operations of sampling and analysis.

5. Sampling of volatile constituents

In this case, to avoid the sources of error, it is better to carry out sampling with large volumes of air.

6. Sampling of particulates

Here errors may be introduced due to agglomeration or breaking up of particulate matter. In order to eliminate the sources of error,

(a) Sampling should be carried out under conditions which are as isokinetic as possible.
(b) A gas stream should be sampled as far as possible in the same direction and at the same speed itself, but never counter-current.
(c) The collecting surface should be as close as possible to the source of the gas stream. To avoid reducing the efficiency of sampling, deposits and condensation should not be allowed to form on the walls of sampling vessels. Hence this condition (c).

7. Sampling of waste gas

Here, difficulties are encountered due to high temperature, lack of uniformity in the composition of gas flow, and difference in speed due to disturbances. Hence to avoid error, the gas stream should be sampled at several points and maximum number of samples should be taken, to get the average value.

8. Sampling in the open air

Here difficulties arise due to,

(a) High dilution of the pollutants dispersed in air
(b) Consequent need to collect large number of samples
(c) Difficulty of sampling under isokinetic conditions

Therefore, to minimise the error of sampling, continuous recording instruments should be used, and samples should be collected at various places.

Also, meteorological data like wind speed and direction, should be collected to know their effects on dilution or concentration of pollutants in the air.

9.3.2 Basic Considerations of Air Sampling

1. The sample collected must be representative in terms of time and location.
2. The sample volume should be large enough to permit accurate analysis.
3. The sampling rate must be such as to provide maximum efficiency of collection.
4. The duration of sampling and frequency of sampling should reflect accurately the occurrence of fluctuations in pollution levels.
5. The contaminants must not be modified or altered in the process of collection.

9.4 Instruments for Sampling Waste Gases and for Atmospheric Sampling

Following are the sampling devices in use:

9.4.1 Devices for General Use

Meters

They are used to determine accurately the volume of the gas collected. They are fitted with manometers and thermometers to indicate the pressure and temperature of the gas stream sampled.

Probes

These are tubes suitable for penetrating into the gas stream and should be constructed of materials which are non-corrosive and which can withstand special temperature conditions. Also, they should be constructed of materials which do not react with the substances to be sampled. Therefore, they should be made of stainless steel or preferably of glass or quartz. A probe should have suitable length and diameter. To ensure isokinetic sampling conditions, the opening of the probe should face the gas stream to be sampled.

Suction Devices

Any suction device which has the required volumetric capacity can be used. Vacuum pumps driven by electric motors are very commonly used.

9.4.2 Devices for Sampling Gases and Vapours

Absorbers

In this process, effluent gases are passed through absorbers (scrubbers) which contain liquid absorbents that remove one or more of the pollutants in the gas stream. The efficiency of this process depends on

(1) amount of surface contact between gas and liquid
(2) contact time
(3) concentration of absorbing medium
(4) speed of reaction between the absorbent and gas.

Absorbents are being used to remove sulphur dioxide, hydrogen sulphide, sulphur trioxide and fluorides and oxides of nitrogen.

The equipment using the principle of absorption for the removal of gaseous pollutants includes: (1) packed tower, (2) plate tower, (3) bubblecap plate tower, (4) spray tower, and (5) liquid jet scrubber absorbers. Selective chromatographic absorption of gases on small pellets may offer much higher rates than those achieved in packed towers.

A gas can be sampled by means of a suitable absorption reagent. For this purpose, U-shaped absorbers are used. These absorbers are filled with a certain measured amount of reagent and fitted with a porous glass partition, so that the air or gas led into them passes through the reagent solution in the form of fine bubbles thus ensuring intimate contact. Sampling by means of such absorbers is usually carried out at an average rate of about 100-150 litres per hour of gas stream. The absorbers may be arranged in series of two or more elements containing the same reagent solution to ensure more complete absorption of the impurity in question, or in series of two or more elements containing two or more different reagent solutions so as to absorb different pollutants successively from the same volume of gas or air sampled. A typical sampling train is shown in Fig. 9.1 comprising an impinger, trap, manometer, flow meter, valve and pump.

By selecting the most suitable absorbent solutions, the gaseous components listed below can be determined in concentrations as low as 0.1 ppm by volume.

1. Oxides of sulphur
2. Oxides of nitrogen
3. Ammonia
4. Hydrogen sulphide
5. Hydrochloric acid
6. Hydrofluoric acid
7. Hydrocyanic acid.

This method can also be used for the determination of ozone, hydrocarbons and organic solvents.

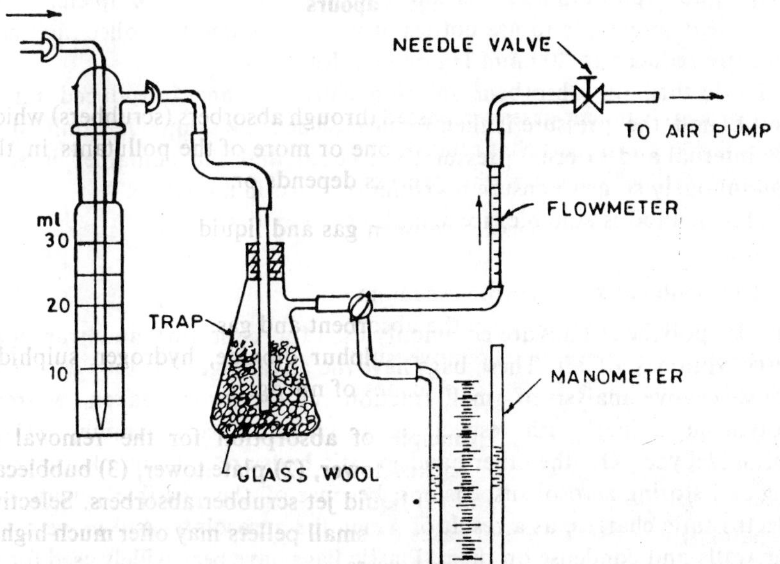

Fig. 9.1 *Sampling train*

Adsorbers

Adsorption is brought about by aspiring the air or gas to be sampled through adsorption columns containing silica gel, activated charcoal or any other suitable agent. After adsorption, the different pollutants can be extracted from the column in various ways. For example, by raising the temperature.

The main difficulty in this method is in selecting a suitable adsorbing medium. This type of sampling is used especially for ozone and light hydrocarbons.

Condensers

Here the gas stream sampled is cooled in suitable containers, thus bringing about the condensation of the volatile substances present. As in the case of absorption devices, here also the condensation traps can be arranged either in series or parallel, at decreasing temperatures. By using various coolants, e.g., ice, liquid air, or liquid nitrogen—the components can be separated by fractional condensation.

This method is used in particular for the sampling of odouriferous substances.

Collectors Under Reduced Pressure

For some substances like nitric acid and aldehydes having a high molecular weight, absorption in aqueous solutions is sometimes incomplete. In such cases, it is preferable to use bottles of known volume for collecting under a pressure reduced to 200 mm Hg or even less.

To do this, the absorbent solution chosen is first introduced into the bottle and the pressure is then reduced. Then the sample is admitted until the internal and external pressures are equal and the container is shaken continuously so as to ensure maximum absorption.

This method is suitable, for sampling the oxides of N_2.

Plastic Containers

Special polythene bags are commonly used for collecting and transporting large volumes of air. These bags have the advantage that they can be used for successive analysis of small fractions of the sample taken. Moreover, polythene is inert with respect to many substances including SO_2 and formaldehyde. On the other hand, plastic bags are not suitable for collecting and storing aerosol suspensions, because of the possible generation of electrostatic charges, as a result of which the aerosols tend to move towards the walls and condense on them. Plastic bags have been widely used for grab sampling and sample storage before analysis.

Samplers for Mass Spectrometric Analysis

Sampling for mass spectrometric analysis can be carried out in various ways. For example, by compressing the gas sample in a pressure flask so as to concentrate a large quantity of gas in a small volume, or by filling evacuated containers.

9.5 Duration of Sampling Period

Two types of sampling are used in studies of air pollution. Short period or 'spot' sampling and continuous sampling for the evaluation of peak and average concentrations over definite time intervals. Spot samples are collected over periods varying from less than 30 minutes to several hours for specific, well defined purposes. The choice of sampling period depends upon the nature of the compound under study and its stability to oxidation, light or other factors such as sensitivity, accuracy and precision of the analytical method to be used for the measurement of pollutants. Short-sampling is useful for the random checking of pollution at many points. Such samples have only limited value because pollution levels fluctuate widely, depending on meteorological conditions, topographical features, and various factors associated with sources of pollution (e.g., mass rates of emission of pollutants from smoke stacks, the temperature, velocity and density of stack gases, the height of smoke stacks, the distribution of sources, and the downwind distance from the sources to the points where the measurements are made). Consequently, spot sampling cannot give adequate data on the nature and magnitude of an air pollution problem. However, by making a series of such observations one can obtain information on the maximum and average concentration of pollutants over a definite time period. Continuous sampling techniques are usually necessary in systematic studies of the nature and extent of air pollution if the data are to be value for epidemiological surveys, for evaluating the potential hazard to man, animal or vegetation, and for control programmes. Continuous sampling may be carried out by the sequential chemical absorption or filtration of measured air volumes, and subsequent analysis of pollutants, by appropriate analytical techniques.

Samples should be collected under all environmental conditions that affect pollution levels, e.g. all weather conditions. This may necessitate sampling continuously round the clock. There is sufficient evidence that pollution is influenced by diurnal variations in wind speed, atmospheric stability, solar heating and cooling, cloud cover and other factors. Biological effects, rates of corrosion, and visibility are greatly influenced by the moisture content of the atmosphere and by interaction of the water vapour with condensation nuclei. Natural air cleaning processes, including precipitation materially affect the levels of pollutants in the atmosphere.

It may be pointed out, that trace element levels in the atmosphere are generally very much lower than those in or near emitting sources. Collection of a given amount of atmospheric material therefore requires longer sampling times or higher air flow rates than is the case of source material. There are several consequences of this.

1. Very sensitive analytical procedures are generally needed to determine trace elements present in collected atmospheric samples.

2. Determination of element volume concentrations is much easier than that of specific concentrations, since the former does not require measurement of the weight of the collected material.

9.6 Location of Sampling Sites

Sampling sites must be carefully selected so as to be representatives of the areas under study. For the study of health effects, the sites must be located so that the collected samples represent air that is actually breathed by the exposed population groups. Since the concentration of pollutants varies with altitude, results obtained on the roofs of tall buildings may differ substantially from conditions at the ground or breathing level. The necessary number of sampling stations, and their location, depend on several factors including the objectives of the programme, the size of the study area, the proximity of the sources of pollution, topographical features and the weather. A representative number of sampling stations for a given area may be established by means of a preliminary survey, whose objectives should be:

1. To gather information on the nature and magnitude of the emission from principal sources of pollution
2. To review the available climatological and meteorological data
3. To gather data on the concentration of pollutants in areas of severe and slight pollution.

9.7 Sampling Methods

Particulate pollutants are classified generally into dust that settles in air and dust that remains suspended as an aerosol. The physical consideration which determines into which class a particle falls, is the particle diameter.

Sedimentation, the ability of particles to settle, is defined by a relationship that involves the terminal settling velocity of the particle falling under the influence of gravity. This relationship, called Stoke's Law, allows one either to predict relative settling velocity (V_t) of various particles knowing their diameter (d) or to compute the diameter knowing the time in which a particular aerosol settles from a known elevation.

Assuming that the particles are spherical and that constant buoyancy is exerted by air, then

$$d^2 = \frac{18\eta V_t}{(d_p - d_a)g} \tag{9.1}$$

where, η = the viscosity of air (poise)
d_p = density of the particle (g/cm^3)
d_a = density of air (g/cm^3)
g = acceleration due to gravity (980 cm/s^2)

Solving the equation for settling rates of particles, gives the results over a range of diameters shown in Table 9.1.

Table 9.1 *Settling Rates for Various Particles at 18°C*

Particle diameter (μ)*	Terminal settling velocity (m/h)
0.01	0.00011
1.0	0.11
5.0	2.77
10.0	10.80

*assuming particles of specific gravity 1.

From Table 9.1, it is obvious that the actual settling ability of very small particles approaching molecular diameter is so small, that for practical purposes, such particles remain suspended in air and are even difficult to remove by air washing, which occurs with rainfall. As a matter of working definition, particles larger than 10 μ diameter are usually thought of as 'settleable' while those of smaller diameter are referred to as 'suspended' or aerosols.

Two general processes are involved in the formation of aerosols or suspended particulates, namely, condensation and dispersion. In condensation processes, clusters of molecules of a substance unite to form particles of colloidal dimensions. The formation of a cloud of liquid droplets when warm air is cooled as it rises into the upper atmosphere, is an example of a condensation process. When volatile substances and solids are heated to a temperature at which they evolve vapours and fumes, small liquid and solid particles may be formed as a result of condensation that occurs on cooling. In the atmosphere, the presence of nuclei and nucleating agents (e.g., ions, polar molecules, fine dust particles, and molecular aggregates of vapour itself) provide the surfaces upon which vapour can condense to produce an aerosol.

In dispersion processes, relatively coarse matter is split up into fine particles by mechanical grinding, abrasion and many other processes. Examples of dispersion are the discharge of fly ash from pulverised fuel furnaces, the emission of dust from foundry shakeouts, the discharge of solids from blast furnaces, the production of dust in the manufacture of cement and the discharge of particulate contaminants by motor vehicles.

The particulate matter in the air may be sampled using equipment based on one or more of the following principles.

9.7.1 Sedimentation

Sedimentation is one of the simplest techniques for the collection of particulate matter. This method is adopted in general for particles whose diameter exceeds about 10 μ. They are known to settle from air and collect on horizontal surfaces and therefore can be sampled merely by placing an open container in an outdoor area that is free from overhead obstructions. This method is widely used in the determination of particle fallout or dust fall in community atmospheres. The collectors are ordinarily constructed of polyethylene, glass or stainless steel, since the inside walls must be inert to atmospheric oxidative flaking, which would contribute to sample weight (Fig. 9.2).

The container is usually more than twice as high as its diameter at the base. This design inhibits the re-entrainment of contained dust because of spurious drafts or spatters due to heavy rain. A typical dust fall container is constructed of plastic and is 20–32 cm high and 10–15 cm in diameter at the base with a slight inward taper of the walls from top to bottom. Occasionally, water containing algicide (1 mg of $CuSO_4$/litre of water) is added up to 500 ml to facilitate retention of particles, where error from loss of particles may occur as a result of gusty winds. After an exposure period, usually a month, the soluble and insoluble matter is determined and the total solids collected are expressed in terms of weight per unit area per 30 days.

In sampling rather large areas such as entire communities, it is common to employ at least one dust fall container for every 2600 hectares. On the other hand, when dust fall sampling is intended to measure the effect of a

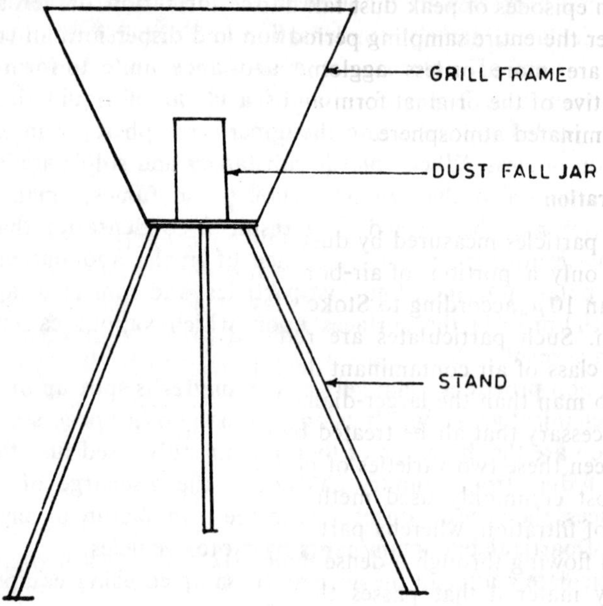

Fig. 9.2 *Dust fall jar*

given industry or industrial complex, containers may be placed as close as a few hundred metres apart. General considerations in site selections are:

1. The site should be free from overhead obstructions and away from interference by local sources such as an incinerator or chimney.
2. The mouth of the dust fall collector should be no less than 2.5 m and no more than 16 m above ground level, with a standard height of 6 m as recommended elevation.
3. When sampling in urban areas, the dust fall collector should be set no less than 10 stack lengths from an operating smoke stack and no closer to a vertical wall than the distance that provides a 30° angle from the sampler to the top of the wall or roof.

The results are affected by the shape and dimensions of the container. Therefore, results for particle fall for different areas cannot be compared unless the dimensions and design of the collectors, their location and other details are rigidly standardised.

The major advantage of dust fall sampling is probably the ease of procurement of 1–5 gram of weightable sample, on which a number of chemical and physical analysis can be performed. In addition, the method is simple and inexpensive and requires no electrical power or moving parts. It facilitates: (a) collection of dust that is representative of a given industry or community; (b) detection of process changes of a given industry; and (c) survey of a community to determine areas of high versus low levels of dust pollution. The disadvantages of the method include lack of precision and inability to distinguish episodes of peak dust fall due to integration of the total sample weight over the entire sampling period (up to 30 days). In addition, particles collected are more or less agglomerated and consequently may not be representative of the original form and size of particulate matter suspended in a contaminated atmosphere.

9.7.2 Filtration

Settleable particles measured by dust fall collectors, as mentioned earlier, represent only a portion of air-borne dust. Particulates having a diameter of less than 10 μ, according to Stoke's law, tend to remain entrained in an air stream. Such particulates are referred to as 'suspended particulates'. Since this class of air contaminant dust is respirable, it is certainly more harmful to man than the larger-diameter settleable particulates. Consequently, it is necessary that air be treated by sampling procedures that discriminate between these two varieties of pollutant dust.

The most commonly used methods of particulate sampling employ the principle of filtration, whereby particles are quantitatively removed from an air stream flowing through a dense material containing sub-micrometre pore sizes. Any material that passes through such a filter is considered, rather arbitrarily, to be non-particulate.

It is widely believed that a particle will be collected by a filter only if its physical size is too great to permit it to pass through the filter pores (direct interception) or if its inertia is sufficiently great for it to impact on the solid structural material of the filter (inertial deposition). However, particle collection is achieved by a combination of five distinct processes, namely, direct interception, internal deposition, diffusional deposition, electrical attraction and gravitational attraction. Only in the first of these processes must the particle be larger than the filter pores. The other four all involve migration of a particle which would otherwise pass through a filter pore, from a stream line to the filter surface under the action of inertial, diffusional, gravitational, or electrostatic forces.

The cooperative effect of these five collection processes gives rise to an important dependence of collection efficiency on particle size in most common filter materials. Thus, larger particles are efficiently removed from the air stream by inertial deposition on filter fibres. They are retained by weak gravitational, electrostatic, and adhesive forces. Similarly, very small particles whose motion is governed by diffusion can move rapidly too and be retained by the filter microstructure. Particles of intermediate size, however, are unable to leave the air stream under the influence of either inertial or diffusional forces. Thus, they pass through the filter. As a result, essentially all filters exhibit a 'window' within which intermediate size range particles are inefficiently collected.

In the selection of a filter medium the following points should be borne in mind.

1. Type of particulates to be collected.

Fine particles collected under conditions of high humidity may absorb surface moisture. Moisture can blind a normally hygroscopic filter, causing decrease inflow rate as well as injury to the air pump. Furthermore, since particulates may be either solid or liquid, high molecular weight organics may deposit as a film, with resultant plugging.

2. Chemical nature of the filter.

Some organic particulates may dissolve certain synthetic fibres instantaneously, while gases such as NO_2 or SO_2 may attack or decompose some membrane filters. Glass-fibre filters should not be used to collect particulates on which analysis for silica is to be performed.

Other factors that lead to collection inefficiency and errors in filtration are particle re-entrainment and vaporisation. Although both these processes have been the subject of considerable speculation, there are no data describing the extent to which they may occur in practical sampling situations. Both loss processes are promoted by high air-flow rates, and it has been pointed that species such as Hg, SeO_2 and As_2O_3 have sufficient vapour pressures at ambient temperatures for substantial losses to occur.

Most air sampling filters collect particles to a lower size limit of 0.1–1.0 μ diameter. The efficiency of filters for collection of smaller particles improves as the sampling progresses because of increasing particulate matting within the filter matrix. Placement of several filters in series may increase the amount of sample collected per cubic metre of air. However, the best use of the technique is to follow a coarser filter with a finer filter so that a wider range of particle size is collected. If a sample of rather high temperature air ($> 110°C$) such as might be found in the vicinity of oxidation or heat creating processes is to be taken, a temperature resistant filter (e.g., glass or polyolefin fibre) should be used in preference to paper or a polyester type synthetic material.

9.7.3 Impingement Methods

Separation of particulates from an air stream by abrupt collision against a flat surface forms the basis of these methods. The apparatus employed may be either a wet or dry type. Wet impingers collect particles by causing them to impinge a surface submerged in a liquid. Dry impingers, usually referred to as impactors collect particles by impaction on a dry surface. In both types of apparatus, collection results as a result of inertial force, as the particles tend to resist a change in direction when the air stream is deflected by a surface or other obstacle. The efficiency of collection by impingement or impaction is high for particles whose diameter is 1 μ or greater. For collection of submicron particles, the impinger should be operated at high velocities.

The well known Greenburg-Smith standard impinger (Fig. 9.3) and midget impinger (Figs. 9.4 and 9.5) are examples of wet collectors. In the former type, the air sample is drawn through an orifice 2.3 mm in diameter

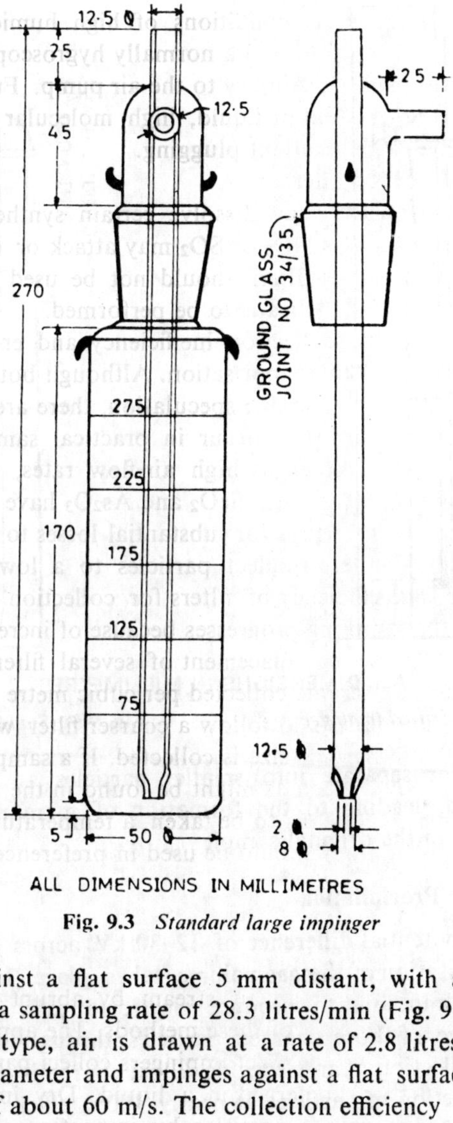

Fig. 9.3 *Standard large impinger*

and impinges against a flat surface 5 mm distant, with a jet velocity of about 113 m/s at a sampling rate of 28.3 litres/min (Fig. 9.3).

In the midget type, air is drawn at a rate of 2.8 litres/min through an orifice 1 mm in diameter and impinges against a flat surface 5 mm distant, at a jet velocity of about 60 m/s. The collection efficiency is low for particles less than 0.7 μ in diameter (Figs. 9.4 and 9.5).

Dry cascade impactors are also available in the market. They are useful for simultaneous collecting and size differentiating of an aerosol. In cascade passes through the impactor, particles whose inertia is too great to enable it to follow the air stream will impact on a collection surface; thus successive impaction stages will collect successively smaller particles. In most cascade impactors, the last stage is followed by a backup filter (usually fibre glass) for collecting the particles that are sufficiently small to pass right through the impactor. One of the principal objections to the use of impingement devices for collection of aerosols is that large particles may

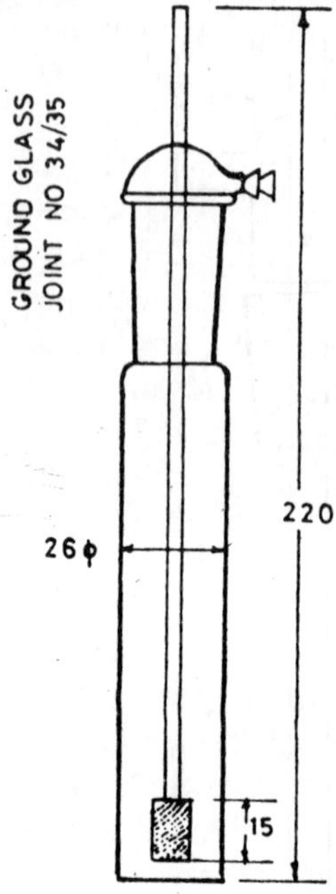

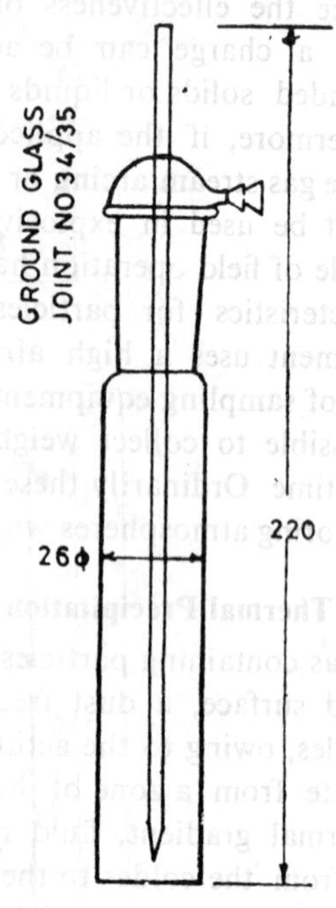

ALL DIMENSIONS IN MILLIMETRES

Fig. 9.4 *Midget fritted impinger*

Fig. 9.5 *Standard midget impinger*

tend to fracture or separate into smaller particles at the high impact velocities attained, leading to the formation of many more fine particles than were present in the original sample.

9.7.4 Electrostatic Precipitation

Application of a potential difference of 12–30 kV across a sample airflow, so that actual ionisation of the gas volume takes place, leads to collisions between particles and ions to produce electrostatically charged particles. Subsequent passage of those charged particles through an electrical field results in their collection at an electrode whose charge is opposite to that of the particle. Collection of particles at nearly 100% efficiency occurs readily over a reasonable length of gas path with particles ranging in diameter from 10–0.2 μ.

The most important mechanism in the charging of particles whose diameter is greater than about 2 μ is bombardment by ions. For particles whose diameter is less than about 0.2 μ, the controlling mechanism is charging by ion diffusion or the motion of ions induced by thermal motion of the surrounding gas molecules. Both mechanisms are effective in the collection of particles whose diameter is 0.2–2.0 μ. Relatively high collection efficiencies (greater than 98%) are possible for many type of dusts and fumes of submicron size.

Since the effectiveness of a precipitator depends upon the extent to which a charge can be acquired by the particles, it will not remove suspended solids or liquids that can volatilise at its operating temperature. Furthermore, if the applied voltage approaches the breakdown voltage for the gas stream arcing or sparking may occur; consequently precipitators cannot be used in explosive atmospheres. Small electrostatic precipitators capable of field operation have been designed and exhibit good separation characteristics for particles having diameters over a wide range. The equipment uses a high air flow rate compared with that of many other types of sampling equipment. Consequently, it has the advantage of making it possible to collect weighable quantities of particles in a comparatively short time. Ordinarily these samplers are employed for source sampling or monitoring atmospheres with relatively high particulate loading.

9.7.5 Thermal Precipitation

If a gas containing particles in suspension is brought into contact with a heated surface, a dust free space develops between the surface and the particles, owing to the action of thermal force that causes the particles to migrate from a zone of high temperature to one of lower temperature. In a thermal gradient, fluid molecules tend to flow around suspended particles from the colder to the warmer region. Reaction to this flow results in the movement of the particles towards the colder region. This behaviour is so subtle as to be unnoticeable in air currents of modest temperatures. However, the effect of thermal repulsion is substantial in cases of temperature gradients of greater than 750°C/cm. Many commercial instruments are available that operate with a thermal gradient greater than 3000°C/cm.

For atmospheric sampling, the collection efficiency of thermal precipitators is almost complete over a wide range of particle sizes from less than 0.01 μ to about 10 μ. Efficiencies approaching 100% are claimed for particles of low specific conductivity (e.g., carbon). Collection by this means is ideal for the direct determination of the number of size distribution of aerosal particles by microscopic examination of the materials deposited on slides. The particles are deposited gently in their natural state without shattering. Disadvantages include necessarily low sampling rates of about 0.02–0.2 lit/min.

9.7.6 Centrifugal Methods

Most centrifugal sampling devices are constructed on the principle of the cyclone. The dust laden gas moving at high velocity, is directed tangentially into a cylindrical chamber, in which it forms a confined vortex. The centrifugal force tends to drive the suspended particles to the wall of the cyclone body, from which they drop into a dust collection chamber. An axial outlet is provided for the clean gas.

The advantages of this sampling device include procurement of a dry, chemically pure sample. The single most important disadvantage is the limit of this method to the particle sizes greater than 5 μ.

9.7.7 Solution Impingers

This consists of a drawn out tube through which air containing particles flows into a solution, where the particles are trapped. The method enables the collection of both particulate and vapour species together, using a suitable solution to trap them. For this purpose, an acidic oxidising solution is best employed. The application of this technique is limited because the collecting solution can be operated only at low flow rates (20 lit/min), is subject to evaporation, and suffers from dependence of collection efficiency on particls size similar to that exhibited by filters. These samplers may be employed in situations where high atmospheric particulate loadings exist, simultaneous collection of particles and gases are required or alternative equipment is not available.

9.8 Sampling Suspended Particulates by High Volume Filtration (the High Volume Sampler)

The high volume sampler is one of the common instruments used for the collection of suspended particulate matter in recent years. Stoke's law offers the basic approach to collection by this technique.

When operated in virtually quiescent air, the geometry of the high volume sampler (Fig. 9.6) employs the sloping roof of the shelter as a means for causing air entering the sampler under the caves of the roof to change direction by at least 90° before entering the horizontal filter. Particles that remained entrained in the air sample prior to horizontal filtration have, in so doing, satisfied the definition of truly suspended dust, i.e. dust that is not subject to settling under the influence of gravitational force.

9.8.1 Air Sampler Operation

The high volume sampler is a vacuum cleaner type motor that is used to draw a sample through a filter area. The filter most commonly used is the 20 cm × 25 cm mat, which allows collection of an air sample at a rate from 40–60 cfm (18,900–28,350 cc/s) over a nominal sampling period of 4–6 h, and a normal sampling period of 24 h. These conditions permit the sampling from 45000–75000 c ft (1260–2100 m^3) of ambient air, with consequent extraction of about 0.5 g of suspended particulate (aerosol). This provides quite a substantial weight of sample, which greatly simplifies subsequent chemical or physical analysis.

The sampler unit may be divided into three parts, the details of which are shown in Fig. 9.7 They are:

(1) the face plate and gasket,
(2) the filter adapter assembly, and
(3) the motor or blower unit.

It is installed inside a casing that is provided with a roof so that the filter is protected from precipitation.

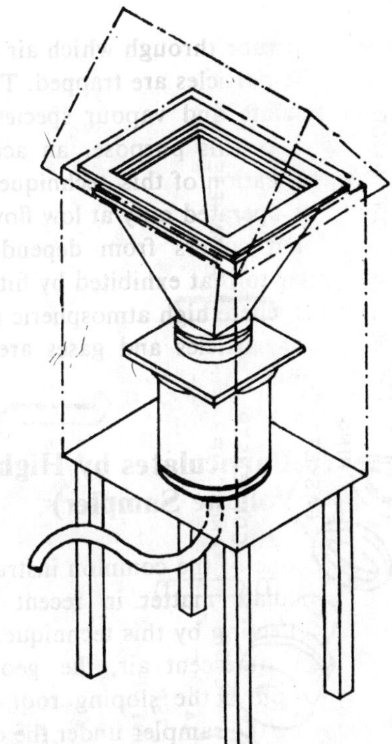

Fig. 9.6 *Assembled sampler and shelter*

The motor, capable of continuous operation for 24 hours with input voltage ranging from 220–230 V, is usually started and stopped by a simple clock timer. The duration of sampling is measured in an elapsed time meter that is placed in series with the blower. Starting and finishing time are at the discretion of the operator, although some recommend starting and finishing from midnight to midnight—24 hours over the same calendar day. The National Air Sampling Network, in the United States, operates such samplers over the entire country on scheduled 24-h intervals every sixth day. This schedule permits a sampling of every day of a week over a period of seven weeks and is the recommended approach for long-term air sampling in a community. On the other hand, short-term studies to determine day-to-day variations in particulate levels may require continuous daily 24-hour sampling. Fulfilment of the aims of such a programme, without manual changing of filters at midnight, is possible by:

1. Using one sampler and manually changing the filter at noon or some other convenient hour rather than midnight.

2. Using two high volume samplers side by side with a timer switch set to start the second motor as the first stops at midnight. The exposed filter can then be changed conveniently during the daylight hours while the second 24-hour sample is being collected.

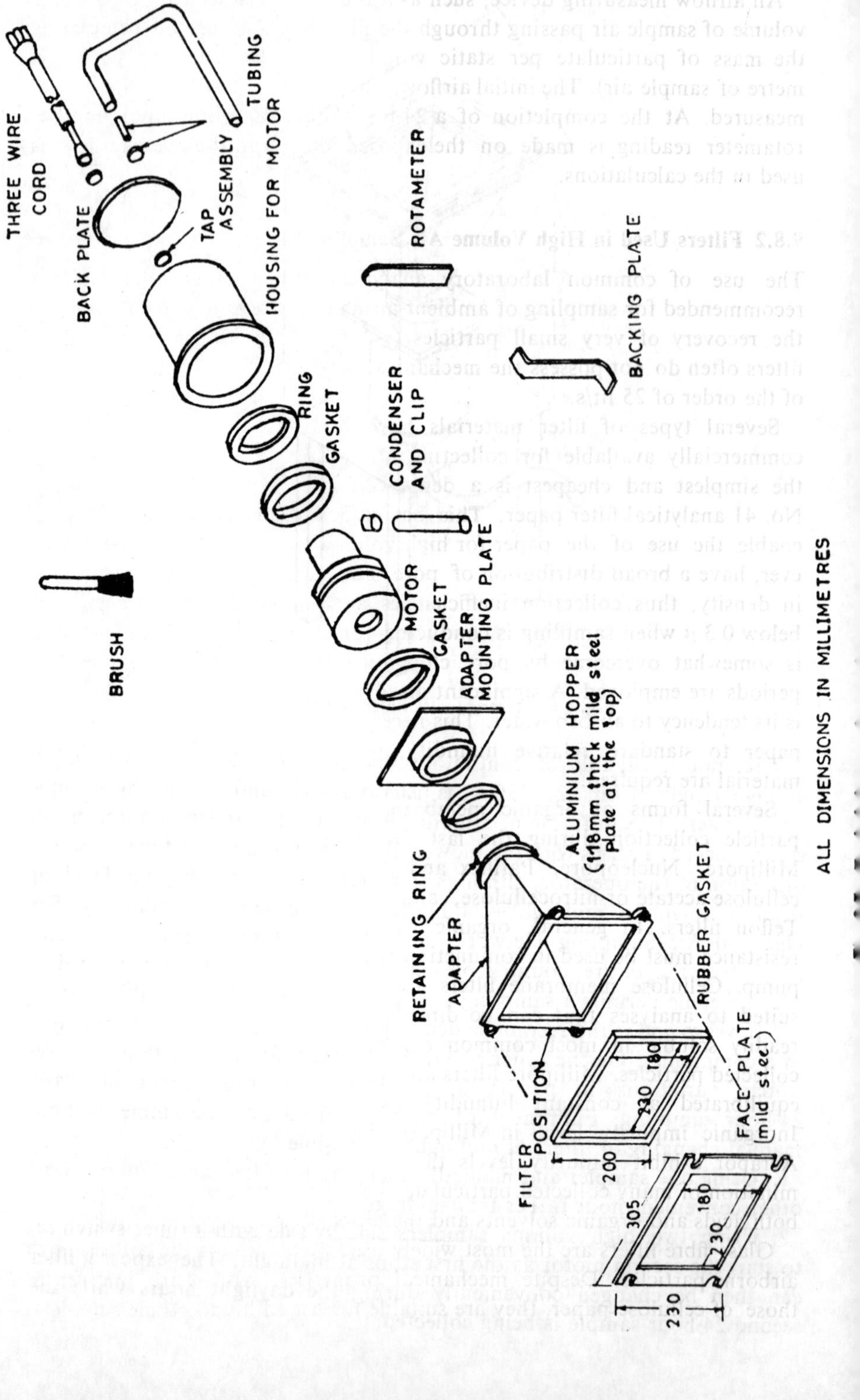
THREE WIRE CORD
BACK PLATE
TAP ASSEMBLY
HOUSING FOR MOTOR
TUBING
RING GASKET
CONDENSER AND CLIP
ROTAMETER
BACKING PLATE
BRUSH
MOTOR GASKET
ADAPTER MOUNTING PLATE
ALUMINIUM HOPPER (118mm thick mild steel plate at the top)
RETAINING RING
ADAPTER
RUBBER GASKET
FILTER POSITION
FACE PLATE (mild steel)
200
230
180
305
230
180
230
ALL DIMENSIONS IN MILLIMETRES

An airflow measuring device, such as a rotameter, is used to record the volume of sample air passing through the filter for later use in calculating the mass of particulate per static volume of air (micrograms per cubic metre of sample air). The initial airflow, through the clean filter mat is first measured. At the completion of a 24-h (or other sampling time), another rotameter reading is made on the exposed filter, and the average flow is used in the calculations.

9.8.2 Filters Used in High Volume Air Sampling

The use of common laboratory analytical grade paper filters is not recommended for sampling of ambient air as they are not satisfactory for the recovery of very small particles ($< 0.5\ \mu$). In addition, such paper filters often do not possess the mechanical strength to withstand airflows of the order of 25 lit/s.

Several types of filter materials have been used and are currently commercially available for collecting atmosphere particulates. Probably the simplest and cheapest is a dense cellulose paper such as Whatman No. 41 analytical filter paper. This has sufficient mechanical stability to enable the use of the paper for high volume sampling. Paper does, however, have a broad distribution of pore sizes and can vary considerably in density, thus collection inefficiencies are observed for particles sizes below 0.3 μ when sampling is conducted for short periods. This problem is somewhat overcome by pore clogging effects when 24-hour sampling periods are employed. A significant disadvantage in the use of filter paper is its tendency to absorb water. This necessitates careful equilibration of the paper to standard relative humidity, if accurate weights of collected material are required.

Several forms of organic membrane filters have come into use for particle collection during the last few years. The most common are Millipore, Nucleopore, Pallflex and Acrapor, which are composed of cellulose acetate or nitrocellulose, and Fluoropor and Miller, which are Teflon filters. In general, organic membrane filters having high flow resistance must be used in conjunction with a relatively pressure insensitive pump. Cellulose membrane filters are mechanically fragile and are best suited to analyses that can be directly performed on the filter. They are readily soluble in most common organic solvents for ease of removal of collected particles. Millipore filters are slightly hygroscopic and should be equilibrated at constant humidity for precise weight determination. Inorganic impurity levels in Millipore filter are generally low. Pallflex and Acrapor exhibit impurity levels that are substantially high for the determination of many collected particulate elements. Teflon filters, are inert to both acids and organic solvents and therefore would permit multiple use.

Glass fibre filters are the most widely used materials for the collection of airborne particles. Despite mechanical properties, which are inferior to those of cellulose paper, they are suitable for use in high volume samplers.

Furthermore, a glass fibre filter is inert to acids (excepting hydrofluoric acid) and to organic solvents. For accurate determination of collected weight, filters must be equilibrated in an atmosphere at constant relative humidity. Unfortunately, glass fibre filters contain substantial amounts of several trace elemental impurities (viz., Cr, Cu, Fe, Mn, Ni, Pb, Sb and Zn). Consequently, in order to reduce the contribution of impurities by the filter it becomes necessary to collect large amount of sample.

Table 9.2 furnishes data on collection efficiency of various filter media which find use for the collection of suspended particulate matter in ambient air.

Table 9.2 *Collection Efficiency of Different Types of Filter Media*

Filter	Composition	Efficiency
W and R Balston, England		
Whatman GF/C	Glass fibre	99.9
Whatman No. 32	Cellulose	99.5
Whatman No. 42	Cellulose	99.2
Whatman No. 44	Cellulose	98.6
Whatman No. 50	Cellulose	97.0
Gelman Instrument Co., USA		
Type VM-4	Poly vinyl chloride	
Acropor, type AN 3000	Acrylonitrile/ Poly vinyl chloride Co-polymer reinforced with nylon	100
Type A	Glass fibre. No organic binder	99.99
Type E	Glass fibre with acrylic binder	99.99
Hollingsworth and Vose, USA		
HV-70,-8-mil	Cellulose and asbestos	99.3
Millipore Filter Corp., USA		
type HA and AA	Cellulose esters	99.9
Mine Safety Appliance Co., USA		
Type 1106-B	Glass fibre	99.99
Type 1106B	Flash fired glass fibre	99.97
Schleicher and Schull, Federal Republic of Germany		
Ultrafilter	Cellulose ester	99.99
Schnelder-poelman. France		
Type 'rose'	Esparto and asbestos fibres	99.98
Type 'jaune'	Esparto and asbestos fibres	97.0
USSR		
FPP-15-4.5		99.995
FPP-15-3.0		99.99
BF		99.9
AFA RMP		99.5
AFA RMA 20		97.0
US Atomic Energy Commission		
AEC 1	Asbestos and vegetable fibres	99.9
CWS 6	Asbestos and cellulose fibres	99.9

9.8.3 Procedure for the Collection of Suspended Particulates

Apparatus

1. High volume sampler with flow meter and elapsed time-meter
2. Barometer capable of recording atmospheric pressure to nearest millimetre of mercury
3. Desiccator (50% humidity) to condition the filter
4. Thermometer to record ambient temperatures
5. Glass fibre filters having a collection efficiency of at least 99% for particles of 0.3 μ diameter

Sampling Procedure

1. Check the fibre glass filter for pinholes, particles or other imperfections. Filters with visible imperfections should not be used.
2. Number the filter and equilibrate the filter in the desiccator for 24 hours and weigh to the nearest milligram.
3. Record tare weight and filter identification number (do not bend or fold the filter before collection of the sample).
4. Open the shelter of the high volume sampler, loosen the wing nuts and remove the face plate from the filter holder.
5. Install a numbered, pre-weighed, glass fibre filter in position (rough side up). Replace the face plate without disturbing the filter and fasten securely (undertightening will allow air leakage and overtightening will damage the rubber gasket). A very light application of talcum powder may be used on the rubber face plate gasket to prevent the filter from sticking.
6. Close the roof of the shelter.
7. Set the on-off timer to start sampling for the prescribed time (24 h), recording the starting time.
8. After five minutes note down the flow rate.
9. At the end of the sampling period, record the length of the sampling period (from elapsed time meter) and the flow rate.
10. Remove the face plate as described before and carefully remove the exposed filter without tearing or touching the collected surface. (High winds or inclement weather occasionally make it necessary for the operator to shield the samples with his body or clothing while transfering the filter).
11. Fold the filter lengthwise so that only surfaces with collected particulates are in contact and place in the special folder.

Expression of Results

The use of a great variety of units to express the results of air sampling and analysis has led to much confusion in the literature. In order to achieve international standardisation and comparability, the results of air pollution measurements are now expressed in CGS units. Using this system, the concentration of pollutants is reported in terms of mass per unit volume at a standard or reference temperature and pressure. The standard pressure of 760 mm Hg and temperature of 0°C should be employed. Although correction to standard or reference gas conditions may not be justified for many local purposes owing to inherent technical errors or the limited accuracy of measurements, it is essential that temperature and pressure conditions be reported in all studies in which comparability and precision are desired.

Normally, the concentration of air pollutants is expressed in terms of milligrams per cubic metre (mg/m^3) or micrograms per cubic metre ($\mu g/m^3$).

Field Data to be Collected

During the sampling period, the data related to sample collection are to be collected and noted down in the field book. A model proforma is given in Table 9.3.

9.8.4 Determination of Mass Concentration

Equilibrate the exposed filter for twenty-four hours in the desiccator and then weigh. After they are weighed save the filters for chemical analysis.

The mass concentration of the suspended particulate is calculated using the formula.

$$C = \frac{(W_2 - W_1)}{V} \times 10^6 \qquad (9.2)$$

where C = mass concentration of suspended particulates in $\mu g/m^3$
W_2 = final weight of the filter in grams
W_1 = initial weight of the filter in grams
V = total volume of air sample in m^3 at STP
10^6 = conversion of gm to μg

$$V = \frac{(Q_1 + Q_2)}{2} \times T \qquad (9.3)$$

where Q_1 = initial flow rate in cu m/min
Q_2 = final flow rate in cu m/min
T = sampling time in min

9.8.5 Example

Average pressure of the day at station level = 712.59 mm of Hg
Average temperature = 30.6 °C or 303.6 °abs.
Actual sampling time = 24 h or 1440 min
Sampling rate = clean filter: 1.6 cu m/min
Filter after exposure = 1.5 cu m/min
Average = 1.55 cu m/min
Therefore volume of air sampled = Sampling rate × sampling time
= 1.55 × 1440
= 2232 cu m

Correction for temperature and pressure:

$$\frac{PV}{T} = \frac{P_1V_1}{T_1}$$

TABLE 9.3 *Field Data for High Volume Air Sample*

Station name : Location :

Technician :

Filter identification No. :

Start date : ________ Time : ________ (use 24 h clock for time)

Finish date : ________ Time : ________

Elapsed time—Final : ________ hrs ________ min Sampling rate—Initial : ________ cu m/min

Initial : ________ hrs ________ min Final : ________ cu m/min

Total time : ________ hrs ________ min Average : ________ cu m/min

Meteorological observations :

Pressure—Initial : ________ mm of Hg Clear sky : Yes/No partial

Final : ________ mm of Hg Cloudy : Yes/No partial

Average : ________ mm of Hg Rain from. . . . . .hrs to. . . . . . .hrs

Temperature(°C) : (hourly recordings)

If the sampling is stopped earlier, specify the reasons.

Signature

$$\frac{760 \times V}{273} = \frac{712.59 \times 2232}{303.6}$$

From which $V = 1881.83$ cu m

Weight of particulate:

Tare weight of filter before exposure $= 3.417$ g

Tare weight of filter after exposure $= 3.925$ g

Weight of particulate $= 0.508$ g

$= 0.508 \times 10^6\ \mu g$

Suspended particulate concentration:

$$= \frac{\text{weight of particulate in } \mu g}{\text{volume of air sampled in m}^3}$$

$$= \frac{508 \times 10^3}{1881.83}\ \mu g$$

$$= 269.95\ \mu g/m^3$$

9.9 Stack Sampling Techniques

Stack sampling or source sampling may be defined as a method of collecting representative samples of pollutant laden air/gases at the place of origin of pollutants to determine the total amount of pollutants emitted into the atmosphere from a given source in a given time. Stack sampling is employed for the assessment of the following:

1. To determine the quantity and quality of the pollutant emitted by the source
2. To measure the efficiency of the control equipment by conducting a survey before and after installation
3. To determine the effect on the emission due to changes in raw materials and processes
4. To compare the efficiency of different control equipments for a given condition
5. To acquire data from an innocuous individual source so as to determine the cumulative effect of many such sources
6. To compare with the emission standards in order to assess the need for local control.

Source sampling is usually carried out in a process ventilation stack to determine the emission rates and/or characteristics of pollutants.

9.9.1 Planning the Study

The success of stack sampling depends on proper initial planning prior to the conduction of a survey. The planning includes the following:

1. Familiarity of the process and operations to determine the time of cyclic operations, peak loading that might cause variations in the characteristics

2. Methods of sampling
3. Methods of analysis of samples
4. Sampling time because certain industries undergo cyclic changes
5. Amount of samples required
6. Sampling frequency.

Representative Sample

It is highly important that the sample collected must truly represent the conditions prevailing inside the stack. The important considerations for accurate representative sample collection include:

1. Accurate measurement of pressure, moisture, humidity and gas composition
2. The selection of suitable locations for sampling
3. Determination of the traverse points required for a velocity and temperature profile across the cross-section of the stack and sampling for particulate matter
4. The measurement of the rate of flow of gas or air through the stack
5. Selection of a suitable sampling train (Fig. 9.8)
6. Accurate isokinetic sampling rate especially for particulate sampling
7. Accurate measurement of weight and volume of samples collected.

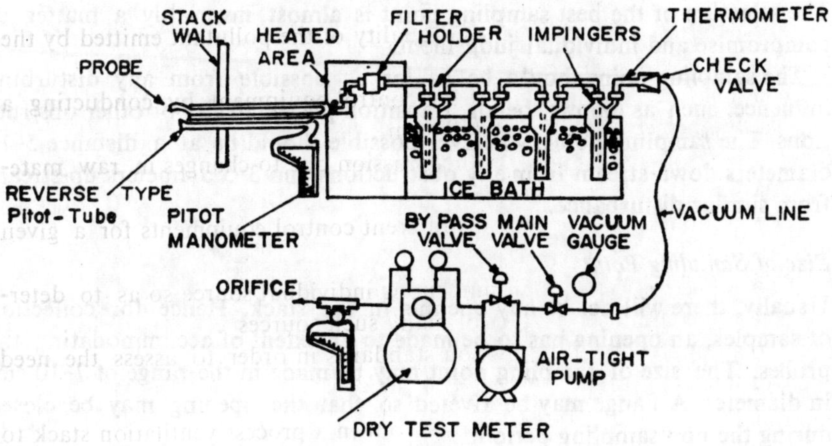

Fig. 9.8 *Particulate sampling train*

Overall Objective

The main tasks involved are to determine the pollutant concentration, stack gas flow rate and pollutant mass emission rate. These terms are related as shown in Eq. (9.4)

$$\overline{PMR_S} = \bar{C}_S \times \bar{Q}_S \tag{9.4}$$

where,

$\overline{PMR_S}$ = is the average pollutant mass emission rate from the stack
$\bar{C}_S$ = is the average stack concentration of the pollutant, and
$\bar{Q}_S$ = is the average volumetric stack gas flow rate.

Hence, the overall objective is to determine $\overline{PMR_S}$ by measuring $\bar{C}_S$ and $\bar{Q}_S$. The average volumetric stack gas flow rate, $\bar{Q}_S$ is determined by measuring the average gas velocity, V_S and the area of the stack A_S.

$$\bar{Q}_S = \bar{V}_S \times \bar{A}_S \tag{9.5}$$

The basic equation to determine the velocity of flow inside the stack is given by Eq. (9.6.)

$$V_S = K_P \times C_P \left\{ \frac{T_S \times \Delta P}{P_S \times M_S} \right\}^{1/2} \tag{9.6}$$

where,

ΔP = is the velocity head, mm H_2O.
T_S = is the stack temperature °K.
P_S = is the absolute stack gas pressure, mm Hg.
M_S = is the molecular weight of the stack gas.
K_P = is the dimension constant.
C_P = is the pitot coefficient.

9.9.2 Selection of Sampling Location

The selection of the best sampling point is almost invariably a matter of compromise and individual judgement.

The sampling point should be as far as possible from any disturbing influence, such as elbows, bends, transition pieces, baffles or other obstructions. The sampling point, wherever possible should be at a distance 5–10 diameters down-stream from any obstructions and 3–5 diameters up-stream from similar disturbance.

Size of Sampling Point

Usually, there will not be any opening in the stack. Hence for collection of samples, an opening has to be made to an extent of accommodating the probes. The size of sampling point may be made in the range of 7–10 cm, in diameter. A flange may be riveted so that the opening may be closed during the non-sampling period.

Traverse Points

For the sample to become representative, it should be collected at various points across the stack. This is essential as there will be changes in velocity and temperature (hence the pollutant concentration) across the cross-section of the stack. Traverse points have to be located to achieve this. These points are to be located at the centre of each of a number of equal areas in the selected cross-section of the stack. The number of traverse points may be selected with reference to Table 9.4.

TABLE 9.4 *Traverse Points*

Cross-section area of stack sq.m	No. of points
0.2	4
0.2 to 2.5	12
2.5 and above	20

In circular stacks, traverse points are located at the centre of equal annular areas across two perpendicular diameters as shown in Fig. 9.9. Table 9.5 may also be referred to for fixing the traverse points and the distances are from the inside wall of the stack.

In case of rectangular stacks, the area may be divided into 12 to 25 equal areas and the centres for each area are fixed. The traverse should be carried out at least on nine hypothetical squares on at least three lines.

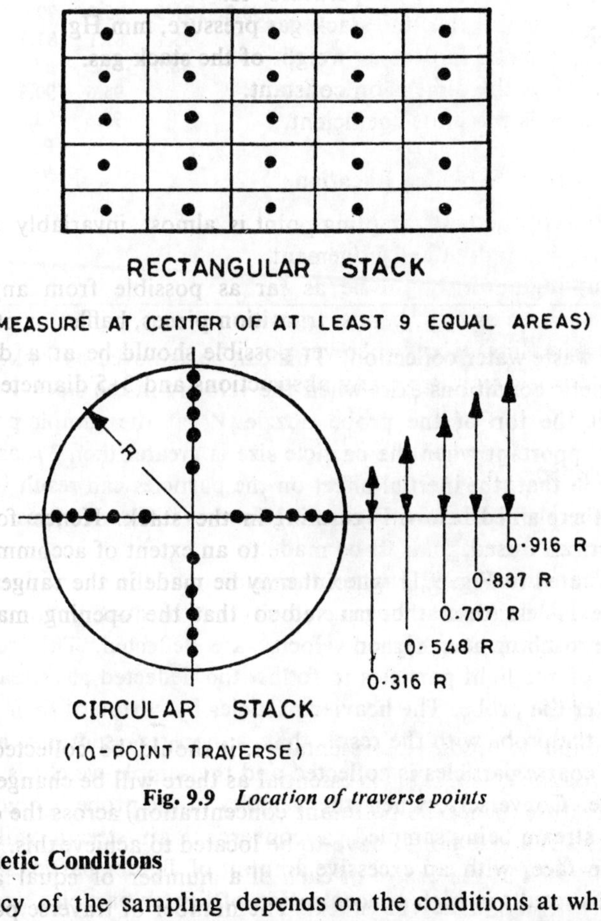

Fig. 9.9 *Location of traverse points*

9.9.3 Isokinetic Conditions

The efficiency of the sampling depends on the conditions at which sampling was carried out. The sample collected must be representative like a

TABLE 9.5 *Location of Traverse Points in Circular Stacks*

Traverse point number on a diameter	Number of traverse points on a diameter									
	6	8	10	12	14	16	18	20	22	24
1	4.4	3.3	2.5	2.1	1.8	1.6	1.4	1.3	1.2	1.1
2	14.7	10.5	8.2	6.7	5.7	4.9	4.4	3.9	3.5	3.2
3	29.5	19.4	14.6	11.8	9.9	8.5	7.5	6.7	6.0	5.5
4	70.5	32.3	22.6	17.7	14.6	12.5	10.9	9.7	8.7	7.9
5	85.3	67.7	34.2	25.0	20.1	16.9	14.6	12.9	11.6	10.5
6	95.6	80.6	65.8	35.5	26.9	22.0	18.8	16.5	14.6	13.2
7		89.5	77.4	64.5	36.6	28.3	23.6	20.4	18.0	16.1
8		96.7	85.4	75.0	63.4	37.5	29.6	25.0	21.8	19.4
9			91.8	82.3	73.1	62.5	38.2	30.6	26.1	23.0
10			97.5	88.2	79.9	71.7	61.8	38.8	31.5	27.2
11				93.3	85.4	78.0	70.4	61.2	39.3	32.3
12				97.9	90.1	83.1	76.4	69.4	60.7	39.8
13					94.3	87.5	81.2	75.0	68.5	60.2
14					98.2	91.5	85.4	79.6	73.9	67.7
15						95.1	89.1	83.5	78.2	72.8
16						98.4	92.5	87.1	82.0	77.0
17							95.6	90.3	85.4	80.6
18							98.6	93.3	88.4	83.9
19								96.1	91.3	86.8
20								98.7	94.0	89.5
21									96.5	92.1
22									98.9	94.5
23										96.8
24										98.9

composite waste water collection. This can be achieved by isokinetic sampling. Isokinetic conditions exist when the velocity in the stack V_S equals the velocity at the top of the probe nozzle V_n at the sample point. This is especially important when the particle size is greater then $3\,\mu$ as it presents a problem in that, the inertial effect on the particles can result in erroneous samples as explained below. Figure 9.10 shows different types of sampling probes normally used.

As illustrated in Fig. 9.11 when the velocity of the gas within the sampling nozzle is less than the gas velocity in the duct, portions of the gas stream approaching at a higher velocity are deflected. This results in the deflection of the light particles to follow the deflected gas stream and they do not enter the probe. The heavier particles by virtue of their inertia, continue into the probe with the result that a non-representative high concentration of coarse particles is collected and the sample weight is in error on higher side. Conversely, when the velocity in the probe is higher than that of the gas stream being sampled, a convergent air stream will develop at the nozzle face, with an excessive amount of lighter particles entering the probe. This results in high concentration of lighter particles and hence the sample weight is on the lower side.

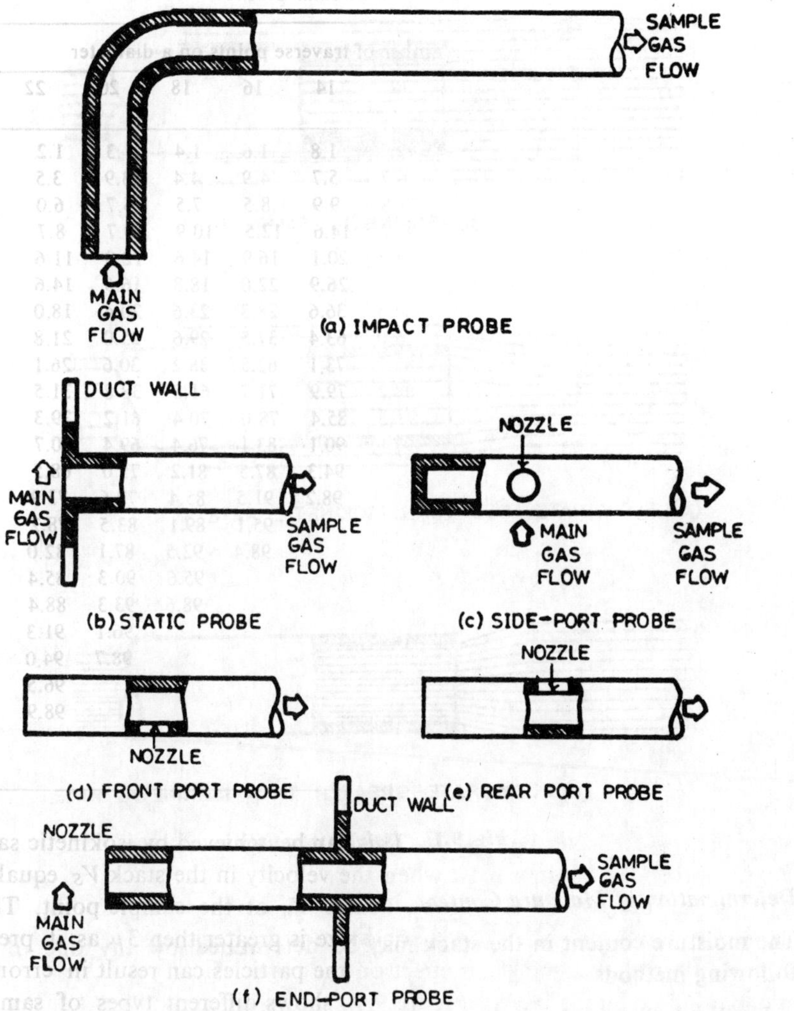

Fig. 9.10 *Sampling probes*

9.9.4 Determination of Gas Composition

The first step in the field work of stack sampling is to determine the gas composition. This can be determined by using Orsat apparatus. The gas is collected in the Orsat apparatus and analysed for the composition of CO_2, O_2 and CO in the same order, and the remaining is assumed to be nitrogen.

$$\text{Molecular weight of gas} = \Sigma M_x . B_x \tag{9.7}$$

where, M_x is the molecular weight of CO_2, O_2, CO and N_2 (44, 32, 28 and 28 respectively) and

B_x represents % of gases.

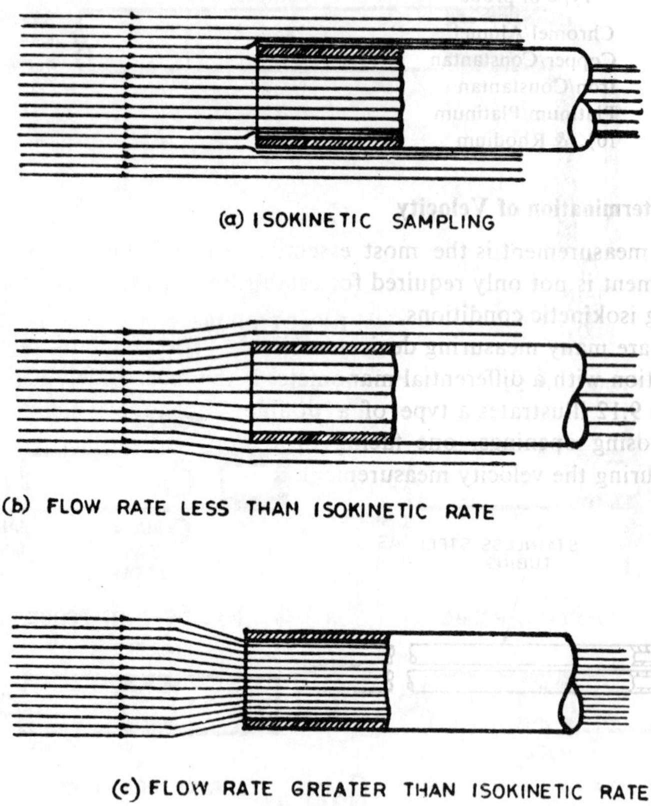

Fig. 9.11 *Isokinetic sampling*

Determination of Moisture Content

The moisture content in the stack may be determined by any one of the following methods:

1. Wet bulb and dry bulb temperature technique
2. Condenser technique
3. Silica gel tube

The wet and dry bulb technique is used when the moisture content is less than 18% and dew point is less than 51°C and cannot be used for acid streams. As the moisture content in the stack is usually less than 18%, the wet bulb and dry bulb technique is being adopted.

9.9.5 Determination of Temperature

The temperature has to be measured across the cross-section of the stack at predetermined traverse points. The temperature probe is inserted into the stack and the readings are taken with the help of a pyrometer. Care should be taken in selecting the probe.

Type of probe	Temperature range
Chromel/Alumel	148.8–1260 °C
Copper/Constantan	148.8–348.9 °C
Iron/Constantan	115.5 °C–1010 °C
Platinum/Platinum 10% & Rhodium	0 °C–1537.7 °C

9.9.6 Determination of Velocity

Velocity measurement is the most essential part of stack sampling. This measurement is not only required for establishing the gas flow rate but also for fixing isokinetic conditions.

There are many measuring devices available. The standard pitot-tube in combination with a differential manometer is the most widely used.

Figure 9.12 illustrates a type of a double Pitot-tube which consists of two opposing openings, one facing upstream and the other facing downstream during the velocity measurement.

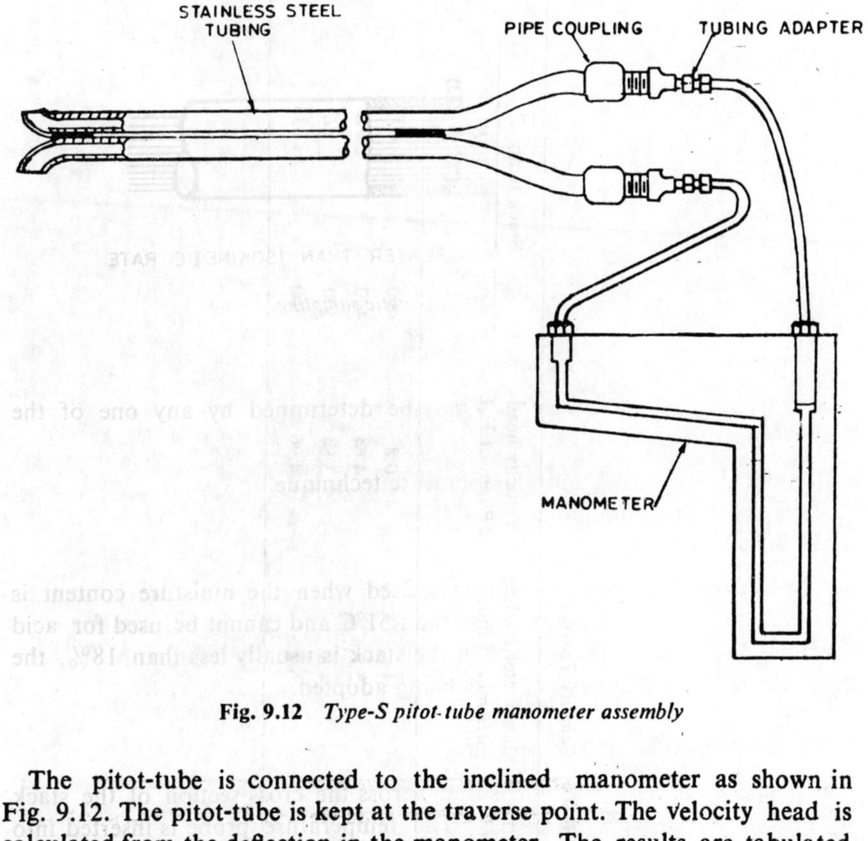

Fig. 9.12 *Type-S pitot-tube manometer assembly*

The pitot-tube is connected to the inclined manometer as shown in Fig. 9.12. The pitot-tube is kept at the traverse point. The velocity head is calculated from the deflection in the manometer. The results are tabulated as shown in Table 9.6.

TABLE 9.6 *Velocity Traverse*

Traverse point	Manometer reading ΔH mm of fluid	Velocity head ΔP mm of H_2O	Stack temp. T		$\sqrt{T \Delta P}$ (col. 3×5)	$\frac{K_P C_P}{\sqrt{P_S M_S}}$	Velocity V_s m/s (col. 6×7)
			C	K (°C + 273)			
1	1	0.8	400	673	23.2034	0.2081	4.8288
2	5.8	4.64	400	673	55.8813		11.6289
3	4.8	3.84	500	773	54.4823		11.3378
4	6.2	4.96	560	833	64.2781		13.3763

The values are substituted in the basic equation (9.6)

$$V_s = K_P C_P \left(\frac{T_s \Delta P}{P_s M_s}\right)^{\frac{1}{2}}$$

Here for a particular stack and pitot-tube, the factors K_P, C_P, P_s and M_s are constant and T_s and ΔP are variables across the cross-section of the stack.

The velocity thus obtained is at stack condition and this has to be corrected to standard condition using the equation

$$\frac{P_1 V_1}{T_1} = \frac{P_2 V_2}{T_2} \tag{9.8}$$

The velocity is multiplied by the cross-section area of the stack to obtain the discharge rate of air from the stack.

Example

Data:

Traverse point	1(col. 1)
Manometer reading	$\Delta H = 1$ mm (col. 2)
Specific gravity of fluid	0.8
Temperature T	400°C
Molecular weight	30
Stack pressure P_s	762 mm. of Hg
Pitot coefficient C_P	0.9

$K_P = 34.96$

Calculation:

$$\text{Velocity head} = \Delta P = \Delta H \times \text{sp. gravity}$$
$$= 1 \times 0.8$$
$$= 0.8 \text{ mm of water (col. 3)}$$

In the equation $K_P\ C_P \dfrac{1}{\sqrt{P_s M_s}}$ is constant for particular stack

$$= \frac{34.96 \times 0.9}{\sqrt{762 \times 30}} = 0.2081 \text{ (col. 7)}$$

$$\sqrt{T \cdot \Delta P} = \sqrt{673 \times 0.8} = 23.2034 \text{ (enter column 6)}$$

velocity at the traverse point $= \text{column } 6 \times \text{column } 7$

$= 23.2034 \times 0.2081$

$= 4.828$ m/s (col. 8)

Add the column-8 and divide by the number of traverse points to get average velocity.

Average velocity = 10.293 m/s (@ stack condition)

9.9.7 Procedure for Particulate Matter Sampling

The general sampling train is shown in Fig. 9.8. The procedure for collection of dust is as given below.

1. Determine the gas composition and correct to moisture content.

2. Determine the temperature and velocity at each traverse point.
3. Determine the empty weight of the thimble (W_1).
4. Mark out the traverse points on the probe. The marks are normally fixed by tying with asbestos thread.
5. Check all points for leakages.
6. Determine the flow rate to be sampled under isokinetic condition.
7. Insert the probe at the traverse point 1, very close to the stack. Start the pump and adjust the flow so that the rotameter reads the predetermined value.
8. Switch off the pump at the end of sampling time.
9. Read the vacuum at the dry gas meter (DGM) and also the temperature.
10. Move the probe to the subsequent traverse points by repeating the steps five to eight.
11. After completion of collection of samples, remove the probe and allow it to cool.
12. Remove the thimble carefully. Some of the dust would have adhered to the nozzle. This should be removed by tapping and transferred to the thimble.
13. Weigh the thimble with the sample. The difference in weight gives the dust collected.
14. The volume of sample collected is either given by the dry gas meter (cu m) or by the sampling rate given by rotameter multiplied by the sampling time.
15. Hence from (13) and (14), the emission rate can be calculated. This will be at DGM conditions. This is to be corrected for temperature and pressure so as to obtain values for standard conditions.

Sample Recovery

After cooling, the outside of probe assembly is cleaned with cotton waste. Disconnect the nozzle. Remove the thimble and keep it in a clean glass beaker. The particulate matter adhered to the inside walls of the nozzle, should be transferred carefully to the thimble. Weigh the thimble with sample (W_2). The difference in weight ($W_2 - W_1$) will give the particulate matter collected.

Calculation

Total volume of gas sampled: V_m @ meter condition.

Total volume of gas V_s sampled $\frac{P_m}{P_s} \times V_m \times \frac{T_s}{T_m}$ @ standard condition

where,

V_m = m^3 as given by DGM
T_m = temperature @ DGM K
P_m = pressure @ DGM mm of Hg
P_s = standard pressure 760 mm of Hg

T_s = standard temperature 298 K

$$F = \text{stack loading} = \frac{\text{Wt. of sample } (W_2 - W_1)}{V_s} \text{ g/m}^3$$

Flow rate = $V_s \times A_s$ cu m/mt @ stack condition
= Q_{std} @ std. condition (applying correction)

Total emission per day $Q_{std} \times 1440$ m³/day

Total particulate matter emission per day

$$= F \times Q_{std} \times 1440 \text{ g/day}$$

$$= \frac{F \times Q_{std} \times 1440}{10^6} \text{ Tonnes/day}$$

Example

(a) To obtain isokinetic condition (columns 1 to 8 same as Table 9.7) for traverse point 1

Velocity of gas stream = 4.8288 m/s (col. 8)

Discharge rate P = Col. 8 × area of nozzle × 60 m³/min.
= $4.8288 \times 0.3166 \times 10^{-4} \times 60$
= 0.0092 m³/min (col. 9)

This is to be corrected to DGM conditions for adjusting the flow rate in the rotameter

The vacuum gauge pressure = 127 mm of Hg (col. 10)

The temperature @ DGM = 32°C = 305 K (col. 11)

$$\text{Temperature correction } Q = \frac{T_m}{T_s} = \frac{305}{673} = 0.4532 \text{ (col. 12)}$$

$$\text{Pressure correction } R = \frac{P_s}{P_s - P_m} = \frac{762}{762 - 127} = 1.2 \text{ (col. 13)}$$

Sampling rate = $P \times Q \times R$
= $0.0092 \times 0.4532 \times 1.2 = 0.005$ m³/min (col. 14)

The rotameter should be adjusted to this sampling rate. The sampling period at each traverse point is entered in col. 15.

(b) For emission rate calculation:

Weight of sample collected $(W_2 - W_1)$ = 1.4842 g

Total volume of gas sampled = Multiply col. 14 and 15 and add
= 0.192 m³, @ DGM condition

$$\text{Vol. of gas sampled at std. condition} = \frac{762 \times V_1}{298} = \frac{635 \times 0.192}{305}$$

(25°C and 762 mm Hg)

$$V_1 = \frac{635}{762} \times 0.192 \times \frac{298}{305} = 0.1563 \text{ m}^3 \text{ @ std. condition}$$

$$\text{Stack loading} = \frac{1.4842}{0.1563} = 9.495 \text{ g/m}^3$$

Stack area = 0.1742 m²

Total discharge per minute = Ave. velocity × area of stack
= $10.2930 \times 0.1742 \times 60$
= 107.582 m³/min.

TABLE 9.7 *Data Sheet of Particulate Matter*

Diameter of nozzle: 6.35 mm

Area of nozzle: 0.3166 sq. cm = 0.00003166 sq. m

P_s = 762 mm of Hg

M_s = 30

Traverse points	Manometer reading ΔH mm of fluid	Velocity head ΔP mm of H_2O	Stack Temp T °C	Stack Temp T K	$\sqrt{T_s \Delta P}$	$\frac{34.96\, C_P}{\sqrt{P_s M_s}}$	Velocity m/s	Discharge $V \times a$ m³/min P	DGM Pressure (abs)	DGM K	$\frac{T_m}{T_s}$ Q	$\frac{P_s}{P_s - P_m}$ R	Sampling Rate PQR cu m/min	Sampling Time in min
1	1	0.8	400	673	23.2034	0.2081	4.8288	0.0092	127	305	0.4532	1.2	0.005	5
2	5.8	4.64	400	673	55.8813		11.6289	0.0220	127	305	0.4532	1.2	0.012	5
3	4.8	3.84	500	773	54.4823		11.3378	0.9215	127	305	0.3946	1.2	0.0102	5
4	6.2	4.96	560	833	64.2781		13.3763	0.0254	127	305	0.3661	1.2	0.0112	5

This is to be corrected to std. condition

$$\frac{762 \times 107.582}{738} = \frac{762 \times V_1}{298}$$

$$Q_{std} = 43.44 \text{ m}^3/\text{min}$$
$$= 43.44 \times 60 \times 24 = 62553.6 \text{ m}^3/\text{day}$$

Total particulate matter emission per day

$$= 9.495 \times 62553.6 \text{ g}$$
$$= 9.495 \times 62553.6 \times 10^{-6} \text{ tonnes/day}$$
$$= 0.594 \text{ metric tonnes/day.}$$

9.9.8 Gaseous Sampling

The procedure adopted for gaseous sampling in a stack differs little from that employed in particulate matter sampling insofar as basic routine is concerned. Isokinetic sampling principles are not required. However, the success of gaseous sampling depends on the proper collecting device, since different gases, mists and vapours exhibit a wide variety of physical and chemical characteristics.

Methods for Collection of Gaseous Samples

Gaseous sampling may be carried out by any one of the following methods:

1. Absorption
2. Adsorption
3. Freezeout/condensation

Whenever the flow rate and the process operation are uniform and whenever analytical equipment of extreme sensitivity is available it is possible to take a grab sample from the stack by any one of the following methods:

1. Use of an evacuated container
2. Purging (displacement of air)
3. Inflation of flexible bag
4. Use of a syringe

Proportional Sampling

Whenever the source conditions, especially flow rate changes, with time, the sampling must be done proportionally so that the sample is representative. This requires the velocity measurement as explained and correction to standard conditions.

$$\text{Concentration of pollutant} = \frac{\text{wt. of gaseous sample}}{\text{volume of gas sample}}$$

For analysis of various gaseous pollutants standard books may be referred.

9.10 Recent Trends in Sampling of Stack Effluents

Two monitors useful for determining the particulate concentration in stack effluents are described below. This recent technology is useful to manufacturers of equipment for on-line sampling of stack effluents.

9.10.1 Piezoelectric Monitor

In this device, particles in the sample stream are electrostatically deposited on to a piezoelectric sensor. The added weight of particulates changes the oscillation frequency of the sensor in a characteristic way. This instrument cannot handle very high particulate loadings found in many stacks without dilution of the sample. The dilution step greatly complicates the sampling problems associated with use of the device since two isokinetic samples are required, one of the stack and another of the diluted stack sample. A piezoelectric microbalance system consists of the components shown in Fig. 9.13. The output signal can be conditioned so that it becomes directly proportional to particulate mass concentration. It can then be recorded either by a digital or analog recorder.

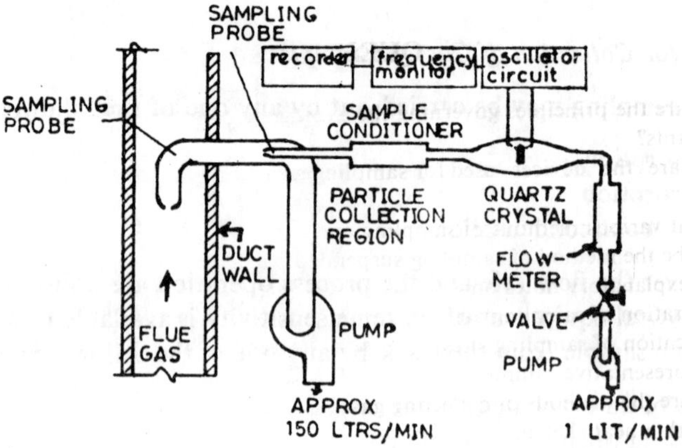

Fig. 9.13 *Piezoelectric microbalance Instrument*

9.10.2 Beta Attenuation Monitor

In this type of device Fig. 9.14, the particulate sample is filtered using a continuous filter tape and the mass of particulates filtered out is determined by measuring its attenuation of beta radiation. Since beta attenuation characteristics are not very different for a wide variety of stack particulate matter compositions, a direct mass measurement is possible. Carbon-14 with a half life of 5,568 years is a typical beta radiation source. Thallium-204, Cerium-137 and Promethium-147 have also been used. Geiger-Muller (GM) counters are the most common detectors, but proportional and scintillation counters have also been used. The difference in count rate of

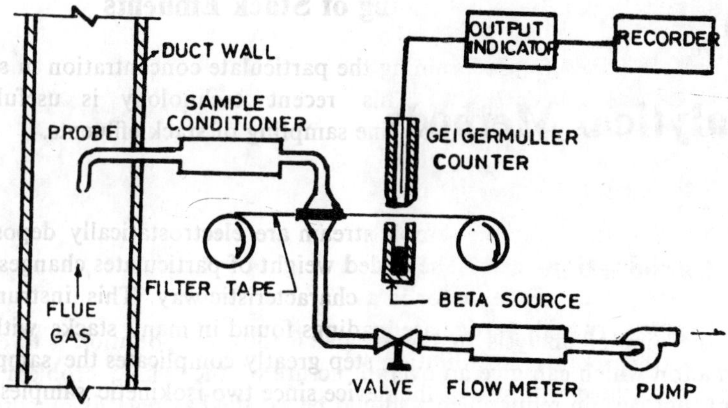

Fig. 9.14 *Beta radiation instrument*

the GM detector before and after the particles are collected is a measure of the mass of particles on the filter.

QUESTIONS

1. What are the principles governing the avoidance of sources of error in sampling pollutants?
2. What are the devices used for sampling gases and vapours? Describe any one in detail.
3. List out various methods of sampling and describe any one in detail.
4. Describe the method of sampling suspended particulates by high volume filtration.
5. Write explanatory notes on:
   (a) Duration of sampling air
   (b) Location of sampling sites
   (c) Representative sample
6. What are the methods of collecting gaseous samples from a stack? Describe any one with a neat sketch.
7. By neat sketches, explain the principle of
   (a) Piezoelectric monitor
   (b) Beta attenuation monitor.
8. Write short notes on:
   (a) Dust fall jar
   (b) High volume air sampler.

10

Analytical Methods

10.1 Introduction

Since the aim of analysis in the special field of air pollution is to obtain information which can give an overall picture of the whole problem, it is evident that certain requirements should be satisfied as regards concordance between sampling and analytical techniques. These requirements are as follows:

1. The efficiency of the sampling device used should be known as accurately as possible
2. The sampling device chosen should cause the minimum of change in the composition of the various substances to be detected
3. The sampling method should supply the analyst with a quantity of pollutants within the quantitative limits called for by the sensitivity of the analytical technique to be used
4. The methods employed for determination should be as sensitive as possible

The various analytical methods require, in addition to a specialized and skilled technical staff, a special high precision equipment.

The methods used for analysis of atmospheric samples can be divided into three basic groups.

1. Chemical methods
2. Instrumental methods
3. Biological methods

10.2 Chemical Methods

(a) Gravimetric method
(b) Volumetric method

    (i) Acidimetric and alkalimetric method
    (ii) Oxidation and reduction method
    (iii) Precipitation method

(c) Colourimetric method

(d) Turbidimetric and nephelometric method
(e) Chromatographic method

10.2.1 Gravimetric Method

In this method the weight of the substance (pollutant) is determined directly (where possible) or indirectly in the form of one of its compounds. From the weight of the compound, the quantity of the substance concerned can be calculated.

The gravimetric method is a very simple analytical technique and can be used for

(1) Determination of a gas-sensitivity: approx. 1 mg. For this very large samples are usually required.
(2) Determination of dusts and soot collected in sedimentation jars following long sampling periods (1 month)—sensitivity: approx. 1 mg.
(3) Determination of particulate matter dispersed in the air, sampled by collection on filter paper by means of the impinger or electrostatic precipitators. Here also, large samples are necessary.

10.2.2 Volumetric Method

In volumetric analysis, the components are not weighed but determined by means of reagents of known concentrations.

Acidimetric and Alkalimetric Method

This is for the determination of acids and bases, using solutions of a base or an acid respectively, in the presence of indicators which undergo a colour change according to the acidity or alkalinity of the medium. The method is very sensitive (0.01 ppm sensitivity)

Oxidation and Reduction Method

These are based on the oxidation or reduction of the compound to be analysed by means of a standardised substance, the quantity of oxygen given up or consumed being measured according to the colour change developed in the reaction, in the presence or absence of an indicator.

Precipitation Method

This is especially suitable for halogens. Here silver nitrate solution of known strength, which precipitates the halogen in the form of silver halides is used. This method is not very sensitive.

10.2.3 Colourimetric Method

This is the most satisfactory method among the various analytical methods now in use. A very large number of elements, radicals and organic compounds form coloured complexes with special reagents and such colourimetric reactions are sometimes sufficiently sensitive to detect even the most minute traces of the compounds which are sought. In other words, the method is very sensitive. The apparatus used for this method is the filter photometer or other types of photometers.

They can be used for the analysis of gas and particulate matter (sensitivity: approx. 0.5 μg), but occasionally they require large samples and difficult separation and concentration techniques for compounds to be examined.

10.2.4 Turbidimetric and Nephelometric Method

Here, precipitates obtained by means of specific reactions are quantitatively analysed by optical measurement of light passing through the corresponding suspensions.

Advantages

1. Suitable for gas and particulate matter
2. Require only small air sample
3. Highly suitable for determination of Cl_2 and SO_3 (sensitivity: 0.005 ppm)

10.2.5 Chromatographic Method

Chromatograpic methods make possible the separation and identification of many substances by means of adsorption on adsorption columns, or on specially sensitised paper, followed by elution of the various compounds adsorbed with selective solvents. In this way, the components of the complex mixtures can be separated without undergoing any chemical change.

Advantages

1. Sensitivity of this method is very high (0.1 μg)
2. Specially suitable for metals and volatile organic compounds.

10.3 Instrumental Methods

1. Emission spectrometric method
2. Absorption spectrometric method (spectrophotometry)
3. X-ray diffraction method
4. Mass spectrometric method
5. Polarographic method (polarography)
6. Methods using microscopy
7. Refractometric method
8. Thermal conductivity method
9. Radioactivity method
10. Sound absorption method
11. Automatic (recording) apparatus (for sampling and analysis)
12. Atomic absorption spectroscopy

10.3.1 Emission Spectrometric Method

Principle

Any element, when in an excited state, has its own special emission spectrum. Spectrometric analysis (quantitative) involves the spectroscopic characterisation (qualitative) of the elements, based on identification of the characteristic lines in their emission spectrum.

Such identification can be made by means of a comparator microscope for the measurement of the distance between the lines themselves, while quantitative determination is based on the consideration that the energy emitted for a given spectral line of an element is proportional to the number of excited atoms and consequently to the concentration of the element in the sample.

Emission spectrometry can be applied to the analysis of more than 70 of the known elements. The alkaline and alkaline-earth elements can be detected with as high an accuracy as 0.1 ppm. Today, it is widely used to determine the composition of mixtures or compounds and is indispensable in the exploratory examination of particulate matter.

Apparatus required

1. Source of radiation
2. A means of splitting up the radiation, which may be either a prism or a grating
3. A receptor, which can be either a photographic plate or a series of photo tubes

Application

In research work, especially for identification of the elements present in particulate matter. This method is particularly suited for Beryllium, Sodium, Potassium, Calcium, Lead, Cadmium and Antimony.

10.3.2 Absorption Spectrometric Method (Spectrophotometry)

Principle

It is based on the absorption of a certain range of electromagnetic radiation in the visible, ultra-violet and infra-red regions of the spectrum.

It is one of the most important of all the instrumental methods of analysis.

The absorption varies with the wavelength of the incident radiation. Therefore, measurements should be carried out either with monochromatic light or with light having a limited spectrum. The wavelength of a beam emerging from the source of radiation is suitably adjusted in this way and then passed through a layer of the substances to be examined.

In the visible region of the spectrum, the concentration of a coloured substance analysed by this method, can be derived from Lambert-Beer law. According to the law, light absorption is proportional to the concentration when the layers analysed are of equal thickness. Further, this method can also be applied to the analysis of colourless substances by studying absorption in the ultra-violet region of the spectrum.

This method is suitable for organic substances, particularly benzene and aromatic compounds. Infra-red spectroscopy is more successful because it can be applied to either solid, liquid, or gaseous substances.

Today this method is mainly used in research, but it has wide application as it has many advantages such as, a simple technique and requirement of only small samples.

But the main disadvantages are:

1. Traces of impurity can be detected only if their absorption is high
2. The sharpness of the spectrum decreases with increasing molecular weight

Further, in the field of spectrophotometric analysis, it is possible to make use of a phenomenon shown by certain substances namely, the re-emission of light energy absorbed in the form of a fluoroscence, whose intensity is measured. By means of measurement of fluoroscence, it is possible to analyse trace amounts of certain atmospheric components.

10.3.3 X-ray Diffraction Method

These are suitable for the qualitative and quantitative analyses of crystalline substances, e.g., mineral dusts.

Principle

It is a well known fact that crystals can serve as diffraction gratings in the study of X-rays and that each crystal has its own particular diffraction pattern, so that it is possible to identify unknown substances by comparing their diffraction patterns with those of known substances. Quantitative determinations are sometimes possible, based on the measurement of the intensity of the diffraction lines after having photographed them on a film.

The X-ray diffraction method also forms part of the spectrometric methods, but while the latter involve measurement ranging from a wavelength of 40 μ (extreme limit of the infra-red) to 0.2 μ (extreme limit of the ultra-violet), X-ray diffraction measurements involve wavelengths of the order of 0.0001 μ. The sensitivity of this method is due to the great penetrating power of X-rays and also because some solid bodies are opaque to any other form of radiation.

However, at present this method is used as an auxillary to other forms of analysis, such as emission spectrography. Moreover, this method requires very large samples and considerable practical experience on the part of the investigator. They are employed principally for the detection of quartz crystals in the air.

10.3.4 Mass Spectrometric Method

It is based on the principle that if a gas is ionised by means of electric discharges, and if the collimated beam of positive ions produced is subjected to the simultaneous action of an electric field and a magnetic field at right angles to each other, then the ions are deflected from their initial straight line direction of propagation and get distributed along a curve, the deflection being proportional to the mass of the ions. The result is a mass spectrum, and the apparatus which records it on a photographic plate is the 'mass spectrograph'.

The mass spectrometer is an apparatus in which the radii of curvature of paths of the various ions can be altered by changing the speed of the ions or the intensity of the magnetic field. In this way, the nature and quantity of the various kinds of ions can be determined by measuring the current flowing in a collecting circuit, when the radius of the paths is changed.

Application

1. For the identification of toxic gases
2. The method is useful in particular for hydrocarbons and compounds of hydrocarbons with Cl_2, N, and O_2

Note: For analysis by this method, the samples need not be very large, but it is essential to isolate and concentrate the fractions to be examined and to conve.` these into gaseous form. This method requires a wide knowledge of mass spectra in order to make comparisons and identifications.

10.3.5 Polarographic Method

This is one of the electrochemical methods of analysis, which makes use of the chemical phenomena which occur in an electrochemical cell, and is based on interpretation of the ratio between current and potential difference. This method is generally referred to as voltametry or polarography, in the particular case where a mercury electrode is used. It is particularly satisfactory for the analysis of dilute solutions.

In the polarograph, the voltage applied to the cell is regulated by means of a potentiometer and the current which passes through the cell is read on a galvanometer. This method can be used to identify a large number of inorganic and organic compounds like Antimony, Arsenic, Benzene, Bismuth, Cadmium, Chromium, Cobalt, Copper, Formaldehyde, Lead, Manganese, Nickel, oxides of N_2, Tin, Titanium, Vanadium and Zinc.

10.3.6 Methods Using Microscopy

The use of microscopy in air pollution research is limited to the examination of particulate substances. (For counting, measuring their size, and optical observation of their surfaces.) In this way, the following can be examined: tarry matter, small drops of oil or water, crystal such as Quartz and ammonium sulphate, nitrate, sulphate, halides and ammonium carbonate.

If an electron microscope is used, it is possible to study even smaller particles present in the atmosphere (sensitivity = 0.005 μ). Further, an electron microscope can give an absolute analysis of particulate matter.

10.3.7 Refractometric Method

It is based on the use of the refractometer for determining the refractive index of the substance under investigation.

This technique is of limited application in air analysis, because of its lack of specificity and sensitivity. It can be used solely under conditions in which only one or two known vapours or gases are to be determined, whose refractive indices are sufficiently different from that of air to permit a certain degree of accuracy.

10.3.8 Thermal Conductivity Method

Principle

It is based on the measurement of thermal conductivity of gases and vapours. The method is of limited use because it cannot be used to identify the gas, but only to evaluate the concentrations of the substances. They give only an overall index of the thermal conductivity and consequently can only be applied to those cases where the conductivity can be measured before and after eliminating a particular constituent, so that the latter can be determined from the difference.

In certain cases, however, it is possible to use a combustion apparatus known as an explosimeter, which gives the concentration of vapour or gas as a percentage with reference to the minimum explosion limit. In this method, a filament is electrically heated in a special apparatus to a temperature at which the gas or vapour becomes oxidised and burns at the surface of the filament. The energy developed during oxidation (combustion) is then measured with a thermocouple. This is the method most used for determining carbon monoxide.

This method has very limited application in air pollution and is limited to the industrial analysis of gaseous effluents.

10.3.9 Radioactivity Methods

Analytical methods based on the measurement of radioactivity have also been used in atmospheric analysis to detect substances and impurities present in traces. The two methods are:

1. Dilution technique
2. Activation technique

In the dilution technique, the element to be detected is diluted with one of its radioactive isotopes and the specific activity of the diluted substance is subsequently measured with a detector. But this method is considerably upset by impurities and by samples which are not well selected.

The activation technique is perhaps the most sensitive of all known analytical methods. It consists of artificial transformation of the substances under examination into their radioisotopes and subsequent measurement of the quantity of disintegration product. This method is not subject to interference because of the specificity of the half-life period of the radioisotopes. But difficulties arise in sampling and preparation of the samples, apart from the problem of obtaining technical personnel with sufficient experience in this field.

10.3.10 Sound Absorption

It is used for measuring particle size.

Principle

It is based on the fact that particles subjected to the influence of a sound field of a particular frequency are affected by those vibrations whose amplitude is related to their size.

Apparatus

1. Sound generator (which can be adjusted for the desired frequency)
2. An observation cell (towards which the sound is directed and into which the sample is introduced)
3. Photographic recording

The photographs obtained show the paths corresponding to the vibratory movements of the particles, and by measuring them the particle sizes can be determined. However, very dilute samples are necessary, since only a few particles should be present in the cell, if clear and interpretable photographs are to be obtained.

It is seen that the same substance can be determined by a variety of methods. Some procedures are suitable for routine work but not practicable for research work and vice versa. Only those procedures should be selected which give the best accuracy, highest sensitivity, highest selectivity, and the best precision. Of course, rapidity of analysis is also an important factor.

It is difficult to find a combination of all these properties in their highest degree in one and the same method. However, it is probable that by combining a wide range of techniques optimum results can be obtained.

10.3.11 Continuous Recording instrumental Methods

In some cases it will be necessary for repeated analysis in order to determine the behaviour of a source of pollution and study its effects with reference to time. In such cases, automatic continuous recording instruments are used. These instruments combine the functions of sampling and analysis in the same apparatus. Although such apparatus is not extremely accurate, nevertheless it furnishes graphs, which are of great help to research workers.

The instruments are based on the usual chemical, electrochemical, optical and other principles. It can be applied for research both on gaseous and particulate matter. For example for determination of sulphurous anhydride (SO_3), instruments known as Thomas Autometer and Titrilog are used.

The Thomas Autometer measures the electrical conductivity of the sulphuric acid solution produced by the oxidation of sulphurous anhydride following absorption in slightly acidified water containing hydrogen peroxide.

There are two types of this apparatus:

1. Accumulating model
2. Instantaneous model

The accumulating type records conductivity continuously for absorption periods of 30 minutes while the instantaneous type records the concentrations at two minute intervals. Both types can be combined into a third or 'mixed' type in which the concentrations are determined at intervals of two minutes as well as the average concentrations over 30 minute intervals. Since, however, many gases or aerosols yield an electrolyte in solution, such apparatus measures the total pollution and not only that specifically due to sulphurous anhydride, although the total is expressed in terms of the latter.

The titrilog also measures sulphurous anhydride (SO_3) to a concentration of 0.02 ppm, by means of oxidation of the gas stream in an electrolytic cell in the presence of bromine. Again, this apparatus is sensitive not only to sulphurous anhydride but also to many other gases which can be oxidised, such as hydrogen sulphide, mercaptans and unsaturated hydrocarbons. With suitable modifications, titrilog can be used to determine H_2S and mercaptans. The autometer can detect any substance which gives an electrolyte in water, either directly or after combustion.

Recording analysers can also be applied for the analysis of oxidants (ozone, chlorine, oxides of nitrogen). Such apparatus measures electrometrically or photometrically the amount of iodine liberated from solutions of potassium iodide by the oxidants.

An automatic instrument, 'microsensor' can detect and measure many kinds of gaseous impurities in amounts as low as 0.1 μg/1. It consists of a colourimetric comparator which measures the colour changes produced on sensitised paper with special reagents.

Recording instruments can also be used to analyse particulate matter. In these instruments, the intensity of light dispersed in the atmosphere owing to the presence of particles is measured and the visibility recorded directly. It is possible to relate these measurements to the particle concentration and size, following suitable calibration of the apparatus. The same principle can also be made use of to count the number of particles present, although sensitivity of such measurements is low. Instruments of this type are sensitive to particles 2.5 μ or more in diameter.

Another example of a sampler and analyser, is a monitor for atmospheric radioactivity, in which the aerosol is deposited electrostatically on the sampling device, connected to a continuous recording radiation counter.

10.3.12 Atomic Absorption Spectroscopy (A.A.S.)

This is a powerful tool for trace metal determinations in air particulate samples. In addition to being a very sensitive method, it provides selectivity, simplicity, and good reproducibility in comparison with most other methods.

Principle

Under most conditions ground-state atoms make up the bulk of an atomic population. These atoms absorb radiation at discrete wavelengths, which are characteristic for each element, and are thus raised to the excited state. The excited state atoms, in turn, emit radiation on returning to the ground state.

The amount of energy thus absorbed and emitted is a function of the number of atoms present and hence may serve as the measure of the concentration of the respective elements in the sample.

Instrument

An atomic absorption spectrophotometer is essentially the same as an emission instrument except for the addition of a radiation source—usually a hollow cathode lamp which has an emitter cathode made of the element to be determined. Electrodeless discharge lamps are also being used in place of hollow cathode lamps, particularly for volatile elements such as Arsenic, Mercury, Cadmium, and Selenium.

The sample to be measured is generally atomised in a flame using conventional atomisers and burners. The radiation from the source transverses the flame into the entrance slit of a monochromator, which separates the desired resonance line from the other lines that may by present. Its intensity is measured by a conventional photomultiplier tube and amplifier.

Usually the flame also emits radiation of the same wavelength as the source. This background radiation is compensated by modulating the radiation from the source, and feeding the signal from the detector to a tuned a.c. amplifier, which rejects the d.c. signal contributed by emission from the flame.

Merits and Demerits

Atomic Absorption Spectroscopy is distinguished from most other analytical methods by its sensitivity, speed, ease of operation, accuracy and relative freedom from physical and chemical interferences. Its principle disadvantage is that it cannot be applied to non-metals since their lines lie in the vacuum ultra-violet region (i.e., below 2000 Å).

10.3.13 Instrumental Methods for the Measurement and Control of Radioactivity

Among the causes of air pollution, those arising from radioactivity are becoming increasingly important, both because of the growing number of industries employing radioisotopes and because of the development of nuclear plants (both for power and research), with the consequent production of fissile materials, radioactive waste, etc.

Ionising radiation causing pollution can be of various types—Alpha rays, Beta rays, Gamma rays, X-rays, proton beams, electron beams, slow neutrons and fast neutrons. Exposure to such radiations may cause considerable harm to living tissue. Therefore, it is becoming increasingly necessary to employ suitable detectors in order to afford some protection against the possibility of harm of this kind. The problem is so important, that particularly in the case of measuring instruments installed near sources of pollution, they are fitted with alarm signals which are set off whenever the radiation level exceeds—in the zone under control—the danger limit which has been established for the various forms of radiation.

The various measuring instruments are

1. Geiger-Muller counter
2. Ionising chamber
3. Scintillation detector
4. Photographic detector
5. Chemico-colorimetric indicator
6. Colorimetric measuring instrument and pocket control and safety detectors for workers

10.4 Biological Methods

By studying effects on plants and animals

It is often possible to identify a pollutant and a source of pollution by analysis of vegetation and of the body fluids of animals bred in the infected area. In certain cases, the tissues and organs of dead animals are very useful for the identification of toxic substances. The visible nature of the harm caused to plants and animals depends on the chemical nature of the pollutant, time of exposure, and climatic conditions.

This method can be used for studying the effect of both gases and particulate matter. Many tests are extremely sensitive and specific. Beryllium can be detected in amounts as low as 0.0003 μg and lead as small as 0.01 μg.

Sensory tests

Some substances can be detected because of their organoleptic properties; odour, lacrimation, and irritant effects on the respiration can be important factors, making it possible to assess the extent of pollution. But sensitivity of such methods is very limited.

QUESTIONS

1. Enumerate the various analytical methods available for monitoring air pollution.
2. Describe the following methods of analysing sample in detail.
   (a) Chemical methods
   (b) X-ray diffraction method
   (c) Mass spectrometric method
   (d) Thermal conductivity method
   (e) Continuous recording instrumental methods
   (f) Atomic absorption spectroscopy
   (g) Biological methods
3. Describe any one method of analysis for
   (a) Suspended particulate matter
   (b) Sulphur dioxide
   (c) Oxides of nitrogen
   (d) Lead
4. Write a note on: Chromatographic method.

11

Air Pollution Due to Automobiles

11.1 Introduction

This chapter discusses the contribution of automotive emissions in aggravating the air pollution menace. Various causes of the genesis and exodus of these pollutants have been identified and methods to control them have been outlined. Certain modifications in the engine design and operating variables are suggested. The possibilities of some alternatives as suitable substitutes for modern automobiles are also explored and future strategy in the Indian context has been proposed.

The three main types of automotive vehicles being used in our country are: (a) passenger cars powered by four stroke gasoline engines, (b) motor cycles, scooters and autorickshaws powered mostly by small two stroke gasoline engines, and (c) large buses and trucks powered mostly by four stroke diesel engines. Emissions from gasoline powered vehicles are generally classified as: (a) Exhaust emissions, (b) Crank-case emissions and, (c) Evaporative emissions. The amount of pollutants, that an automobile emits depends on a number of factors, including the design and operation (idle, acceleration, etc.). Of the hydrocarbons emitted by a car with no controls, the exhaust gases account for roughly 65%, evaporation from the fuel tank and carburettor for roughly 15% and blowby or crank-case emission (gases that escape around the piston rings) for about 20%. Carbon monoxide, nitrogen oxides and lead compounds are emitted almost exclusively in the exhaust gases. Effect of engine operating conditions on the composition of auto exhaust is shown in Table 11.1.

Diesel powered vehicles create relatively minor pollution problems compared to gasoline powered ones. The diesel engine exhausts only about a tenth of the amout of carbon monoxide exhausted by a gasoline engine, although its hydrocarbon emissions may approach those of the gasoline engine. Blowby is negligible in the diesel, since the cylinders contain only air on the compression stroke. Evaporative emissions are also low because the diesel engine uses a closed injection fuel system and because the fuel is less volatile than gasoline. The major problems of diesel engines are smoke and odour.

TABLE 11.1 *Effect of Engine Operating Conditions on the Composition of Auto Exhaust*

	Idle	Acceleration	Cruising	Deceleration
Air fuel ratio	11 : 1-12.5 : 1	11 : 1-13 : 1	13 : 1-15.1	11 : 1-12.5 : 1
Exhaust analysis				
CO%	4-6	0-6	1-4	2-4
NO, ml/m³	10-50	1000-4000	1000-3000	10-50
Hydrocarbons, ml/m³	500-1000	50-500	200-300	4000-12000
Unburnt fuel as % supplied fuel	4-6	2-4	2-4	20-60

1 ml/m³ = 1 ppm

11.2 Exhaust Emissions

The important exhaust emissions from a gasoline engine are carbon monoxide, unburnt hydrocarbons, nitrogen oxides and particulates containing lead compounds. These emissions vary with the air-fuel ratio, spark timings and the engine operating conditions.

To meet the exhaust emission standards for carbon monoxide and hydrocarbons, automobile manufacturers have used two basic methods. The first is to inject air into the exhaust manifold near the exhaust valves, where exhaust gas temperature is highest, thus inducing further oxidation of unoxidised or partially oxidised substances. The second basic method is to design cylinders and adjust the fuel-air ratio, spark timing and other variables to reduce the amounts of hydrocarbons and carbon monoxide in the exhaust to the point where air injection is not required.

Devices and methods to control hydrocarbon emissions fall into three classes (a) devices that modify engine operating conditions such as intake manifold vacuum breakers, carburation mixture improvers, throttle retarders, etc., (b) devices that 'treat' exhaust gases such as after burners, catalytic converters, absorbers and adsorbers and filters, and (c) use of modified or alternate fuels.

11.3 Crank-Case Emissions

Crank-case emissions consist of engine blowby which leaks past the piston mainly during the compression stroke, and of oil vapours generated into the crank-case. The quality of blowby depends on engine design and operating conditions. Worn out piston rings and cylinder liner may greatly increase blowby. These gases mainly contain hydrocarbons and account nearly for 25% of the total hydrocarbon emissions from a passenger car.

Emissions of hydrocarbons from the crank-case of automobiles can be largely eliminated by the positive crank-case ventilation (PCV) system. These systems recycle crank-case ventilation air and blowby gases to the engine intake instead of venting them to the atmosphere.

11.4 Evaporative Emissions

Through a short term experimental determination at the Indian Institute of Petroleum, it has been estimated that an average Indian passenger car would emit about 20 kg of hydrocarbons through evaporation, annually. For controlling evaporation of fuel from the carburettor and fuel system. systems are being developed that store fuel vapours in the crank-case or in a charcoal canister that absorbs hydrocarbons for recycling to the engine. Evaporative emissions might also be dealt with by changing the properties of gasoline such as reducing the volatility of fuel and replacing the C_4 and C_5 olefinic hydrocarbons in the fuel with the less reactive C_4 and C_5 paraffinic hydrocarbons. Mechanical methods can also be used to control evaporative emissions.

The exhaust gas. pollutants comprise of hydrocarbons, carbon monoxide, nitrogen oxides and lead compounds. Evaporative emissions essentially constitute the fuel evaporation from the fuel tank and carburettor and consist of hydrocarbons alone. The chief constituents of crank-case emissions are also hydrocarbons. Thus exhaust emissions contribute 100% to CO, $(NO)_x$ and lead compounds. Regarding hydrocarbons, the typical break up is exhaust 55%, evaporative emissions 20% and crank-case emission 25%. The exhaust and crank-case emissions also depend upon engine load conditions.

11.5 Formation of Photochemical Smog

The deleterious effects of automotive pollutants include toxic effects of CO and lead compounds and the formation of photochemical smog. The chief culprits in the 'smog' dilemma are the volumetrically lower concentrations of unburnt or partially burnt hydrocarbons and nitrogen oxides. The relative concentration of either pollutant varies with the engine operation.

The chemical reactions involved in smog production are extremely complex and not fully understood yet. The necessary conditions for smog formation are:

1. Sufficient quantity and concentration of unburnt hydrocarbons and nitrogen oxides in the atmosphere
2. Stagnant atmospheric conditions produced by meteorological thermal inversions
3. Strong sunlight

The formation of photochemical smog is ascribed to the following reaction:

Hydrocarbons + nitric oxide + Sunlight → photochemical smog.

Figure 11.1, gives the several reactions which may be involved in the overall reaction indicated. It is to be noted, that the 13 reactions occur simultaneously and in many a case the products of one reaction furnish the reactants for another.

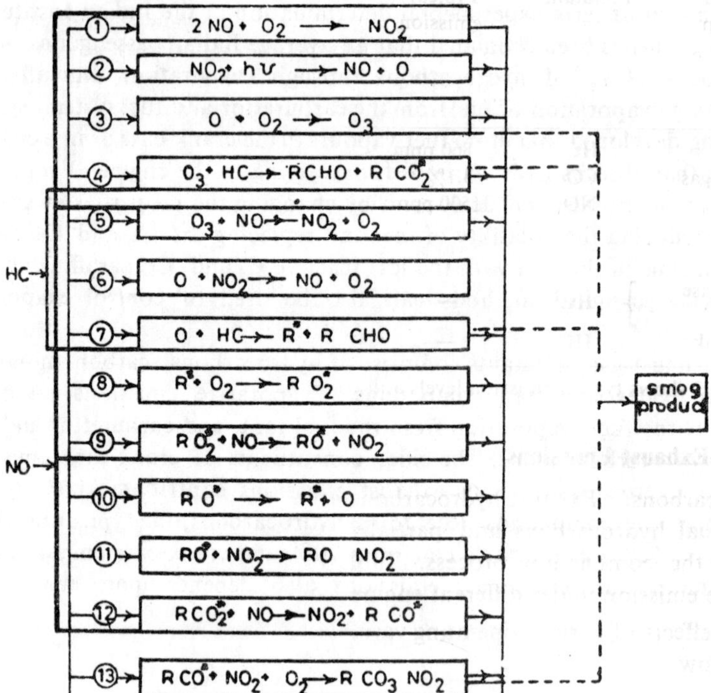

NO	NITRIC OXIDE	HC	HYDROCARBON
NO_2	NITROGEN DIOXIDE	O_2	OXYGEN
$h\nu$	SUNLIGHT ENERGY	O	ATOMIC OXYGEN
O_3	OZONE	R^*	ALKYL RADICAL
RO^*	OXYALKYL RADICAL	RO_2^*	PEROXYALKYL RADICAL
RCO^*	ACYL RADICAL	$RONO_2$	ALKYL NITRATE

RCO_3NO_2. COMPOUND X-ACYL PEROXYNITRATE

RCHO ALDEHYDE

Fig. 11.1 *Reactions leading to photochemical smog formation*

Formaldehyde and per-oxyacyl nitrate (PAN) are eye irritants whereas PAN and ozone cause plant damage. The oxidation reactions accompanied by formation of aerosols or hazes also result in eye irritation and plant damage.

Olefins are the major contributors to the 'smog forming potential' (SFP) of the exhaust gas. Test reported by Caplan, show that the olefins contribute nearly 86% to the reactivity of the exhaust emission, 84% to the reactivity of the carburettor hot soak evaporative emissions and 74% to the reactivity of tank evaporative emissions.

TABLE 11.2 *Typical Emissions Levels and Control Standards*

Source of emission	Pollutant	Typical emission	Emission as % wt of supplied fuel	Estimated pollution/ car/day	US national standards for control equipment 1970
	HC	860 ppm	6	337 g	180 ppm
Exhaust gas	CO	3.1%	—	2150 g	1%
	NO_x	1500 ppm	—	114 g	600–800 ppm or 350 ppm as NO
Crank-case Blowby	HC	—	2.5	164 g	—
Fuel Tank	HC	—	—	114 g	—

Table 11.2 shows typical emission levels and control standards

11.5.1 Exhaust Emissions

Hydrocarbons: Exhaust hydrocarbons are composed of a large number of individual hydrocarbons and partially oxidised hydrocarbons produced during the combustion process. Table 11.3, shows the variation in the average emission under different engine loadings.

The effects of various operating variables on exhaust emissions are discussed below.

TABLE 11.3 *Percentage of Total HC Contribution by Driving Modes*

Mode	Driving time		Exhaust volume		HC	A×B×C	HC
	Total %	Time ratio (A)	cu m/min	Volume ratio (B)	Concentration ppm (C)		contributed %
Idling	19	1.0	0.224	1.0	800	800	5
Accelerating	19	1.0	1.68	7.5	540	4050	26
Cruising	37	2.0	0.98	4.4	485	4270	27
Decelerating	25	1.3	0.224	1.0	5000	6500	42

11.6 Air-fuel Ratio

A decrease in the AF ratio increases the HC content (expressed as wt % of supplied fuel) in the exhausts of passenger cars at idle, but does not have any effect at part throttle.

On the basis of experiments conducted on a single cylinder engine operating at full throttle on propane, Daniel reported that 'relative HC concentration' measured with a dispersive infra-red analyser decreased with an increasing AF ratio and reached a minimum at an AF ratio leaner than stoichiometric. Methane and acetylene are the two hydrocarbons most greatly affected by the AF ratio.

11.6.1 Spark Timing

The HC emission generally decreases as the spark is retarded at constant power. A 10° spark retard from the optimum economy value causes 7–18% reduction as measured by a flame-ionisation analyser.

11.6.2 Combined Effects of AF Ratio and Spark Timing

The reductions in HC emission due to leaner AF ratios and due to retarded spark timing are additive, but while the former improves fuel economy, the latter impairs it. However, these fuel economy effects tend to balance each other when both methods are employed.

11.7 Combustion Chamber Surface-Volume Ratio

The effects of combustion chamber S/V ratio on the exhaust emissions have been investigated by Scheffler. Engines with S/V covering a wide range were installed in cars and tested according to the California schedule. The results are depicted by curve 1 in Fig. 11.2. Curve 2 in Fig. 11.2, gives results for a second group of cars having modified carburettors and distributors (combination of lean AF ratio and spark retard) to minimize HC emission. Figure 11.2 shows that engines ....ing low S/V ratios give lower HC ..ssions, but S/V ratio has no effect on CO concentration.

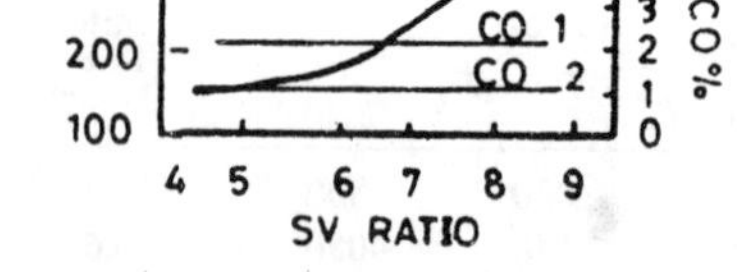

Fig. 11.2 *Effect of surface/volume ratio*

As shown in Table 11.4, the S/V ratio can be changed by altering the compression ratio (CR), number of cylinders, stroke-bore ratio or the displacement. It has been found that when S/V is decreased by decreasing CR, a greater reduction in HC emission is obtained than when S/V is decreased by changing other variables.

TABLE 11.4 *Variables Affecting S/V Ratio*

Variable	Effect on S/V ratio	Modern design practice limits
Increase in CR	Increases	7-10
Increase in no. of cylinders	Increases	4-8
Increase in stroke-bore ratio	Decreases	0.8-1.4
Increase in displacement	Decreases rapidly	2458-7374 cu cm

11.7.1 Compression Ratio

The exhaust HC emission decreases with a decrease in CR. A low CR engine has a larger clearance volume which traps in the cylinder a larger quantity of the HC-rich last part of exhaust gas as residual gas. This results in a lesser HC concentration in the tail pipe. The lower compression engine also has a higher exhaust temperature which may promote oxidation of the HC in the burned mixture and thereby reduce HC concentration in the exhaust. Also, due to low CR the fuel-air mixture will be less packed into stagnant pockets of the combustion chamber where it is likely to escape ignition. The lower CR engine is, however, less efficient and results in higher fuel consumption, though the effect is small. The net effect with reduced CR is a substantial reduction in both concentration and total volume of exhaust hydrocarbons.

11.8 Combustion Chamber Deposits

HC exhaust emissions are significantly enhanced with accumulation of chamber deposits. The quantity of deposits affects the CR of the engine by lowering exhaust temperatures and reducing clearances volume and, exhaust emission increases. Gases near the chamber wall are the last to be expelled out of the cylinder because of viscous resistance. This includes the gases in the clearance volume and forms what is called a 'quench zone'. Experimental evidence shows that residual gases contain about 11 times more HC concentrations as compared to exhaust gases. Thus, as the CR increases with accumulation of more deposits, the clearance volume becomes small and more residual gases with very high HC concentration are expelled out of the exhaust manifold. This greatly increases HC emission levels.

The volume of combustion chamber deposits rather than their weight seems to be more important, in that, it is the porosity of the deposits (void space) and degree of pore interconnections which affect exhaust emissions. During combustion, these pores and their interconnecting passages are filled with unburnt HC and escape burning. These are discharged back to the atmosphere during exhaust in the same state, increasing emission levels considerably. Uniform distribution of deposits is likely to provide more void space for the absorption of unburnt HC during combustion thereby raising exhaust emission levels.

11.8.1 Coolant Temperature

The data shows significant reduction in HC emission when tap water is changed to boiling water but further temperature increase of the coolant produces no further reduction in emission. Higher emission with tap water is probably due to condensation of fuel on some of the colder spots in the chamber.

11.8.2 Other Operating Variables

Existing literature shows that the concentration of exhaust gas HC decreases to a certain extent with increase in engine load, speed, exhaust back pressure, and decrease in valve-over-lap.

11.9 Control of Exhaust Emissions

Two main approaches to minimize exhaust emissions are:

1. Modifications in the engine design and operating variables
2. Treatment of exhaust gases after emission from the engine

1. The following modifications may help in cleaner exhaust.

(a) Use of leaner idle mixtures

(b) Use of leanest possible mixture and maximum spark retard compatible with good power output and drivability.

It is well known that burning leaner mixtures greatly reduces HC and CO emissions (Fig. 11.3). However, this reduction will result only if there is good mixture formation and distribution.

One of the methods of producing a uniform mixture and refined fuel-metering is the adoption of a three barrel or triple venturi carburettor incorporating a single primary barrel for idle and light load operation and two large secondary barrels having the air capacity required for good performance.

It has also been observed that a dual manifold system ensures better fuel distribution.

(c) Use of minimum valve-over-lap necessary.

(d) Pre-treatment of the mixture to improve vaporisation and mixing of fuel with air. Among the methods employed for the purpose are:

(i) Use of narrower venturies to produce higher air speeds and better fuel atomisation,

(ii) Provision of exhaust-heated hot spots to vaporise the heavier liquid droplets in the air stream,

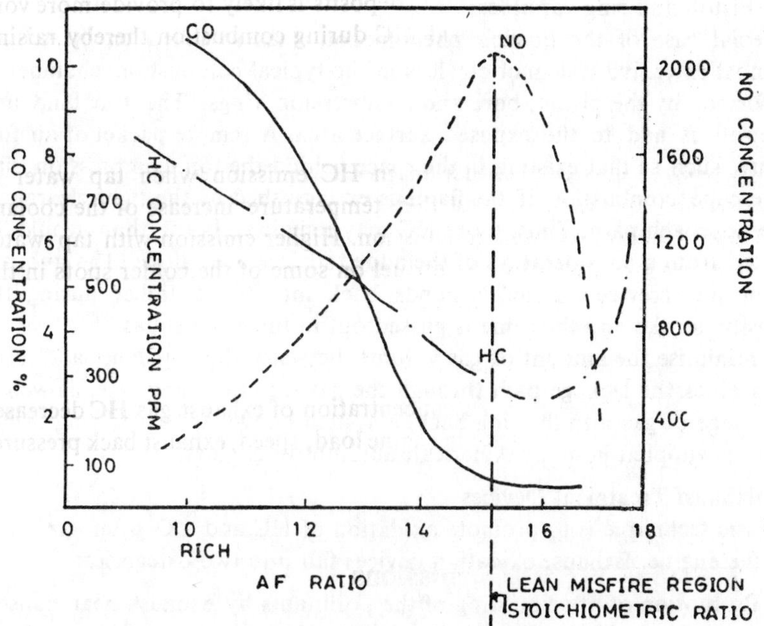

Fig. 11.3 *Effect of AF Ratio*

(iii) Use of exhaust heat to pre-heat the mixture at part-loads,

(iv) Use of automatic transmission (because manual transmission produces a high intake vacuum during deceleration as the engine is motored at high speeds, thus giving high HC emissions and momentary deceleration during manual shifting of gears as well as the subsequent action of acceleration of the pump which increases HC concentration),

(v) Special devices for reducing or cutting off fuel supply during deceleration.

(e) Low quench combustion chambers

The combustion of petrol-air mixture within any engine is by no means an instantaneous event. The flame is propagated from the ignition point at varying rates depending among other things, on turbulence, the presence of petrol vapour and even the size of the fuel particles suspended in the air. The flame is, in fact, completely quenched as it approaches the chamber walls; thus leaving a thin layer of unburned HC, part of which mixes with and escapes with the products of combustion. Certain areas in the chamber may possibly be predisposed to causing this phenomenon. HC emissions are reported to increase when flame quenching occurs. Flame photography and sampling techniques corroborate the above hypothesis.

The surface-volume ratio is a significant parameter to be considered in designing the chamber, for a low S/V ratio is conducive for low emissions. The wall and mixture temperature, as also the amount of deposits on the walls influence HC emission concentrations.

(f) Piston and ring variables

A special case of the quench phenomenon is the so called 'crevice-effect'. The most extensive system of crevices in the typical combustion chamber is that formed by the piston, bore and compression rings. The top land and ring grooves add to the exposed surface area. A remote packet of air fuel mixture, such as that existing in the space behind the top compression ring may escape combustion if the flame is extinguished by the tight clearances in the approach path. Thus, more HC may remain unburnt than would be expected from a consideration of the added surface area alone. The unburnt fuel in the 'crevice volume' expands back into the chamber during the expansion stroke and the same is passed out in the exhaust gas.

To minimise the amount of gas volume between the top land and bore and to block the leakage path through the gas into the crank case, as well as the passage of gas into the ring back, a system known as 'sealed ring-orifice system' is adopted in many American automotive engines.

2. Exhaust Treatment Devices.

The basic technique is to promote oxidation of HC and CO after emission from the engine. Exhaust oxidation devices fall into two categories:

1. Promotion of after burning of the pollutants by exhaust heat conservation, introduction of additional air and by providing sufficient volume to ensure adequate reaction time,
2. Use of catalytic converters

   In after-burners air from an engine driven blower is injected into the exhaust stream just after the exhaust valve. Sometimes an additional source of ignition is provided in the exhaust passage to initiate the after burning under certain engine conditions.

Catalytic converters (Fig. 11.4) depend on the action of a catalyst containing certain exotic chemicals to convert HC and CO emissions to their oxidised products. Extra air is introduced by an engine driven blower. Vanadium pentoxide (V_2O_5) is one of the successful catalysts used so far. Although non-flame after-burners are reported to reduce exhaust emissions, they require replacement of spent catalyst with fresh material at regular intervals (one year) and entail about 7% fuel economy loss.

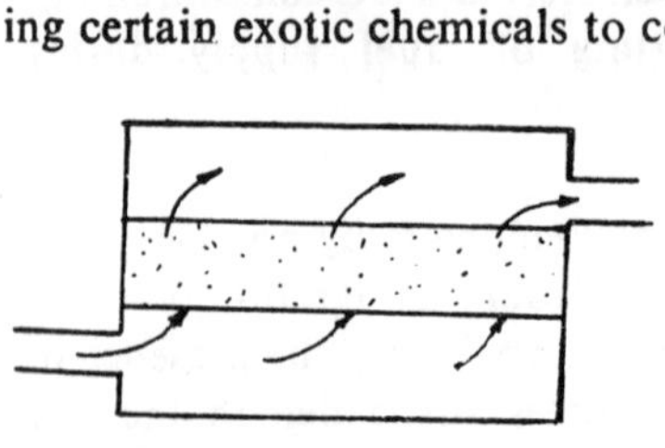

Fig. 11.4 *Catalytic burning*

Catalytic converters hold great promise for the future, and their wide acceptance seems to be around the corner. However, they have certain limitations, viz.

1. 'Poisoning' of catalyst by lead compounds in fuel
2. Deterioration with time
3. Pressure loss and noise problems

11.10 Other Methods

(a) Petrol Injection—The petrol injection system provides the flexibility needed to meet the engine requirements and eliminates the problem of mixture distribution altogether and hence has lower levels of exhaust emissions and better specific fuel consumption. Hydrocarbon emission is controlled by the ability to maintain optimum mean AF ratio at all operating conditions. Nitrogen oxides are also controlled because of the relatively lower peak temperatures obtained.

The factors against its wide adoption are the comparatively high cost (about 6-8 times that of a conventional carburettor) and the difficulty of servicing.

(b) Stratified charge engine—It operates with very lean air-petrol mixtures, depending on a localised rich mixture region near the spark plug to initiate combustion. The emission levels of all the three pollutants CO, HC, and NO_x are very much lower with this engine at part throttle due to the leanness of the mixture. Mixtures as lean as 30 : 1 can be successfully used.

11.11 Nitrogen Oxides in Exhaust Gas

The concentration of NO_x in exhaust gas is found to depend mainly on the peak temperature and pressure and the supplied air-fuel mixture composition.

The effect of AF ratio is shown in Fig. 11.3. The NO_x concentration rises rapidly as the mixture is made leaner around the stoichiometric ratio of 14.6. As the mixture is made much leaner, the nitrogen oxides again drop sharply. Cruising and part throttle acceleration conditions contribute nearly all of the NO_x emissions.

Table 11.5 gives the variables influencing NO_x concentration in exhaust gas.

TABLE 11.5 *Engine Variables Affecting Nitrogen Oxides Concentration in Exhaust Gas*

Variable	Influence
Leaning mixture ratio	Increases (peaks at 15–18 AF ratio)
Advancing spark timing	Increases
Increasing engine load	Increases
Increasing CR	Increases
Increasing engine speed	Increases or decreases depending on mixture ratio

Reduction of NO concentration by changes in the variables listed in Table 11.5 is almost always associated with corresponding increases in fuel consumption. This approach is therefore not attractive. Also, the use of richer mixtures to reduce NO emission will increase CO emission. The following methods have been tried to reduce the concentration of oxides of nitrogen.

1. Exhaust gas recycling
2. Water injection into inlet manifold
3. Exhaust manifold reactor with recycling
4. Catalytic reduction
5. Lean operation

11.11.1 Exhaust Gas Recycling

The exhaust re-circulation systems utilise up to 15% of the exhaust stream to recycle into the intake manifold with the fresh air-fuel mixture. The exhaust gas dilutes the fresh charge and thus lowers the flame temperature and consequently the nitrogen oxides. The fuel-air equivalence ratio has a marked influence on the effectiveness of exhaust recycling.

The estimated effect of exhaust gas recycling on NO concentration for various equivalence ratios and recycle percentages is shown in Fig. 11.5.

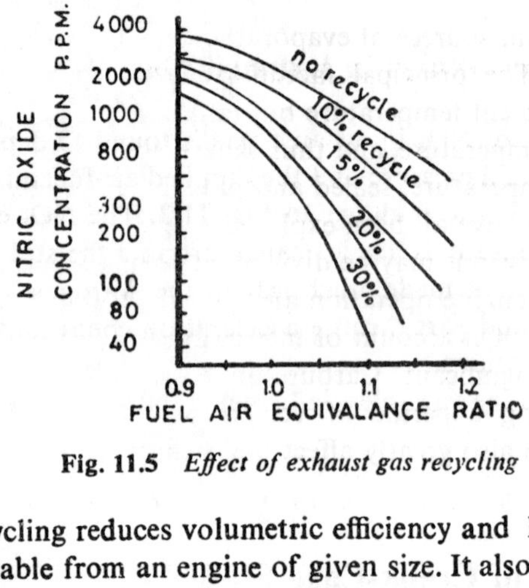

Fig. 11.5 *Effect of exhaust gas recycling*

Exhaust gas recycling reduces volumetric efficiency and hence maximum power output available from an engine of given size. It also adversely affects the vehicle drivability.

11.11.2 Water Injection

The water injection, by lowering the flame temperature leads to reductions in nitrogen oxides. At the leaner fuel-air ratios, water injection is somewhat less effective in reducing NO_x than at the richer F-A ratios.

11.11.3 Exhaust Manifold Reactor with Recycling

Installing a modified recycling system on a vehicle coupled with various exhaust manifold reactors will result in substantially reduced emissions. Drivability problems may perhaps occasionally arise. Air injection has great influence on reactor performance. The fraction between 0.15 and 0.4 ensures good oxidation.

11.11.4 Catalytic Reduction

The laboratory treatment of engine exhaust with a copper-cobalt-alumina catalyst can reduce NO by 80%. Approximately 1.5% CO in the exhaust is necessary to supply the reducing agent. This requires rich setting of the carburettor which lowers fuel economy.

11.11.5 Lean Operation

This approach merely supplies more air as a diluent rather than exhaust gas. Severe drivability problems due to momentary leanness during transient conditions causing misfires may be overcome by supplying a homogeneous mixture under transient conditions.

11.12 Control of Evaporative Emissions

There are two main sources of evaporative emissions viz., the fuel tank and the carburettor. The principal factors governing tank emissions are fuel volatility and ambient temperature but tank design and location can also influence tank temperature and thus tank emissions. Insulation of the fuel tank to reduce temperature, sealed and pressurised fuel systems, and vapour collection systems have all been explored to reduce tank emissions.

Carburettor emissions may be divided into two categories; 'running' losses occurring during engine operation and 'hot soak' losses occurring when the vehicle is parked. On account of internal venting of carburettors the running losses are insignificant. Carburettor losses are substantial only during hot soak following a period of vehicle operation. The fuel volatility and carburettor design also greatly affect the carburettor emissions.

11.13 Control of Crank-Case Emissions

These consist of engine blowby gases, ventilation air and crank-case lubricant fumes. For air pollution blowby is most important and the principal constituents in blowby gases are hydrocarbons. The quantity of blowby depends on many engine design and operating variables and it may be enhanced by poor engine maintenance and by use of crankcase oils.

Designers are shifting crank-case exhaust vents from simple open ending to a feed back. New engines equipped with this Positive Crank-case Ventilation (PCV) system return crank-case vapours through a vacuum valve, back to downstream side of the carburettor. Recycling burns hydrocarbons in the cylinders, dropping overall pollution by 25% (Fig. 11.6).

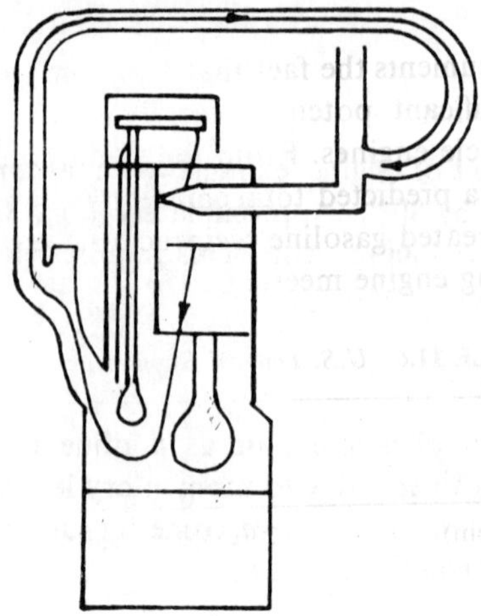

Fig. 11.6 *Crank-case fume feedback*

Starting with 1963 models, all gasoline passenger cars and trucks sold in the USA are equipped with PCV systems to control crank-case emissions.

11.14 Alternatives

1. *Electric Car*

Contrary to the general impression that an electric car holds great promise as a pollution-free automobile, its wide use for vehicular application may accentuate and aggravate the problem. Power generating capacity would have to be stepped up—a Herculean task indeed if not an impossible one. Furthermore, more fuel will be burnt at the power plant on an equivalent basis, to supply necessary electrical energy to charge the batteries of the electric car. Also, the increased stack emissions like oxides of nitrogen, particulates and sulphur dioxide (a much more severe air pollutant than CO) will defeat the very purpose.

2. *Natural Gas*

Compressed natural gas can essentially eliminate the pollutants but supply and proved reserves are limited.

3. *Wankel Engine*

This engine being compact has more space available for emission-control-equipment, can operate on fuel of low octane rating and what is more important, NO_x emissions are 30% of those of piston engines. However, HC emissions are decidedly higher and CO emissions equal. The temperature of exhaust gas being higher by 150°C, an exhaust reactor can be employed most effectively.

4. *Gas Turbine*

Technical literature documents the fact that a properly designed automative gas turbine offers significant potential for alleviating air pollution caused by conventional otto-cycle engines. From the stand point of emissions this potential is projected by a predicted total emission level of only 2–4% of that characteristic of an untreated gasoline reciprocating engine and only 50% of that of a reciprocating engine meeting 1975 standards (Table 11.6).

TABLE 11.6 *U.S. Federal Regulations*

Year	HC_x g/mile	CO g/mile	NO_x g/mile
1973	2.2(50 ppm)	23(0.5% by vol)	3.0(250 ppm)
1975	0.5	23(0.5% by vol)	0.9
1976	0.41	3.4	0.4

Because of high air-fuel ratios associated with gas turbines, CO and HC_x emissions are usually negligible. NO_x concentration is 100 ppm with 200% excess air in a gas turbine as compared to 1200 ppm in a SI engine. However, NO_x emissions expressed as g NO_x/hphr are comparable in magnitude. NO_x emissions can be reduced substantially by lowering the flame temperature by primary zoneleaning. It is possible to reduce NO_x levels up to 85% by direct water injection in a combustion primary zone which will also reduce its part-load fuel consumption.

In fact, the standard production passenger car (80 hp) with a WR 26 turbine engine installed in it, developed by Williams Research Corporation and the other standard production model Dodge Dart with the latest EPA-baseline Chrysler Gas turbine engine installed for low emission passenger cars (130 hp approx) in the United States, show the encouraging prospect. But prohibitive manufacturing costs of its cost-critical components preclude the possibility of general acceptance of a gas turbine as an automobile in the foreseeable future.

5. *Ammonia-fueled SI Engine*

The use of ammonia will eliminate two main pollutants, viz., CO and HC_x. The emission of NH_3 in engine exhaust however, is to be avoided because of its irritating odour and toxic effect. This can be minimized by adding hydrogen in small quantities—about 2% (obtained by dissociation of a part of ammonia fuel in the vehicle) which will act as a combustion promoter in accelerating the burning of ammonia.

6. *Unleaded-gasoline Powered SI Engine*

The use of leaded gasoline to improve antiknock characteristics of the fuel has ironically proved to thwart attempts of modern anti-pollution devices by causing deterioration in both their life and efficiency. Lead compounds are toxic and conductive to more exhaust HC emissions. The most immediate benefit accruing from the removal of lead from gasoline is a 20 ppm reduction in HC emissions from new cars. This reduction is effected by the process of natural oxidation of the hydrocarbons in the exhaust gas which is apparently inhibited by the presence of lead. The lead deposits are also responsible for spark plug fouling with consequent misfires and accompanying drastic increase in HC emissions. Fouling is the basic cause of spark plug replacement in service. Without lead deposits spark plug life would be improved and the anti-foul phosphorous additives used in some gasoline could be eliminated. Engine rusting due to chlorine and bromine containing scavengers will be reduced. Counter-balancing these potential benefits however, is a potentially serious problem aggravated by unleaded fuel-exhaust valve recession which is serious under high speed and/or high load conditions.

11.15 Air Pollution Due to Automobiles in Calcutta

NEERI carried out a short-term survey in Calcutta to collect information on the level of carbon monoxide pollution due to automobile exhausts.

High traffic density streets in Calcutta were studied by measuring carbon monoxide levels during different phases of traffic: Heavy traffic, traffic jams with slow movement, normal and light traffic.

It was observed that the level of carbon monoxide due to auto exhausts at street level varied from as high as 35 ppm to less than 10 ppm at all selected streets depending upon traffic conditions. The levels of carbon monoxide recorded during the observations at Calcutta are comparable with that in the large cities of the world such as New York, Chicago, London, etc., as can be seen from Table 11.7.

TABLE 11.7 *Level of Carbon Monoxide*

City	Maximum 1 hr values—ppm
London	58
Chicago	46
Los Angeles	43
Washington	41
New York	27
Calcutta	35

The Calcutta data shows that although the number of vehicles in Indian cities is smaller compared to Western cities, yet the intensity of pollution caused by them is apparently the same, if not more. This can be attributed to old vehicles with poor engine performance.

There is no simple and inexpensive solutions to the complex problem of auto-exhaust emissions. A beginning could however be made on new Indian Petrol engine cars by adjusting carburettors and air-fuel ratios to reduce carbon monoxide emissions by about 25% in the first instance. Further reductions would need major engine modifications which may be kept in view for future models.

11.16 Conclusion

Extensive research conducted during the last 15 years has given valuable insight into and complete understanding of the causes and control of automative engine emissions. The cost, complexity and maintenance of emission-controlled piston engines in passenger cars are all increasing. Future strategy should be to go in for unleaded gasoline even at the expense of a 2–4% increase in cost that it may entail.

The various alternatives for automative application discussed above are no doubt plausible but problematical today. While efforts should be made to make them more amenable to vehicular use, the gasoline powered SI engine with emission-control devices seems to be the best compromise in the national interest.

We must go the wholehog to forestall the devastating effects of air pollution due to automative vehicles by incorporating methods to control emissions. A uniform regulation is also necessary for automobile pollution control in our country.

The automobile industry in India must increase the scope and efficiency of its research. It is inevitable that as traffic density increases, legislation concerning safety, noise, and exhaust emissions will become imperative and mandatory.

QUESTIONS

1. How are the emissions from gasoline-powered vehicles classified?
2. What is the effect of engine operating conditions on the composition of auto exhaust?
3. Write short notes in 15 sentences on
   (a) Exhaust emissions
   (b) Crank-case emissions
   (c) Evaporative emissions
4. What are the various approaches to minimize exhaust emissions?
5. Suggest suitable methods of minimizing crank-case emissions and evaporative emissions.
6. Discuss the recent developments in the automobile industry to reduce air pollution.
7. Write brief notes on:
   (a) Air-fuel ratio
   (b) Compression ratio.

12

Photochemical Air Pollution

12.1 Introduction

Photochemical air pollution occurs predominantly in highly motorised areas and where inversion conditions prevail in the atmosphere. For e.g., Los Angeles. It is due to the action of sunlight on the two pollutants—hydrocarbons and nitrogen oxides. The major source of these two pollutants are the exhaust gases from the automobiles. Air pollution due to photochemical smog is characterized by intense eye irritation, unusual odour, and reduction in atmospheric visibility. It also causes damage to plants. Other undesirable effects of photochemical smog are cracking of rubber and fading of dyes.

12.2 Theory of Formation

Photochemical smog is formed due to photochemical oxidation of hydrocarbons and nitrogen oxides. Among the theories proposed to account for the formation of smog in the atmosphere of Los Angeles the theory given by Prof. A.J. Haagen-Smit is found to be generally acceptable as it is very comprehensive.

It has been found that at the time of photochemical smog formation, there is a considerable increase in the amount of ozone and oxidant material. The ozone is not found in appreciable amounts at night, but only during the day. It begins to form simultaneously throughout the Los Angeles basin shortly after dawn. These facts clearly indicate that photochemical formation of ozone or oxidant from impurities, takes place due to the action of sunlight.

12.2.1 Nitrogen Dioxide Photochemical Reaction

In the beginning, ultra-violet light energy is absorbed by NO_2 (Eq. 12.1). The highly energized molecule (NO_2^*) then decomposes into nitric oxide and atomic oxygen (Eq. 12.2). The atomic oxygen reacts quickly with molecular oxygen to form ozone (O_3). However, unless some other energy—absorbent molecule is present, the ozone will decompose rapidly. But if a third body (X) is present a stable ozone molecule is formed (Eq. 12.3). If nitric oxide is present it reacts with ozone to form NO_2 and an oxygen molecule (Eq. 12.4).

$$NO_2 + h\nu \rightarrow NO_2^* \quad (12.1)$$

$$NO_2^* \rightarrow NO + O \quad (12.2)$$

$$O + O_2 + X \rightarrow O_3 + X \quad (12.3)$$

$$O_3 + NO \rightarrow NO_2 + O_2 \quad (12.4)$$

In the above reactions, NO_2 behaves like a catalyst.

Further, in the presence of certain hydrocarbons, other reactions take place. Some of the atomic oxygen, ozone, and nitric oxide react with hydrocarbons to form a variety of products and intermediates with which even further reaction can probably take place. Some of the products formed are aldehydes like formaldehyde and acrolein, peroxides, and peroxy acetyl nitrate (PAN).

If nitric oxide is also present in the air, following reaction can take place

$$NO + O_3 \rightarrow \dot{NO_2} + O_2 \quad (12.5)$$

If ozone is present in excess, then

$$2NO_2 + O_3 \rightarrow N_2O_5 + O_2 \quad (12.6)$$

In presence of water vapour, the following reaction can take place

$$N_2O_5 + H_2O \rightarrow 2\ HNO_3 \quad (12.7)$$

Thus nitric acid may be formed.

12.2.2 Sulphur Dioxide Photochemical Reaction

Ozone may be formed in the atmosphere as a byproduct during the photochemical oxidation of sulphur dioxide to sulphuric acid. The reaction takes place as follows:

$$SO_2 + h\nu \rightarrow SO_2^* \quad (12.8)$$

$$SO_2^* + O_2 \rightarrow SO_4 \quad (12.9)$$

$$SO_4 + O_2 \rightarrow SO_3 + O_3 \quad (12.10)$$

$$SO_3 + H_2O \rightarrow H_2SO_4 \quad (12.11)$$

The quantum efficiency for this process is low.

12.2.3 Aldehydes Photochemical Reactions

The photochemical oxidation of aldehydes can be represented as follows:

$$R\ CHO + h\nu \rightarrow R\ CHO^* \quad (12.12)$$

$$R\ CHO^* + O_2 \rightarrow R\ CO_3H \quad (12.13)$$

$$R\ CO_3H + O_2 \rightarrow R\ COOH + O_3 \quad (12.14)$$

The primary photochemical decomposition of aldehydes produces organic free radicals. For example, acetaldehyde decomposes into methyl and formyl radicals.

$$CH_3CHO + h\nu \rightarrow CH_3 + HCO \quad (12.15)$$

Similarly, acetone decomposes into methyl and acetyl radicals.

$$(CH_3)_2CO + h\nu \rightarrow CH_3 + CH_3CO \quad (12.16)$$

Some of these reactions have been shown to occur in the Los Angeles atmosphere and in experimental smog chambers.

12.3 Factors Affecting Photochemical Reactions

The nature of the photochemical reactions that take place in the atmosphere depends on a number of factors. Some of these factors are light intensity, hydrocarbon reactivity, ratio of hydrocarbons to nitric oxide, presence of light absorbers, and meteorological variables. The height and intensity of atmospheric inversion is a prime factor. Other things being equal, the lower the inversion base, the more intense will be the smog.

12.4 Measurement of Photochemical Smog

As mentioned earlier, photochemical smog is a complex mixture of several compounds. Among its various constituents, ozone and PAN (Peroxy acetyl nitrate) are significant. Even though ozone is the major oxidising agent in smog, there are small amounts of other agents which vary in quantity and interfere with ozone determination. On account of this, all oxidising compounds in smog with oxidising power more than molecular oxygen are combined and known as 'oxidants'.

Generally, 'oxidants' are measured by passing the air stream containing the oxidant into a neutral solution of potassium iodide. The oxidant reacts to release iodine and the tri-iodide ion, which can be measured colorimetrically. The two interfering gases during the measurement are NO_2 and SO_2. PAN may be measured either by gas chromatography or infrared spectroscopy.

12.5 Effects of Photochemical Smog

The effects of photochemical smog on human beings, plants, and materials have been studied in places like Los Angeles, where it usually occurs. Additional information has also been obtained by simulating photochemical smog in environmental chambers. Following are the important effects of photochemical smog.

1. *Eye Irritation*

Probably, the compounds responsible for eye irritation are formaldehyde, acrolein, PAN, and peroxy benzoyl nitrate (PB_zN).

2. *Vegetation Damage*

The effects observed are silvering and bronzing of the underside of leaves, followed by collapse of cells, and necrosis. Growth retardation has also been reported. The three principal phytotoxicants are ozone, nitrogen dioxide, and PAN. This has resulted in economic loss. The details are given in Chapter 7.

3. *Visibility Reduction*

This is perhaps the most commonly observed effect of photochemical smog. The aerosol particles causing the photochemical smog contain compounds of carbon, oxygen, hydrogen, nitrogen, sulphur, and halides. The size of the particles is about 0.3 μ. The liquid phase is largely made up of organic matter. Attempts are now being made to establish a quantitative relationship between visibility and oxidant index.

4. *Cracking of Rubber*

This is primarily due to the ozone constituent of photochemical smog. An important economic effect of smog in Los Angeles, is the deterioration of the side walls of automobile tyres. To overcome this problem, an anti-ozonant is being used. The other rubber products that are affected are gaskets, hoses, and wire insulation. This has been discussed in Sec. 8.4.

5. *Fading of Dyes*

This is another important economic effect of photochemical smog, which has been discussed in Sec. 8.4.

QUESTIONS

1. Explain the theory of formation of photochemical smog.
2. What are the factors affecting photochemical reactions?
3. Explain the term 'oxidants'. How are they measured?
4. Explain the effects of photochemical smog on human beings, plants, and materials.
5. Write short notes on:
   (a) Photochemical smog
   (b) Oxidants
6. "Photochemical air pollution occurs predominantly in highly motorised areas and where inversion conditions prevail in the atmosphere." Substantiate the statement with special reference to Los Angeles.

13

Control of Air Pollution by Equipment

13.1 Introduction

Basically, four means are available for the control of effluent discharges into the atmosphere and thereby control of their detrimental effects. They are:

1. Reduction of pollutant discharge at the source by the application of control equipment
2. Reduction at the source through raw material changes, operational changes, or modification, or replacement of process equipment
3. Dilution of the source discharge by the use of tall stacks
4. Dispersion of source locations through allocation of land usage, i.e., proper planning and zoning of industrial areas

But the most effective methods are reduction at the source by the application of control equipment and process control. If air pollution problems are properly considered when an industry is designed and built, real economy can be effected. But in most cases air pollution control is an afterthought and ways and means have to be devised to treat the polluted effluents.

To remove the particulate matter from gas streams, various types of control equipment are available. But to select the required equipment, certain basic data must be available:

1. Quantity of gas to be treated and its variation with time
2. Nature and concentration of the particulate matter to be removed
3. Temperature and pressure of the gas stream
4. Nature of the gas phase (for solubility and corrosive effects)
5. Desired quality of the treated effluent, i.e., efficiency of removal of particulates required

13.2 Objectives of Using Control Equipment

1. Pr vention of nuisance
2. Prevention of physical damage to property
3. Elimination of health hazards to plant personnel and to the general population

4. Recovery of valuable waste products
5. Minimisation of economic losses through the reduction of plant maintenance
6. Improvement of product quality

13.3 Types of Collection Equipment

A list of common types of collection equipment for aerosols (particulates) is as follows:

1. Settling chambers
2. Inertial separators (other than cyclones)
3. Cyclones
4. Filters
5. Electrostatic precipitators
6. Scrubbers or wet collectors

This classification is somewhat arbitrary since, for example, the installation of water sprays in a cyclone may change the collection method to scrubbing, while the insertion of baffles changes a settling chamber into an inertial separator.

13.4 Efficiency of a Separating Device

The commonly used method of defining efficiency η is as follows:

$$\eta = \frac{\text{Quantity of particulates collected from the gas}}{\text{Quantity of particulates initially present in the gas}}$$

Efficiency is generally expressed as a percentage.

13.5 Use of Collector in Series

Air pollution control systems frequently incorporate two or more contaminant collectors in series. This has several advantages.

1. The primary collector is used to remove the larger particles from the gas stream to enhance collection of finer particles in the secondary collector. Thus, the primary collector also reduces the load on the secondary collector.

2. The secondary collector(s) may act as a reserve in the event of malfunction in the upstream collector.

3. Series collectors may be used to separate materials in a mixture, i.e., one collector may remove a reusable or soluble material, the other a waste material.

4. The primary collector may act as a fire protection device by preventing a source of ignition such as incandescent particles, or readily combustible material such as, cotton, linen from entering the secondary collector.

5. A primary collector may act as a concentrator so that there are two exit gas streams, one with a small gas flow and high particle loading, the other with a large gas flow and low particle loading. Either or both of the exit gas streams may go to the secondary chambers.

6. A primary collector may act as a plenum chamber from which several secondary collectors may be operated in parallel.

13.6 Settling Chambers

The settling chamber is the simplest type of equipment used for the collection of solid particulates. It consists of a chamber in which the carrier gas velocity is reduced so as to allow the particulates to settle out of the moving stream under the action of gravity.

The most common form is a long box like structure, with an inlet at one end and an outlet at the other, set horizontally, often on the ground. It can be constructed from brick and concrete. The carrier gas is made to pass at low velocities. The solid particulates having higher density than the surrounding gas, settle under the influence of gravity on the base of the chamber, from where they are removed through hoppers. A typical settling chamber is shown in the Fig. 13.1.

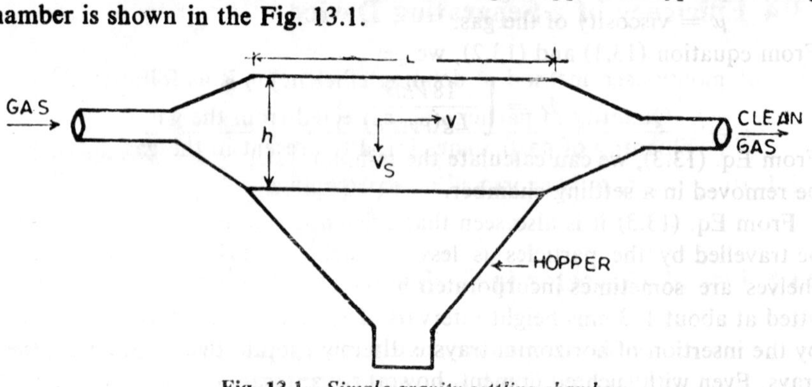

Fig. 13.1 *Simple gravity settling chamber*

The gas velocity must be sufficiently low (less than about 3 m/s to prevent re-entrainment of the settled particles; less than 0.5 m/s for good results. To minimise turbulence and ensure uniform velocity, curtains, rods, and wire mesh screens may be suspended in the chamber. The pressure drop through the settling chamber is usually small and consists mostly of entrance and exit losses. Installation costs are low because of the simple structure.

A simplified analysis of the gravity settling chamber assumes that the solids move along the chamber with the velocity of the gas and also settle with Stoke's velocity. A particle entering the chamber at the top will be collected by the chamber, if its settling time is the same (or less than) the time the gas takes to pass through the chamber.

Thus, $$\frac{h}{V_s} = \frac{L}{V}$$

or $$V_s = \frac{hV}{L} \tag{13.1}$$

where,

L = length of the chamber
V = horizontal velocity of the carrier gas
V_s = settling velocity of particulates
h = height through which the particulates travel before settling down.

By Stoke's Law

$$V_s = \frac{g(\rho_p - \rho)D^2}{18\,\mu} \tag{13.2}$$

where,

D = Diameter of the particle
g = accelaration due to gravity
ρ_p = density of the particle
ρ = density of the gas
μ = viscosity of the gas.

From equation (13.1) and (13.2), we get

$$D = \left[\frac{18 Vh\mu}{Lg(\rho_p - \rho)}\right]^{1/2} \tag{13.3}$$

From Eq. (13.3), we can calculate the minimum size of the particle that can be removed in a settling chamber.

From Eq. (13.3) it is also seen that efficiency is improved if the height to be travelled by the particles is less. To achieve this, horizontal trays or shelves are sometimes incorporated in the chambers. The trays (plates) are fitted at about 1–3 cms height intervals. The increase in efficiency obtained by the insertion of horizontal trays is directly proportional to the number of trays. Even with such equipment, however, the minimum particle size which can be removed in practice is about 10 μ. Also, the use of this modified settling chamber is limited by difficulties in cleaning the closely spaced trays and by their tendency to warp during high temperature operation.

13.6.1 Advantages

1. Low initial cost
2. Simple construction
3. Low maintenance cost
4. Low pressure drop
5. Dry and continuous disposal of solid particulates

6. It can be constructed out of almost any material
7. Temperature and pressure limitations imposed only by materials of construction used

13.6.2 Disadvantages

1. Large space requirements
2. Only comparatively large particles (definitely not less than 10 μ, if very dense and 40 μ, if of low density) can be collected.

13.6.3 Applications

Industrial application of this equipment is limited. Settling chambers are used generally to remove particulates above 40 μ in diameter. However, fine materials such as carbon black and various metallurgical fumes form agglomerates which have enough mass to permit collection in settling chambers. Settling chambers are used widely for the removal of large solid particulates from natural draft furnaces, kilns, etc. They are also, sometimes used in the process industries, particularly the food and metallurgical industries, as a first step in dust control.

Because of simplicity of construction and low maintenance costs, gravity settling chambers have found quite widespread application as pre-cleaners for high efficiency collectors. This reduces the inlet dust loadings to the second stage cleaner and can remove large highly abrasive materials, thus reducing maintenance costs of high efficiency equipment which is more subject to abrasive deterioration.

13.7 Inertial Separators

This is another type of pollution control equipment.

13.7.1 Principle

This type of control equipment includes all collectors which utilise the relatively greater inertia of the dispersoid to effect the particulate-gas separation.

Two types of equipment utilise this fundamental principle. They are inertial or impact separators and cyclonic separators.

Inertial or Impact Separators

They employ incremental changes of direction of the carrier gas stream to exert the greater inertial effects of the dispersoid.

Cyclonic Separators

They produce a continuous centrifugal force as a means of exerting the greater inertial effects of the dispersoid. The Cyclonic Separator is an important and popular type of dust removal equipment and is explained in detail in Sec. 13.8.

13.7.2 Types of Inertial or Impact Separators

Here there are three fundamental types:

1. Baffle type separator
2. Louvre type separator
3. Dust traps

Baffle Type Separator

This is the simplest type of impingement separator. Here the gas stream is made to follow a tortuous flow path, which is obtained by the insertion of staggered plates into the path of the gas stream. As a result, the gas is subjected to a series of sudden changes of direction, with the resulting impaction of the particles on solid surfaces.

This device is suitable for removing particles larger than 20 μ in diameter. Frequently, commercial units are equipped with some machanism for cleaning the impaction surfaces. For example, rappers are provided to remove the collected materials. Other units employ a continuous or intermittent flowing film of water over the collecting surfaces.

There are many different baffle designs as shown in Fig. 13.2. They can be in the form of louvres, scrolls, or venturis, etc. Scrolls as shown in Fig. 13.2(d) have air foil surfaces that minimise the formation of down stream turbulence. The air-foil design of the scrolls promotes a continuous trajectory of the particles without boundary layer separation and loss due to turbulence. This promotes particle collection on the next scroll. If it gets past that one, perhaps the particle will be collected on the next scroll, etc. Figure 13.2(c) shows venturi type baffles. In this, the carrier gas must traverse a steadily narrowing path. The venturis are rectangular in cross section. An impingement baffle is usually kept opposite the venturi exit to improve the collection of particles in 20–50 μ range.

The baffle type separator, is simple to operate. It has no moving parts. It is slightly more expensive to construct than a simple open settling chamber, because of the interior work involved in fabricating and installing the baffles. But owing to high impaction velocity, abrasion is often severe. The efficiency is totally dependent on the degree of impaction which occurs. Baffled chamber efficiency is a function of four variables: the number, length, spacing, and configuration of the baffles.

These devices are widely used for particulate removal in power plants and rotary kilns.

Louvre Type Separator

Here there is a series of louvres or impingement elements set at an angle to the carrier gas stream so as to cause a rapid reversal of the gas flow direction and thereby cause the particulates to impinge on the louvres. The

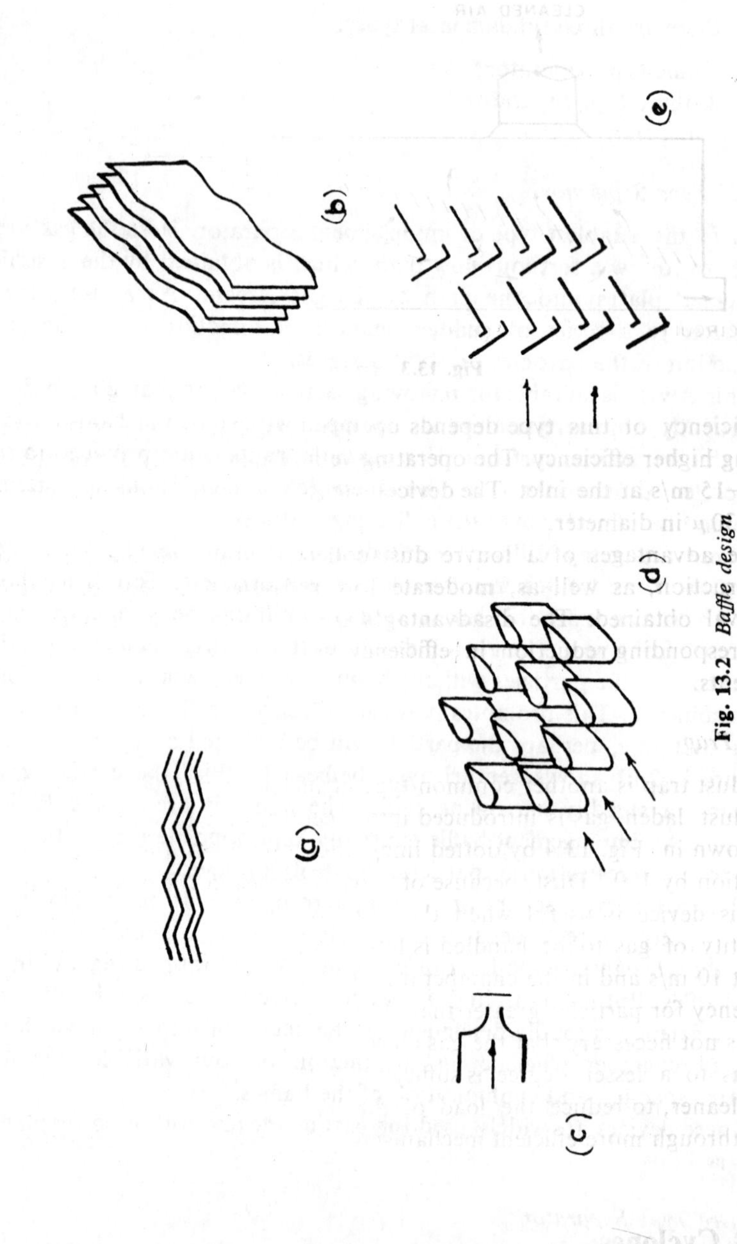

Fig. 13.2 *Baffle design*

particles thus impinged rebound back into the moving gas stream in the inlet chamber and are removed from the collector by a secondary air circuit (Fig. 13.3).

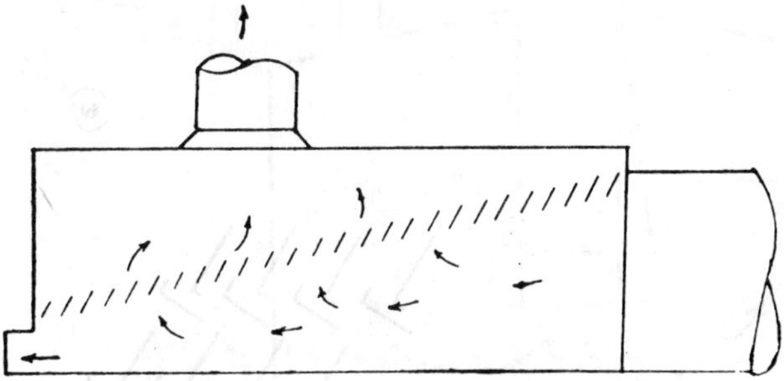

Fig. 13.3 *Louvre dust collector*

Efficiency of this type depends upon louvre spacing, closer spacing producing higher efficiency. The operating velocities normally are of the order of 12–15 m/s at the inlet. The device is suitable for removing particles larger than 30 μ in diameter.

The advantages of a louvre dust collector are, simple and low cost of construction, as well as, moderate low pressure drop for the degree of removal obtained. The disadvantages are, clogging of the louvre grid with a corresponding reduction in efficiency and excessive abrasion of louvre elements.

Dust Trap

The dust trap is another common type of inertial separator. In this device, the dust laden gas is introduced into a central pipe (cylindrical or tapered as shown in Fig. 13.4 by dotted line), and is made to undergo a change in direction by 180°. Dust, because of inertia, settles in the conical chamber.

This device is useful when the dust loading is high (> 100 g/m^3) and quantity of gas to be handled is low. Normally gas velocity in the inlet is about 10 m/s and in the chamber it is reduced to about 1 m/s. The collection efficiency for particles greater than 30 μ is about 70%.

It is not necessary that the gas must reverse its flow. Sometimes, deflection of gas to a lesser degree is sufficient. A dust trap, is sometimes used as a pre-cleaner, to reduce the load of the larger diameter particles so as to pass through more efficient mechanisms, for removing smaller particulates.

13.8 Cyclones

This class of separators is the most common of a general group of separators

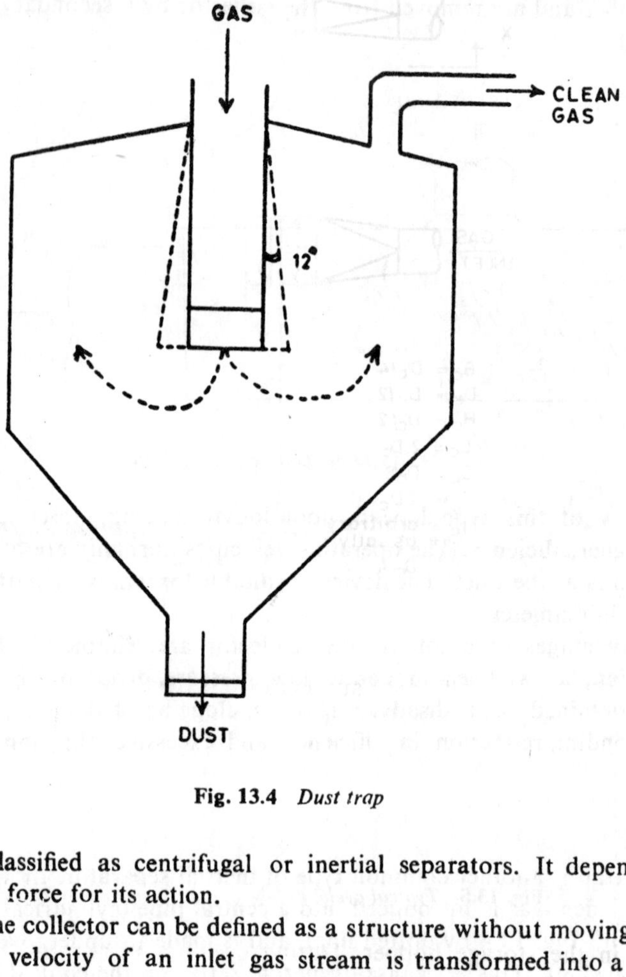

Fig. 13.4 *Dust trap*

that are classified as centrifugal or inertial separators. It depends upon centrifugal force for its action.

A cyclone collector can be defined as a structure without moving parts in which the velocity of an inlet gas stream is transformed into a confined vortex from which centrifugal forces tend to drive the suspended particles to the wall of the cyclone body.

It consists of a vertically placed cylinder which has an inverted cone attached to its base. The particulate-laden gas stream enters tangentially at the inlet point into the cylinder. The outlet pipe for the purified gas is a central cylindrical opening at the top. The dust particulates are collected at the bottom in a storage hopper (Fig. 13.5).

The gas path generally follows a double-vortex. First, the gas spirals downwards at the outer periphery of the cylindrical portion continues through the conical portion and reaches the bottom. The gas stream then moves upwards in a narrower inner spiral, concentric with the first and leaves through the outlet pipe. Due to the rapid spiralling movement of the gas, the dispersoids are projected towards the wall by the centrifugal force and then they drop by gravity to the bottom of the body, where they are

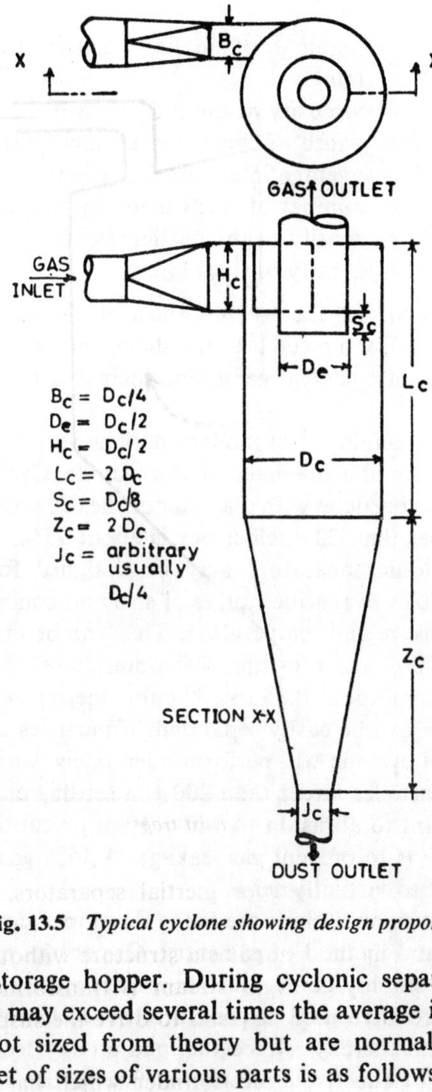

Fig. 13.5 *Typical cyclone showing design proportions*

collected in the storage hopper. During cyclonic separation, carrier gas rotational velocity may exceed several times the average inlet gas velocity.

Cyclones are not sized from theory but are normally designed by set procedures. One set of sizes of various parts is as follows:

$$D_e = D_c/2$$
$$H_c = D_c/2$$
$$S_c = D_c/8$$
$$L_c = 2D_c$$
$$Z_c = 2D_c$$
$$B_c = D_c/4$$
$$J_c = D_c/4$$

The minimum diameter of a particle which can theoretically be completely separated from the gas stream is given by,

$$D_p, \text{min.} = \left[\frac{9\,\mu\,\text{B}}{\pi\,V\,N_t(\rho_p - \rho)}\right]^{1/2} \tag{13.4}$$

where,

D_p, min = diameter of smallest particle that can be removed (cm)
μ = viscosity of the fluid (poise)
B = width of cyclone inlet duct (cm)
V = average inlet velocity (cm/s)
N_t = number of turns made by gas stream in cyclone
ρ_p = density of the particle (g/cc)
ρ = density of fluid (g/cc)

The design factor having the greatest effect on the collection efficiency is the cyclone diameter. For a given pressure drop, smaller the diameter, higher is the efficiency, because centrifugal action increases with decreasing radius of rotation.

Centrifugal forces employed in modern designs vary from 5 to 2500 times gravity depending on the diameter of the cyclone. Cyclone efficiencies are greater then 90% for particles with diameter of the order of 10 μ. For particles with diameter higher than 20 μ, efficiency is about 95%.

In practice, cyclonic separators may be designed for the satisfactory collection of particles over wide ranges of size and concentration, and over wide ranges of pressure and temperature. They can be operated at temperatures as high as 1000°C and pressures 500 atmospheres. They can handle gas volumes ranging from about 0.85 to 700 cubic metres per minute. Particles of diameter 5–10 μ can be easily separated. If particles are large (5–200 μ), a properly designed cyclone will perform adequately with moderate power requirement. For particles larger than 200 μ, a settling chamber is desirable, as it is more resistant to abrasion. An important precaution to be taken in operating a cyclone is to prevent gas leakage. A 15% gas leakage can bring down the efficiency to virtually zero.

Today, various cyclone designs are being developed covering a wide range in size, geometry, and method of gas entry. Gas entries may be tangential, or involute or may employ guide vanes to impart a spiral motion to the gas stream. Cyclones are often used as pre-collectors to remove the bulk of the particulate mass and thereby reduce the load on any secondry control unit specially designed for effective collection of fine particles.

13.8.1 Efficiency

Cyclones are generally, divided into two classes, 'conventional' and 'high efficiency'. High efficiency cyclones merely have a smaller body diameter to create greater separating forces, and there is no sharp dividing line between the two groups. High efficiency cyclones are generally considered to be those with body diameters up to about 0.25 m.

One particular cyclone efficiency problem is the formation of eddies at the top of the unit where the dirty gas is introduced. The turbulence in the eddies causes some of the incoming dirty gas to be mixed with the outgoing clean gas stream. The effect of this problem can be minimised by removing the exit gas stream at a point below the zone of maximum turbulence. This is done by adding a central tube called a vortex finder which projects into the cyclone body below the turbulent entry region to confine the rising inner gas spiral.

In general, increase in collection efficiency will result if there is an increase in any of the following: dust particle size, dust particle density, gas inlet velocity, inlet dust loading, cyclone body length (number of gas revolutions), and ratio of body diameter to gas outlet tube diameter.

On the contrary, collector efficiency will decrease if there is an increase in gas viscosity or density, cyclone diameter, gas outlet diameter, inlet width and inlet area.

To get increased efficiency, especially for the collection of smaller sized particles, a small diameter, long taper cyclone should be used.

Table 13.1 indicates the efficiency range of cyclones.

TABLE 13.1 *Efficiency Range of Cyclones*

Particle size range (μ)	Efficiency percentage	
	Conventional	High efficiency
Less than 5	Less than 50	50–80
5–20	50–80	80–95
15–40	80–95	95–99
Greater than 40	95–99	95–99

13.8.2 Tangential Inlet and Involute Inlet—A Comparison

Inlets are of two types—tangential and involute. A straight tangential entry creates quite a bit of turbulence which will lead to back mixing and loss of efficiency even when a vortex finder is included in the cyclone design. On the other hand, the involute design (Fig. 13.6) brings in the gas parallel to the outer edge of the cyclone (tangent at that point) and leads it around a spiral for 180° to enter the top section with minimum turbulence. The other advantages of the involute design are better particle projection to the wall and a decrease in the loss of finer particles. The higher velocity in the central core can cause a slight increase in the Bernoulli effect, drawing more fine particles from the wall towards the central core. However, fines losses at the top and the pressure loss are much less and the efficiency much higher for the involute design than for the tangential entry design.

13.8.3 Operating Problems

There are three important operating problems associated with cyclones. They are erosion, corrosion and material build-up.

Erosion Heavy, hard, sharp-edged particles, in a high concentration, moving at high velocity in the cyclone, continuously scrape against the wall and can erode the metallic surface unless suitable materials are used.

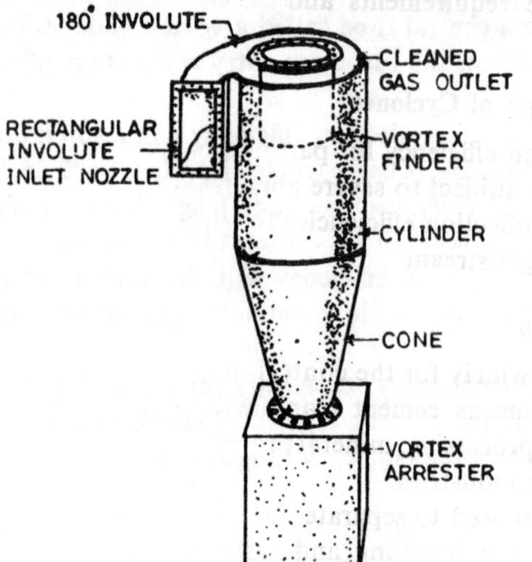

Fig. 13.6 *Involute cyclone*

Corrosion It is a problem if the cyclone is operating below the condensation point when reactive gases are present in the effluent stream. The best solution to any corrosion problem in a cyclone is to keep the product above the dew point. If the gas and dust are corrosive at low temperatures then perhaps the only alternative is to use a stainless alloy.

Build-up Build-up of dust cake on the cyclone walls, especially around the vortex finder, at the ends of any internal vanes, and opposite the entry can become a severe problem. It occurs most frequently with hygroscopic dusts.

In case the dust builds up on the wall of the cyclone cone, a simple solution is to pound on the cone with a sledge hammer. Another solution is to hang chains inside; this works, but reduces efficiency. A better solution is to flange a section between the dust collecting hopper and the cyclone body. It can be removed periodically and scraped.

13.8.4 Advantages of Cyclones

1. Low initial cost
2. Simple construction and operation
3. Low pressure drop
4. Low maintenance requirements
5. It has no moving parts
6. Continuous disposal of solid particulates
7. They can be constructed of any material which will meet the temperature

and pressure requirements and the corrosion potential of the carrier gas stream.

13.8.5 Disadvantages of Cyclones

1. Low collection efficiency for particles below 5–10 μ in diameter
2. Equipment is subject to severe abrasive deterioration
3. Decreasing collection efficiencies for decreasing dispersoid concentrations in the gas stream

13.8.6 Applications

Cyclones are used widely for the control of gas-borne particulates in such industrial operations as cement manufacture, feed and grain processing, food and beverage processing, mineral processing, paper and textile industries, and wood working industries.

Cyclones are also used to separate dust in disintegration operations, such as rock crushing, ore handling and sand conditioning in industries. They are also used in the recovery of catalyst dusts in the petroleum industry, and in the reduction of fly ash emissions.

13.8.7 Multiple Cyclones

For high efficiency at reasonable capacity, a battery of smaller cyclones operating in parallel is used in preference to a large single unit. This battery cf smaller cyclones is known as 'multiple cyclones'. It is also referred to as multi-clones, tubular cyclones, tubular collectors and multiple collectors.

Multiple cyclones consist of rows of parallel tubes, about 25 cms in diameter, with a common inlet chamber, a common outlet plenum, and a common dust collection system as shown in the Fig. 13.7. These chambers must be designed for a constant pressure drop in each, to avoid any channeling of the dirty gas to any particular single cyclone or group of cyclones.

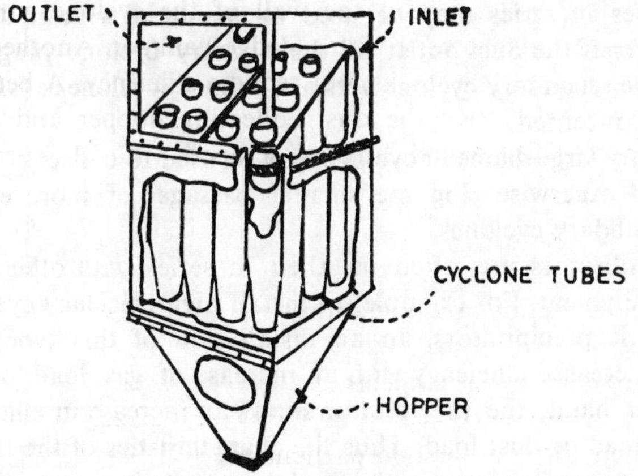

Fig. 13.7 *Multiple cyclone*

Such a situation leads immediately to overloading which increases the pressure drop in that section and causes further over-loading in other parts of the system. The hopper (dust bin) must be designed with more care than that designed for a single cyclone because the rate of dust collection will be generally much larger (because of greater centrifugal action due to smaller diameter of the cyclones). The velocities are much higher in the smaller cyclones which can cause re-entrainment problems to be more severe than in larger units. Continuous dust bin emptying, and very often a deeper hopper are required to reduce the potential for dust re-entrainment.

Multiple cyclones have good abrasion resistance, are compact and have convenient inlet and outlet arrangements. They have low pressure drops and are highly efficient in collecting heavy particles. But their main disadvantage is they plug (materials built-up).

Multiple cyclones are used as collectors for cement clinkers, steel mill sinter and stone dust in quarry and asphalt operations.

13.8.8 Cyclone in Series

The efficiency of two cyclone dust collectors operating in series is given by:

$$\eta = \eta_p + \eta_s (100 - \eta_p) \qquad (13.5)$$

where,

η = efficiency of the combination of both cyclones

η_p = efficiency of the primary cyclone

η_s = efficiency of the secondary cyclone (based on the inlet dust load to it)

The second cyclone serves to remove particles which were not collected in the first cyclone owing to statistical distribution across the inlet, accidental re-entrainment due to eddy currents, and re-entrainment in the vortex core. The efficiency of the second cyclone will be less than that of the first cyclone.

Cyclones operated in series are advantageous under the following circumstances.

1. Cyclones in series may be used to maintain a large degree of dust collection even if the dust outlet of the primary cyclone plugs. In such a condition, the secondary cyclone acts as a primary cyclone as far as collection efficiency is concerned.

2. A primary large diameter cyclone may be used to collect coarse material which would otherwise clog the smaller passages of more efficient small diameter secondary cyclones.

Cyclone collectors are often installed in series with other type of dust collecting equipment. For example, to install high efficiency cyclones ahead of electrostatic precipitators. In an installation of this type, the cyclone exhibits an increased efficiency with an increase of gas load or dust load. On the other hand, the precipitator shows an increase in efficiency with a reduced gas load or dust load. Thus the characteristics of the two types of equipment compensate for each other and this results in maintaining efficiency over a wide range of gas flow and dust loading.

In general, if the particle-size distribution is such that most of the particulates can be removed in cyclones, then in such a case, very high overall efficiencies can be obtained by operating cyclones in series.

13.9 Filters

Filteration is one of the most reliable, efficient and economic methods by which particulate matter can be removed from gases. Filters can be broadly divided into the following two types.

1. Fabric or cloth filters
2. Fibrous or deep-bed filters

In cloth filters, the filter is in the form of a fabric bag arrangement—tubular bags or as cloth envelopes, and is suitable for a dust loading of the order of 1 g/cu m. In case of deep-bed filters, a fibrous medium like mats of wool, cellulose, etc., acts as a separator and the collection takes place in the interstices of the bed, and is suitable for light dust loads of the order of 1 mg/cu m.

13.9.1 Fabric or Cloth Filters

The most common type of fabric collector is the tubular type, consisting of tubular bags. A bag house or bag filter consists of numerous vertical bags 120–400 mm diameter and 2–10 m long. They are suspended with open ends attached to a manifold. The hopper at the bottom serves as a collector for the dust (Fig. 13.8).

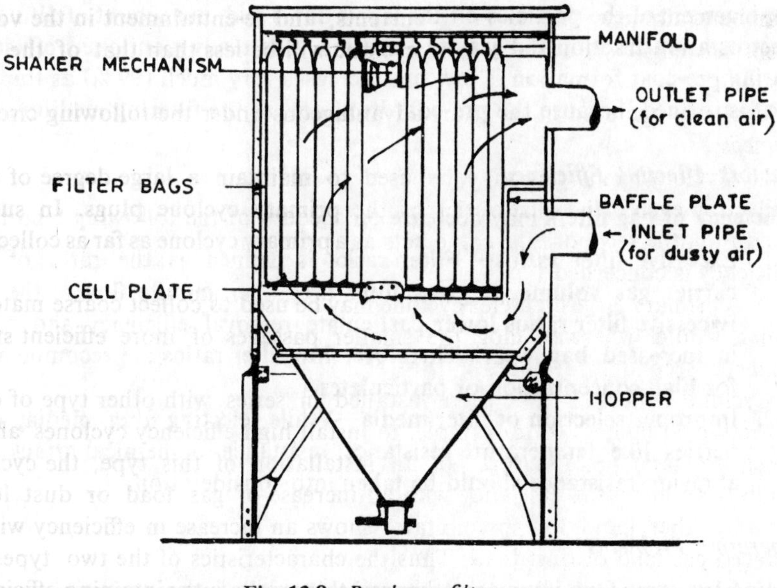

Fig. 13.8 *Bag house filter*

The gas entering through the inlet pipe strikes a baffle plate, which causes the larger particles to fall into a hopper due to gravity. The carrier gas then flows upward into the tubes and then outward through the fabric leaving the particulate matter as a 'cake' on the inside of the bags.

Efficiency during pre-coat formation is low, but increases as the pre-coat (cake) is formed, until a final efficiency of over 99% is obtained. Once formed, the pre-coat forms part of the filtering medium, which helps in further removal of the particulates. Thus dust becomes the actual filtering medium! The bags, in effect, act primarily as a matrix to support the dust cake. The cake is usually formed in a matter of minutes, or sometimes even seconds.

The accumulation of dust increases the air resistance of the filter and therefore filter bags have to be periodically cleaned. They can be cleaned by rapping, shaking or vibration, or by reverse air flow, causing the filter cake to be loosened and to fall into the hopper below.

The normal velocities at which the gas is passed through the bags is 0.4–1 m/min. There are many types of 'filter bags' depending on bag shape. type of housing and method of cleaning the fabric.

Theory of Separation

The theory of removal of dispersoids by a fabric filter has not been fully developed. Separation of the particulates from the carrier gas stream is not a simple filtration or sieving since the pores of the fabric employed in fabric filters are normally many times the size of the particles separated.

It is postulated that the initial deposition of the particulates, prior to the formation of the cake (pre-coat), takes place through interception and impingement of the particles on the fabric fibres and on account of diffusion, electrostatic attraction and gravity—settling. As already mentioned, efficiency during pre-coat formation is low but increases very much (99%) as the pre-coat is formed, because the pre-coat itself acts as the filtering medium.

Factors Affecting Efficiency

Efficiency of bag filters may decrease on account of the following factors:

1. Excessive filter ratios—'Filter ratio' is defined as the ratio of the carrier gas volume to gross filter area, per minute flow of the gas. Excessive filter ratios lower particulate removal efficiency and result in increased bag wear. Therefore, low filter ratios are recommended for high concentration of particulates.
2. Improper selection of filter media—While selecting filter media, properties like temperature resistance, resistance to chemical attack and abrasion resistance should be taken into consideration.

Operating Problems

1. Cleaning—One important operating problem is the periodic cleaning

PLATE 1

DUSTACLONE - TA®
Totally accessible mechanical collector
(Courtesy: THERMAX LTD.)

DUSTACLONE - S®
Standard or conventional multiclone
(Courtesy: THERMAX LTD.)

DUSTOMAT®
Pulse jet bag filter
(Courtesy: THERMAX LTD.)

Side view of ESP showing inlet and outlet cones and TR sets / Rappers located on ESP top. (Courtesy: VOLTAS LTD.)

Cement kiln ESP under erection - to the right are the gas conditioning towers. (Courtesy: VOLTAS LTD.)

Top view of an ESP showing outlet ducting to the stack, TR sets and IIT bus duct and rappers. (Courtesty: VOLTAS LTD.)

of the bags. At intervals they get so clogged up with a covering of dust particles that the gas can no longer pass through them. Then the bags have to be cleaned by rapping, shaking, or by reverse air flow or by a pulse jet.

2. Rupture of the cloth—The greatest problem inherent in cloth filters is rupture of the cloth, which results from shaking. It is often difficult to locate ruptures, and when they are found the replacement time is often considerable.
3. Temperature—Fabric filters will not perform properly if a gross temperature overload occurs. In case the gas temperature is expected to fluctuate, a fibre material that will withstand the upper limit of that fluctuation must be selected. Generally, the upper temperature limit for fabric filters is about 260 to 290°C. Filters under development in Germany and USSR, reportedly can operate at 370°C. Research work is going on to find coatings for fibre glass fabrics so that they can withstand temperature up to 650°C.

   Another temperature related problem occurs whenever the effluent contains a reactive gas such as sulphur dioxide which can form an acid if the temperature in the bag house falls below the dew point. This can be prevented by having the bag house fairly well insulated. Sometimes, it may even be necessary to provide an auxiliary heater for use during the winter season to make sure that the bag house temperature does not fall below the acid gas dew point.
4. Bleeding—Bleeding is penetration of the fabric by the fine particles, which we come across occasionally in fabric filtration. Bleeding can occur if the weave is too open or if the superficial filtration velocity (filter ratio) is too high. Generally, only the smaller particles penetrate through the fabrics; but they are the ones that cause the primary human health hazard. The solution to the bleeding problem is either to use a double layer material or a thick woven fabric.
5. Humidity—Humidity control is a common and important problem, especially if a hygroscopic dust is involved. The problem can be overcome by taking suitable precautions.
6. Chemical attack—Another problem associated with fabric filters is the possibility of chemical attack due to corrosive chemicals present in the process effluent.

Filter Cleaning

Following are some of the common methods of filter cleaning:

1. Rapping
2. Shaking
3. Reverse air flow (back wash)
4. Pulse jet

The shaking mechanism is probably the oldest, and bag house units can range in size from those with only half a dozen bags that are shaken manually up to a very large compartmented system containing a number of bags that are shaken mechanically. Generally each row of bags is joined to the shaker framework by a rod; and each shaker assembly consists of an electric motor, connecting arms and bearings. Shaker cleaners hold the top of the bag still and shake the entire tube sheath (Fig. 13.9). Periodic or intermittent cleaning can be done depending on process requirement. The required specifications are the shaking duration, amplitude and frequency. This method is not used for sticky dusts because the forces needed to remove the collected dust would probably cause tearing and ripping in the bags very quickly.

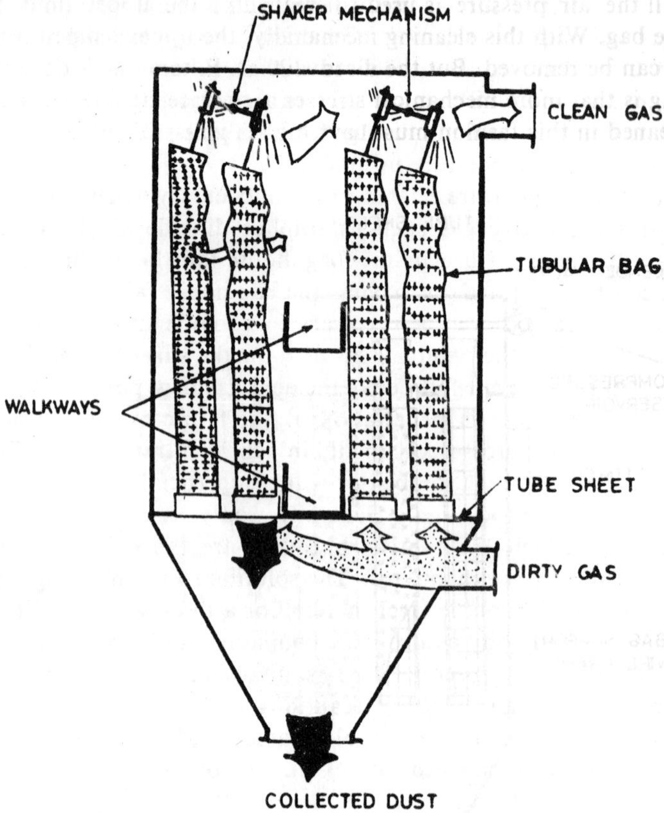

Fig. 13.9 *Shaker mechanism in a bag house*

Traversing annular rings (reverse jets) involve reversing the air flow. It is explained in detail in the section on 'Reverse Jet Filter' on page 173.

In a pulse jet bag house, periodically a jet of high pressure air is blasted down the inside of the bag which is supported interiorly by a wire frame. During the filtering operation, the bag is collapsed on the frame because of the pressure of the gas being cleaned on the outside. During the filter cleaning, the air jet inflates the bag momentarily. When the bag is inflated, the dust cake is loosened and falls into the hopper below. The two important advantages of this method are, there are no moving parts and continuous cleaning is possible. It is not necessary to isolate an entire row, or a compartment from service for cleaning as done for shaker cleaning. Also, duration of the air jet cleaning is insignificant compared to the length of filtering time between cleaning intervals. Moderate to high pressures are involved.

An even better approach is high pressure cleaning with improved pulse jets. Here instead of just having a nozzle or a simple opening for compressed air above the bag, the compressed air is introduced into the top of the bag through a venturi. The venturi is situated on the top of the bag (Fig. 13.10), so that all the air pressure is used to create a pressure wave down the inside of the bag. With this cleaning mechanism, even deposits of hygroscopic materials can be removed. But the disadvantage of high pressure methods of cleaning is that more mechanical stresses are imposed on the bag. Hence, fabrics cleaned in this fashion must have high tensile strength.

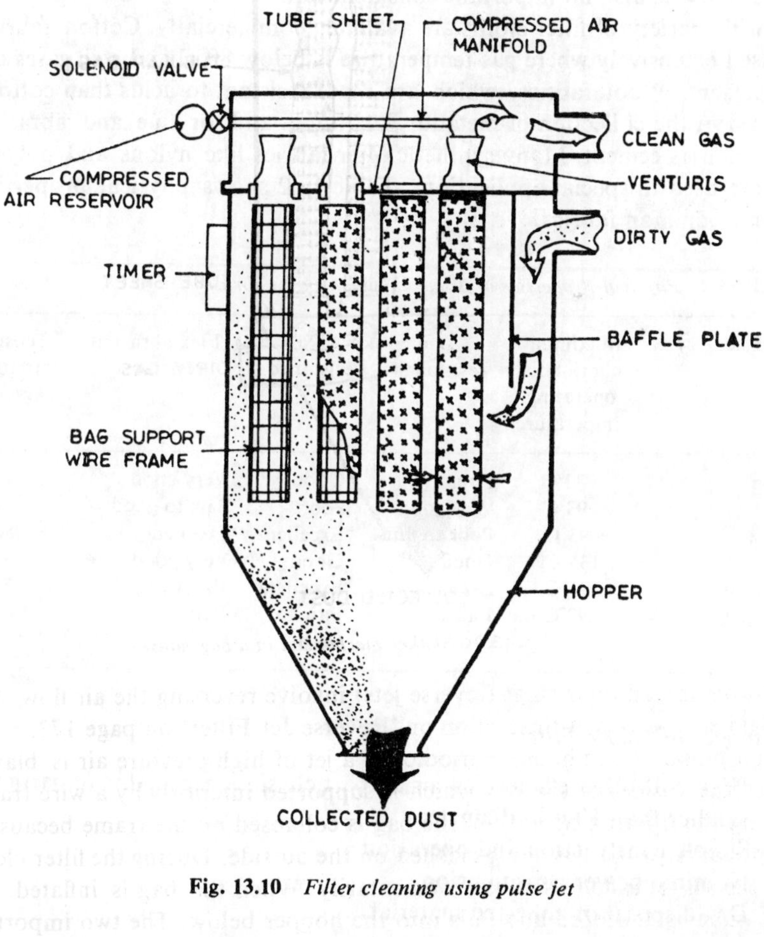

Fig. 13.10 *Filter cleaning using pulse jet*

In general, while selecting the specific method of filter cleaning, we have to consider the cleaning mechanism, frequency, duration of cleaning, and any undesirable effects on the bags.

Filter Medium

While selecting the filter medium for bag houses, we must consider the characteristics and properties of the carrier gas and dust particles. Following are the properties to be noted.

1. Carrier gas temperature
2. Carrier gas composition
3. Gas flow rate
4. Size and shape of dust particles and its concentration

As for the fabric concerned, its abrasion resistance, chemical resistance, tensile strength, and permeability should be considered. Of course, the cost of the fabric is also an important consideration.

A wide variety of filter cloths are available commercially. Cotton fabrics are used extensively where gas temperature is below 80°C and acid gases are not present. Wool fabrics, which are more resistant to acids than cotton, are used in the collection of metallurgical fumes and for fine and abrasive dusts such as cement. Many synthetic filter fabrics like nylons and polyesters, are used in special applications. Table 13.2 shows physical properties of some common fabrics.

TABLE 13.2 *Physical Properties of Some Common Bag Fabrics*

Fabric	Maximum continuous operating temperature	Acid resistance	Alkali resistance	Flex abrasion resistance	Tensile strength kg/cm²
Cotton	82°C	Poor	Good	Very good	4920
Wool	93°C	Very good	Poor	Fair to good	1755
Nylon	93°C	Poor to fair	Excellent	Excellent	5625
Dacron	135°C	Good	Good	Very good	5625
Polypropylene	93°C	Excellent	Excellent	Excellent	7730
Fibre glass	290°C	Fair to good	Fair to good	Fair	14,060

Advantages of Fabric Filters

1. High collection efficiencies for all particle sizes, especially for particles smaller than 10 μ in diameter.
2. Simple construction and operation
3. Nominal power consumption
4. Dry disposal of collected material

Disadvantages of Fabric Filters

1. Operating limits are imposed by high carrier gas temperatures, high humidity, etc.
2. High maintenance and fabric replacement costs (bag houses are difficult to maintain because of the difficulty in finding and replacing even a single leaking bag. Also, as a general rule, about one-fourth of the bags will need replacement every year).
3. Large size of equipment.
4. Problems in handling dusts which may abrade, corrode, or blind the cloths.

Applications

Fabric filters find extensive applications in the following industries and operations.

1. Metallurgical industry
2. Foundries
3. Cement industry
4. Chalk and lime plants
5. Brick works
6. Ceramic industry
7. Flour Mills

Reverse Jet Filter

Recently, bag filters cleaned by an air jet from a traversing ring have been developed. This unit differs basically from the usual tubular type unit with regard to the method of cleaning and the type of filter medium employed. The tubular filters in this unit are cleaned by a high velocity air jet discharged from the inner side of a traversing ring which moves on the outside of the filter tube (Fig. 13.11). The air jet passes through the fabric in a direction reverse to the normal flow and removes cake continuously from the filter surface. Current designs employ felt as the filtration medium.

Following are the advantages and disadvantages of a reverse jet filter.

1. Requires no shut downs or programming for cleaning. Blow rings are a continuous cleaning mechanism in that some part of each bag is being cleaned at all times.
2. Filter works at higher velocity of 3–6 m/min as compared to 0.5–2 m/min for the usual cloth filters.
3. The cloth resistance can be maintained at a nearly constant value in contrast to other cloth filters in which the resistance builds up to the design value.

But the main disadvantages of blow rings are the mechanical linkages and the individual air hose attachments required for each bag. The degree of maintenance required is also high.

The bag house specifications for a reverse jet filter must include linear

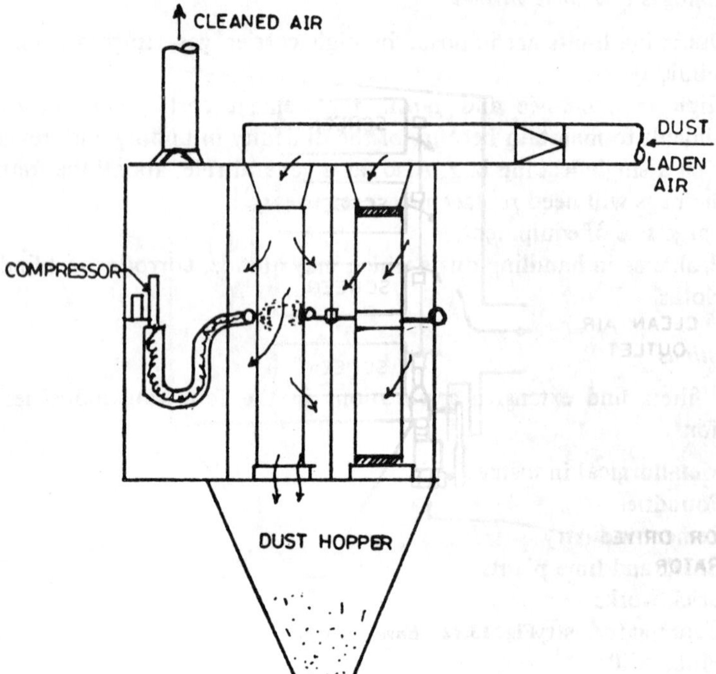

Fig. 13.11 *Reverse jet filter*

ring travel rate, number of rings per bag, and the flow rate and pressure of the cleaning air.

Envelope Type Fabric Filter

This type of a filter is in the form of a cloth envelope supported on a wire-screen frame. The individual units range from 0.6–1 m wide, and 3–5 cm thick. In contrast to the bag filter, gas is introduced on the outside of each envelope and passes through the fabric into the frame of the unit and then out of the collector. The filters are prevented from collapsing due to the inner frame work. As in bag filters, the envelope panels are mounted in multiples (Fig. 13.12).

The fabric is cleaned normally by rapping action on the supporting frame. It can also be cleaned using a shaking device fitted to screen frames. For very fine particulates this equipment is not suitable, since the efficiency of cleaning here, is not as effective as that in the tubular type.

Multi-Compartment Type Bag House

If the requirements of the process being controlled are such that continuous operation is necessary, the bag filter must be of a multi-compartment type to allow individual units of the bag filter to be successively off-stream during

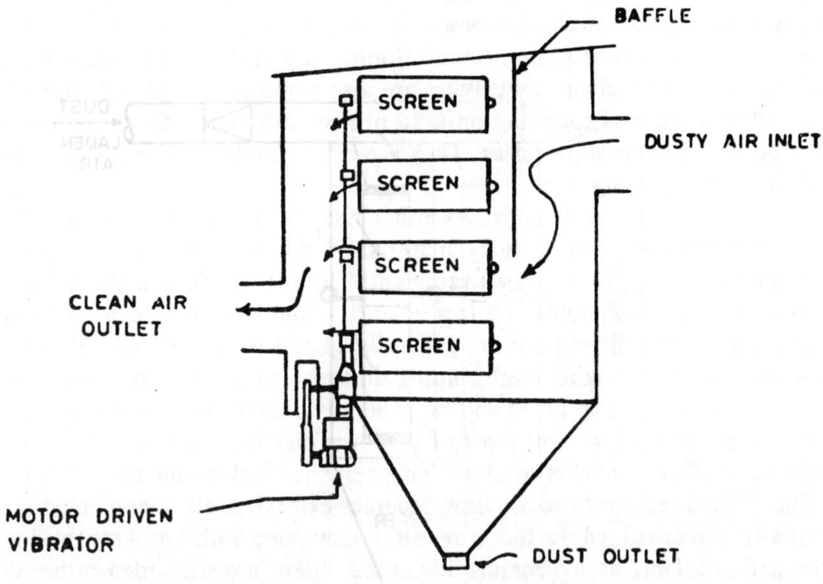

Fig. 13.12 *Envelope type fabric filter*

shaking. This is achieved either manually in small units or by programming control in large, fully automatic units. In this case sufficient cloth area must be provided to ensure that filtering capacity will not be reduced during shaking periods when any one unit of the filter is off-stream (Fig. 13.13).

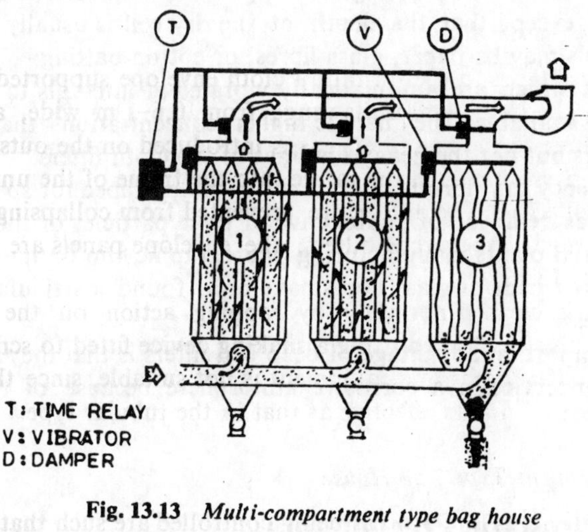

Fig. 13.13 *Multi-compartment type bag house*

13.9.2 Fibrous or Deep-bed Filters

In the deep-bed filters, a fibrous medium acts as the separator and the collection takes place in the interstices of the bed. Fibrous filters may be composed of mats of wool, asbestos, cellulose, glass or iron fibres. They find their most extensive use in air conditioning and heating and ventilating systems, but high efficiency mat filters are also employed as aftercleaners and less efficient ones as roughing units to protect still more efficient equipment for pollution control purposes. They are most suitable for light dust loads, of the order of 1 mg/cu m.

The filters can be classified as either viscous or dry. In the viscous type, the filter is coated with a sticky material or 'adhesive' to help in catching the particles and prevent re-entrainment. The adhesive is usually an oil or grease of high flash point and low volatility, and it should be a good wetting agent. The filter medium, generally glass wool, is placed on metal or cardboard with a wire mesh, approximately 50 cm square and several centimetres deep. The fibres will be generally packed with increasing density from front to back so that most of the large particles are removed before the more efficient part is reached. This helps in prolonging the life of the filter. When resistance to air flow becomes excessive, glass wool media are usually discarded while the wire mesh is washed with hot water and then reoiled. In fact, as a general rule, used filters are discarded rather than cleaned and reused.

Automatic viscous filters are also available. They can be used for handling large quantities of atmospheric air containing high concentration of dust (about 2 grains per 1000 cu ft). The fibre is formed into an endless belt, which continuously moves through an oil bath at the bottom of the housing. They can be operated at high air velocities of the order of 90–150 m/min.

Dry filters for air conditioning are supplied in units similar in size to the viscous type except that the depth of the dry cell is usually greater. The filter materials may be paper, glass fibres, or cotton batting.

Dry filters which are automatically vibrated at intervals to dislodge the dust are also available. They handle higher concentrations than the usual dry-cell filters but not the heavy loadings of the cloth filters.

High efficiency dry fibrous filters have been developed for special applications, such as removal of radioactive or toxic particles or the cleaning of air in industrial plants manufacturing photographic film or fine instruments. Loosely packed pads of glass fibres have been found particularly suitable for such cases.

Another important advantage of deep bed filters is that they are generally very good for service in a corrosive atmosphere because of their 'throw-away' nature.

13.10 Electrostatic Precipitators

Electrostatic precipitators (ESP) are particulate collection devices that utilise electrical energy directly to assist in the removal of the particulate matter. They have been successfully used for removal of fine dusts from all kinds of waste gases with very high efficiency. Particles as small as a tenth of a micron can be removed. The principle on which this equipment operates is that, when a gas containing aerosols is passed between two electrodes that are electrically insulated from each other and between which there is a considerable difference in electric potential, aerosol particles precipitate on the low potential electrode.

There are various types of electrostatic precipitators. They are mainly used for industrial purposes. They can also be used for air cleaning in public buildings, theatres, railway cars, etc.

13.10.1 Description of General System

Basically, an electrostatic precipitator consists of six major components:

1. A source of high voltage
2. Discharge electrodes and collecting electrodes
3. Inlet and outlet for the gas
4. A means (generally a hopper) for disposal of the collected material
5. An electronic cleaning system
6. An outer casing to form an enclosure around the electrodes (precipitator shell)

The gases entering an electrostatic precipitator may or may not have pre-treatment before entering the unit. Pre-treatment could consist of removing part of the dust loading by use of certain mechanical collectors, or by adding chemicals to the gas to change the physical properties of the gas to enhance precipitator action.

The entire electrostatic precipitator is enclosed in a casing. The actual geometric configuration may be either rectangular or cylindrical. In all cases, there is an inlet and outlet connection for gases, hoppers to collect the precipitated matter and the necessary discharge electrode and collector surfaces. Usually, there will be a penthouse on the precipitator. This is a weatherproof gastight enclosure over the precipitator that houses the high voltage insulators.

Precipitators are usually built with a number of auxiliary components, which include access doors, dampers, safety devices, and gas distribution systems. The access provisions can be either a door or a plate which are closed and bolted under normal operating conditions but which can be opened when necessary for inspection and maintenance. Dampers are the means by which the quantity of gas is controlled. They may be either a guillotine, a louvre or some similar device which opens and closes to adjust gas flow. In addition, the safety grounding system is extremely important and must always be in place during operation and especially during periods of inspection. The most common grounding system consists of a conductor, one end of which is grounded to the casing, and the other end is attached to the high voltage system by an insulated operating lever.

The collecting system consists of the collecting surfaces where the particulate matter is deposited. Also, a device for rapping or vibrating the collecting surfaces is included so that the deposited particles can be dislodged.

The high voltage system includes the discharge electrode and the insulators. Electrical insulation must be done for safety purposes.

The precipitator hopper is integral with the precipitator shell and is made from the same materials. The function of the hopper is to collect the precipitated material for final disposal.

Electrostatic precipitators require a high voltage direct current source of energy for operation. Therefore, transformers are required to step up the normal service voltages to high voltages. Rectifiers convert the alternating current to unidirectional current.

13.10.2 Principle of Electrostatic Precipitation

In electrostatic precipitators the gas stream is passed between two electrodes, across which a high potential difference is maintained. Out of the two electrodes, one is a discharging electrode and the other a collecting elecrode. Because of a high potential difference and the discharge system, a powerful ionising field is formed. Potentials as high as 100 kilo volts (but usually 40–60 kV) are used. Consequently, ionisation creates an active glow zone (blue electric discharge) called the 'corona' or 'corona glow'. Gas ionisation is the dissociation of gas molecules into free ions.

As the particulates in the carrier gas pass through this field, they get charged and migrate to the oppositely charged collecting electrode. The particles, once deposited on the collecting electrode, lose their charge and are removed mechanically by rapping, vibration or washing to a hopper below.

Thus, the four steps in the process are:

1. Place the charge on the particle to be collected
2. Migrate the particle to the collector
3. Neutralise the charge at the collector
4. Remove the collected particle

13.10.3 Single-stage and Two-stage Precipitators

Electrostatic precipitators can be either single-stage or two-stage in design. In a single stage precipitator, gas ionisation and particulate collection are combined in a single step. Whereas in the case of a two-stage precipitator, particles are ionised in the first chamber and collected in the second chamber.

Almost all industrial precipitators are of the single-stage design. Usually, the two-stage precipitator is used for lightly loaded gases and the single-stage, for more heavily loaded industrial gas streams.

13.10.4 Pipe-type and Plate-type Precipitators

There are two basic designs of the single-stage precipitator. They are the pipe-type precipitator and the plate-type precipitator.

Pipe-type Precipitator

In the pipe-type precipitator, the collecting electrodes are formed by a nest of parallel pipes which may be round, square or octagonal. Generally, the pipe is about 30 cm in diameter or less. The most common form of a discharge electrode is a wire with a small radius of curvature, suspended along the axis of each pipe. They must be weighted or supported to retain proper physical tension and location, electrically insulated from the support grid, and strong enough to withstand rapping or vibrating for cleaning purpose. The gas flow is axial from bottom to top. Figure 13.14, represents the schematic diagram of a pipe-type precipitator.

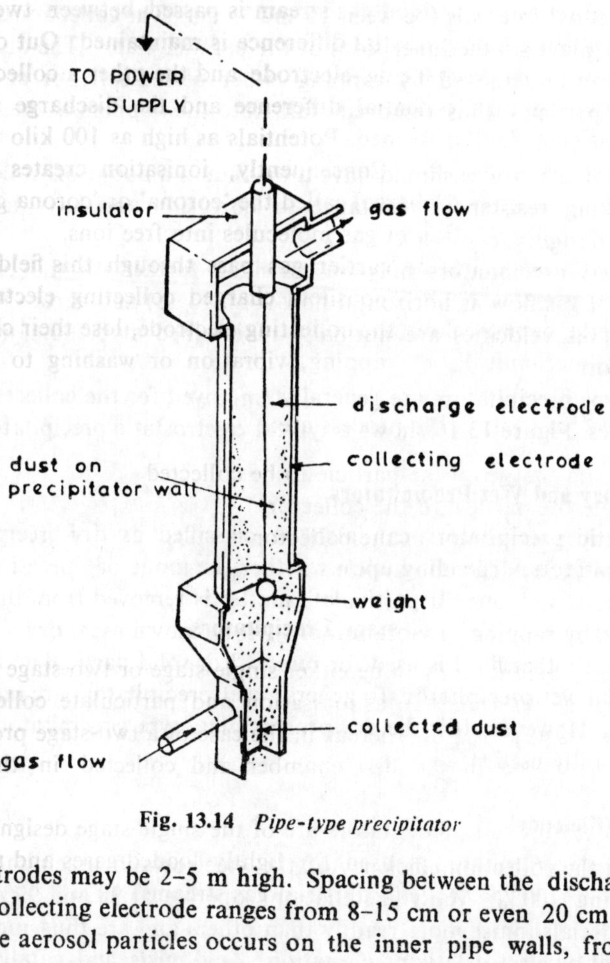

Fig. 13.14 *Pipe-type precipitator*

Pipe electrodes may be 2–5 m high. Spacing between the discharge electrode and collecting electrode ranges from 8–15 cm or even 20 cm. Precipitation of the aerosol particles occurs on the inner pipe walls, from which the material can be removed by periodic rapping of the pipes or flushing with water. Generally, the pipe-type precipitator is used for the removal of liquid particulates.

Plate-type Precipitator

In the plate-type precipitator, the collecting electrodes consist of parallel plates. The discharge electrodes are similar to those used in pipe-type precipitators, i.e., they are wires with a small curvature. Sometimes square rods (approximately 4.8 mm) and twisted square rods (from 3.2–6.4 mm) are used. The wires are suspended midway between the parallel collecting electrodes and usually hang free with a weight suspended at the bottom to keep them straight. Discharge electrodes are made from non-corrosive materials like tungsten, and alloys of steel and copper. The gas flow is parallel to the plates.

Plates may be 1–2 m wide and 3–6 m high. The parallel plates should be spaced at equal intervals (between 15 and 35 cm). The collection of the aerosols takes place on the inner sides of the parallel plates. The dust material collected can be removed by rapping either periodically or continuously. The dust particles thus removed, fall, into the hoppers at the base of the precipitator.

Collection electrodes should have a maximum amount of collection surface, bulking resistance, resistance to corrosion and have a consistent economic design.

Plate-type precipitators may be classified further on the basis of the direction of gas flow as horizontal-flow (Fig. 13.15) and vertical-flow precipitators. Gas velocities are normally maintained at 0.5–0.6 m/s in these precipitators.

Plate-type precipitators are generally employed for the collection of solid particulates Figure 13.16 shows a typical electrostatic precipitator.

13.10.5 Dry and Wet Precipitators

Electrostatic precipitators can also be classified as dry precipitators and wet precipitators, depending upon whether the mode of operation is a dry process or a wet one. If particulate matter is removed from the collecting electrodes, by rapping or vibration only, it is known as a dry precipitator. If water or other fluid is used for removal of solid particulate matter, it is known as a wet precipitator. In general, wet precipitators are more efficient collectors. However, it is the dry process plate type precipitators which are predominantly used.

13.10.6 Efficiency

Generally the collection efficiency of the electrostatic precipitator is high, approaching 100%. Many installations operate at 98 and 99% efficiency. Some materials ionise more readily than others and are thus more adapted to removal by electrostatic precipitation. Acid mists and catalyst recovery

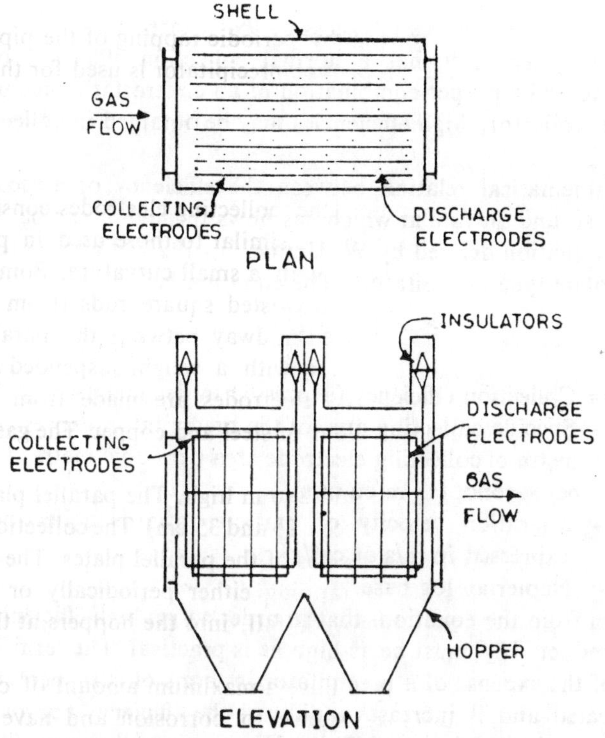

Fig. 13.15 *Horizontal flow precipitator*

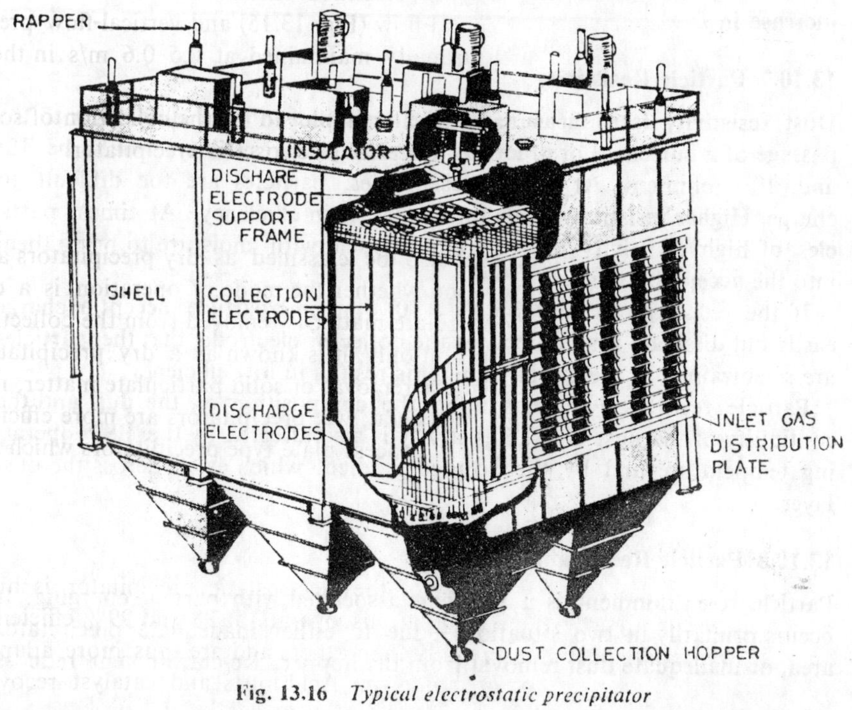

Fig. 13.16 *Typical electrostatic precipitator*

units often have efficiencies in excess of 99%; carbon black, because of its agglomerating tendency, has a normal collection efficiency of less than 35%. However by proper combination of an electrostatic precipitator with a cyclonic collector, high efficiencies may be obtained in collecting carbon black.

The mathematical relation between the efficiency of a precipitator, its size and cost, and the rate at which gas flows through it can be best expressed by an equation derived by W. Deutsch in 1922. It applies to both pipe-type and plate-type precipitators. The equation is

$$E = 1 - e^{-U_t f} \tag{13.6}$$

where,

E = Collection efficiency (expressed in decimals).

f = Specific collecting area of the precipitator, expressed as square metre of collecting electrode area per cubic metre of gas handled per second, (sq m/cu m/s).

U_t = Migration velocity of particle towards collection electrode, expressed in m/s or cm/s.

e = Napierian log base.

It is seen from the equation, that in order to get high efficiency, the value of the product $U_t f$ must be as high as is practical. The term f is a rough measure of the expense of a precipitator in terms of cost per cubic m/min of gas treated and it increases rapidly as the efficiency approaches 100%, where f theoretically becomes infinite. For a given efficiency, any decrease in the migration velocity U_t must be compensated by a corresponding increase in f.

13.10.7 Particle Resistivity

Dust resistivity is a measure of the resistance of the dust layer to the passage of a current. For practical operation, the resistivity should be 10^7 and 10^{11} ohm-cm. At higher resistivities, particles are too difficult to charge. Higher resistivity leads to a decrease in efficiency. At times, particles of high resistivity may be conditioned with moisture to bring them into the acceptable range.

If the resistivity is too low ($< 10^5$ ohm-cm) particles accept a charge easily but dissipate it so quickly at the collector electrode, that the particles are re-entrained in the gas stream. This results in low efficiency.

Particle resistivity depends upon the composition of the dust and the continuity of the dust layer. Resistivity is also affected by the ESP operating temperature and by the voltage gradient which exists across the dust layer.

13.10.8 Particle Re-entrainment

Particle re-entrainment is a problem associated with particle charging. It occurs primarily in two situations—due to either inadequate precipitator area, or inadequate dust removal from the hopper. Re-entrainment reduces

ESP performance because of the necessity of recollecting dust which had been previously removed from the carrier gas. The problem can be overcome by a proper design of the ESP.

13.10.9 Design Parameters

The design parameters relating to the carrier gas are volumetric flow rate, composition, temperature and dew point. The design parameters in relation to the dust particles are concentration, size distribution, resistivity, bulk density, composition, hygroscopicity, and tendency to agglomerate. Other important factors affecting the efficiency of the ESP are velocity distribution at the entrance of the precipitator, design of duct work, collection electrode area, ionisation potential, etc.

13.10.10 Advantages of Electrostatic Precipitators

1. High collection efficiency.
2. Particles as small as 0.1 μ m can be removed.
3. Low maintenance and operating costs.
4. Low pressure drop (0.25–1.25 cm of water).
5. Satisfactory handling of a large volume of high temperature gas.
6. Treatment time is negligible (0.1–10 s).
7. Cleaning is easy by removing units of the precipitator from operation.
8. There is no limit to solid, liquid or corrosive chemical usage.

13.10.11 Disadvantages of Electrostatic Precipitators

1. High initial cost.
2. Space requirement is more because of the large size of the equipment
3. Possible explosion hazards during collection of combustible gases or particulates.
4. Precautions are necessary to maintain safety during operation. Proper gas flow distribution, gas resistivity, particulate conductivity, and corona sparkover rate must be maintained carefully.
5. The poisonous gas, ozone, is produced by the negatively charged discharge electrodes during gas ionisation.

13.10.12 Applications of Industrial Precipitators

The important applications of electrostatic precipitators in industries are given in the following paragraphs.

1. Cement factories

   a) Cleaning the flue gas from cement kilns
   b) Recovery of cement dust from kilns

2. Pulp and paper mills

   a) Soda-fume recovery in Kraft pulp mills

3. Steel plants
   a) Cleaning blast furnace gas to use it as a fuel
   b) Removing tars from coke oven gases
   c) Cleaning open hearth and electric furnace gases
4. Non-ferrous metals industry
   a) Recovering valuable material from flue gases.
   b) Collecting acid mist.
5. Chemical industry
   a) Collection of sulphuric and phosphoric acid mist
   b) Cleaning various types of gas, such as hydrogen, CO_2 and SO_2
   c) Removing the dust from elemental phosphorus in the vapour state
6. Petroleum industry
   a) Recovery of catalyst dust
7. Carbon black industry
   a) Agglomeration and collection of carbon black
8. Electric power industry
   a) Collecting fly ash from coal-fired boilers

13.10.13 Electrostatic Precipitators for Air Cleaning

Air cleaning precipitators are used for cleaning air in public buildings, theatres, ships, railway cars, and in some private homes and club houses. The object of using air cleaning precipitators is to make the air more healthy or pleasant to breathe by removing tobacco smoke, pollen, etc., or to prevent dust from interfering with delicate industrial operations such as manufacture of watches or electronic tubes, or to prevent soiling of walls, paintings, etc.

13.11 Scrubbers or Wet Collectors

These are devices which utilise a liquid to assist in the removal of particulates from the carrier gas stream. Generally, water is used as the scrubbing liquid. In a wet collector, the dust is agglomerated with water and then separated from the gas together with the water.

Nature has used wet scrubbing as a method of cleaning the atmosphere since the world began. Every rainstorm scrubs the air producing that fresh air sensation we all experience. In industrial applications, it is necessary to force the stack gases to be cleaned by contact with the liquid which is going to do the scrubbing. This requires a lot of energy. As such, scrubbers are basically, 'cheap' to install but 'expensive' to operate.

The operational range for particle removal by scrubbers includes material less than 0.2 μ in diameter to large particles which can be suspended in the gas phase.

Four major steps are involved in collecting particles by wet scrubbing. The first of these is transport. The particles must be moved to the vicinity of the water droplets which are usually 10 to 1000 times larger. The second step is collision. The particle must collide with the droplet. The third step is adhesion. Adherence is promoted directly by the property surface tension. The fourth step is precipitation, or removal of the droplet containing the dust particle from the gas phase.

In addition to removing entrained particulate matter, scrubbers can remove gases by absorption and or adsorption. This capability of scrubbers is not possessed by other types of particulate control equipment.

13.11.1 Principle of Scrubbers (Collection Mechanisms)

In general, particulates are removed from the gas stream by one or a combination of the following mechanisms.

Impingement

When gas containing dust is swept through an area containing liquid droplets, dust particles will impinge upon the droplets and if they adhere they will be collected by them. In general, the efficiency of collection is more when the liquid droplet is approximately 100 to 300 times the size of the dust particle, in order to increase the number of inelastic collisions.

Interception

Particles that move with the gas stream may not impinge on the droplets but can captured because they brush against the droplet and adhere there. This is known as interception.

Diffusion

Diffusion of the dispersoid on to the liquid medium helps in removal of the particulate matter.

Condensation

Condensation of the liquid medium vapours on the particulates increases the size and weight of particles. Consequently, it helps in the easy removal of particulates.

In all probability, the efficiency of a scrubber is dependent to a great extent upon impingement and Brownian diffusion.

13.11.2 Types of Scrubbers

Following are the common and important types of scrubbers.

1. Spray towers

2. Venturi scrubbers
3. Cyclone scrubbers
4. Packed scrubbers
5. Mechanical scrubbers

Spray Towers

A spray tower is the simplest type of gas scrubber. It can be either round or rectangular. Gas is passed counter-current to falling drops of liquid (usually water) from a bank of spray nozzles. A simple spray tower is shown in Fig. 13.17.

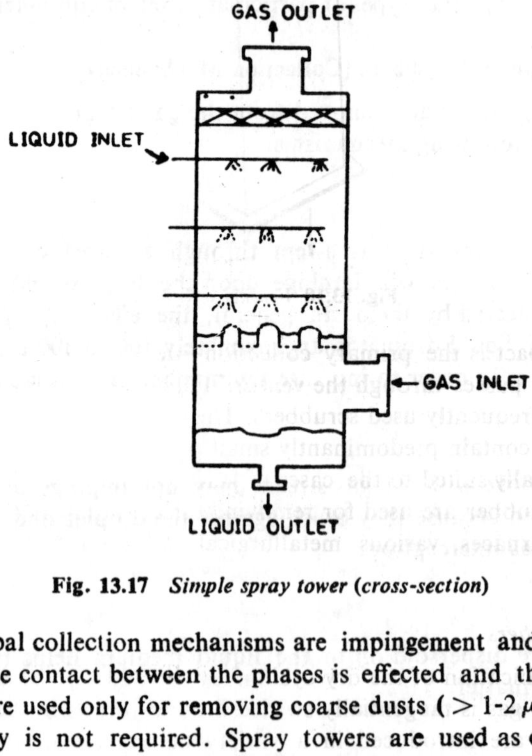

Fig. 13.17 *Simple spray tower (cross-section)*

The principal collection mechanisms are impingement and interception. Only moderate contact between the phases is effected and therefore, units of this type are used only for removing coarse dusts ($>$ 1-2 μ range) where high efficiency is not required. Spray towers are used as coolers and as primary cleaners in treating blast furnace gas and for fly ash and cinder removal.

A slight modification of the simple spray tower can be made by fitting a series of internal baffles. This helps in providing additional impingement surfaces.

Venturi Scrubbers

It is a high energy wet scrubber with a good efficiency. In this the particulate laden stream is directed through a venturi tube at throat velocity of 60 to 100 metres per second. Water sprays are introduced just ahead of the venturi throat (Fig. 13.18).

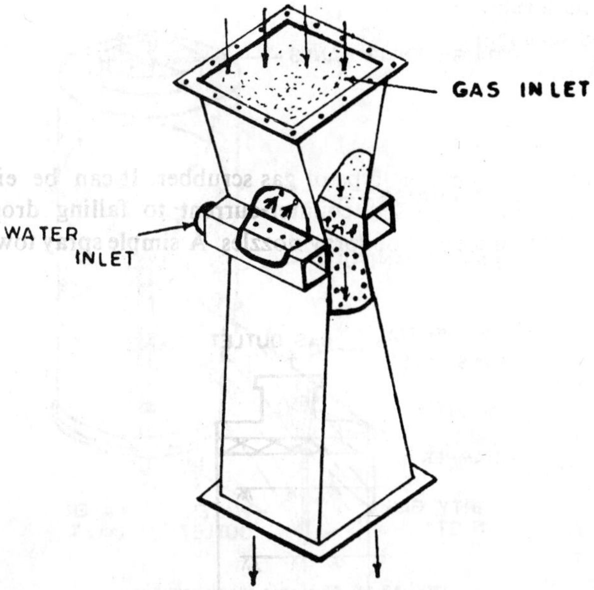

Fig. 13.18 *Venturi scrubber*

Inertial impact is the primary collection mechanism; consequently, the faster the gas passes through the venturi, the higher the efficiency. Venturis are the most frequently used scrubbers. They are used routinely for small gas rates that contain predominantly small particles. Control devices of this type are specially suited to the cases of variable and intermittent gas flows.

Venturi scrubber are used for removing mists and dusts from gases from Kraft mill furnaces, various metallurgical furnaces, and sulphuric acid concentrators.

Cyclone Scrubbers

This is a modification of the dry cyclone by the addition of a liquid phase. Here also, the gas is tangentially swirled around just as in the dry cyclone. Water sprays are introduced in a variefy of ways, e.g., either across the cyclone from the outside wall to the centre line or down the cyclone from the top. These sprays assist in the collection of the dispersoid and tend to prevent re-entrainment. Impingement and inertial separation are the primary collection mechanisms.

One type of cyclone scrubber is shown in Fig. 13.19.

As it prevents re-entrainment, the wet cyclone has higher efficiency than the dry cyclone. However, sludge removal is more difficult than dry particulate removal.

Packed Scrubbers

The packed tower uses fibre glass (fine glass filaments) or other packing.

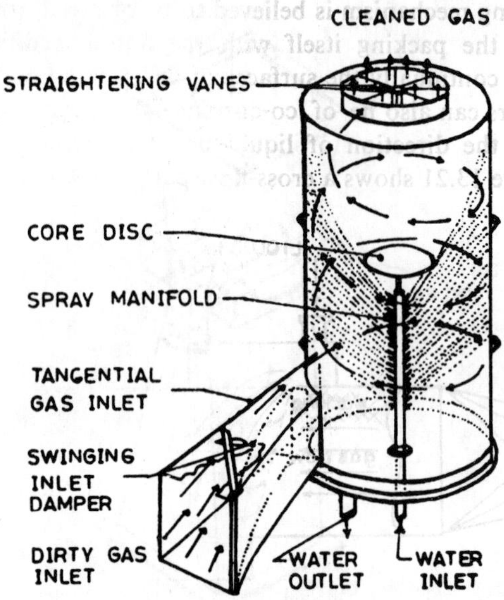

Fig. 13.19 *Cyclonic spray scrubber*

Normally, gas flow is counter-current to liquid flow in this type of equipment (Fig. 13.20).

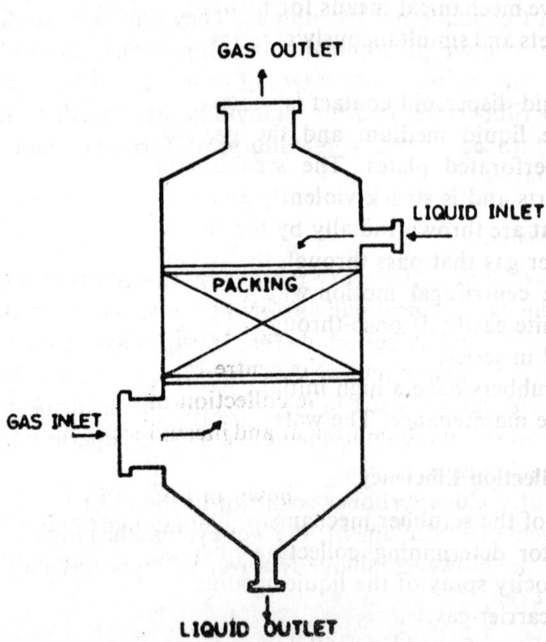

Figr 13.20 *Counter-current packed tower*

The separating mechanism is believed to be primarily impingement of the dispersoid on the packing itself with the liquid medium (water) merely acting to clean continually the surfaces of the packing material.

Packed towers can also be of co-current type or cross-flow type design depending on the direction of liquid flow with respect to the direction of gas flow. Figure 13.21 shows a cross-flow packed tower.

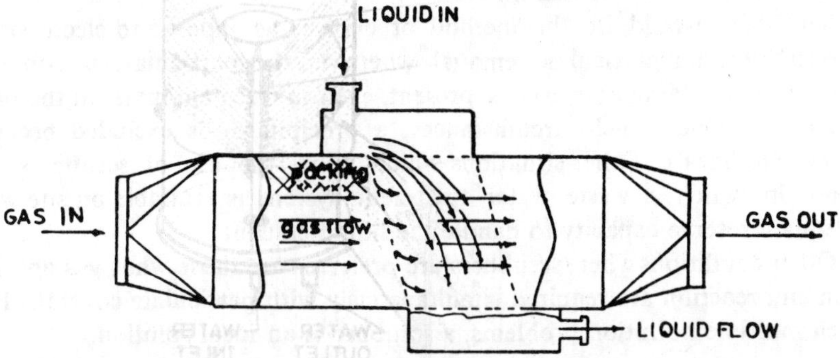

Fig. 13.21 *Cross-flow packed tower*

Mechanical Scrubbers

A mechanical scrubber is a mechanically aided unit which has an internal rotating mechanical part. It is the only scrubber other than the centrifugal types to have mechanical means for breaking up the scrubbing liquid into small droplets and simultaneously creating turbulence. These are high energy scrubbers.

Here liquid-dispersoid contact is achieved by the simultaneous introduction of the liquid medium and the gas stream on to the rotating discs, blades or perforated plates. The scrubbing liquid dribbles down on the rotating parts and is struck violently and shattered (disintegrated) into fine droplets that are thrown radially by the centrifugal force. The dust particles in the carrier gas that pass through the droplet curtain are thrown to the wall by the centrifugal motion which the droplets already possess and are removed quite easily. If once-through contact is insufficient, several stages can be used in series.

These scrubbers have a high initial cost, high operating cost, and require considerable maintenance. The water use rate is also high.

13.11.3 Collection Efficiency

Regardless of the scrubber mechanism involved, power consumption is the critical factor determining collection efficiency. Energy is consumed either by high velocity spray of the liquid medium or high velocity of the dispersion-laden carrier gas.

Simpler types of scrubbers with low energy inputs are effective in collecting particles above 5–10 μ in diameter, while the more efficient, high energy input scrubbers will perform efficiently for collection of particles as small as 1–2 μ in diameter.

13.11.4 Suitability of Scrubbers for Particulate Control

Wet scrubbers are particularly useful in the case of a hot gas that must be cooled for some reason. For some effluents, adding a liquid to the gas would be harmless or even beneficial.

Scrubbers would be the method of choice, as opposed to electrostatic precipitators for particulate removal whenever the particulate is combustible or any flammable gas is present, even in trace amounts, in the bulk gas-phase. Under such circumstances, a precipitator is excluded because of the fire hazard. Other situations which allow the use of scrubbers are those in which a waste water treatment system is available on site with adequate reserve capacity to handle the liquid effluent.

Other conditions where scrubbers are preferred are those when gas absorption and reaction are required simultaneously with particulate control. For such multiple pollution problems, a scrubber is an ideal solution.

13.11.5 Advantages of Scrubbers

1. Low initial cost
2. Moderately high collection efficiency for small particles
3. Applicable for high temperature installations
4. They can simultaneously remove particulates and gases
5. There is no particle re-entrainment

13.11.6 Disadvantages of Scrubbers

1. High power consumption for higher efficiency
2. Moderate to high maintenance costs owing to corrosion and abrasion
3. Wet disposal of the collected material.

13.12 Choice of Equipment

While selecting a particulate collector from various equipment available, the following factors must be taken into consideration.

1. Particulate size, shape, and density
2. Particulate loading, often expressed in milligrams per cubic metre (mg/m^3)
3. Efficiency required
4. Properties of the carrier gas
   (a) Composition
   (b) Temperature
   (c) Pressure

(d) Viscosity
(e) Density
(f) Humidity
(g) Combustibility
(h) Reactivity
(i) Toxicity
(j) Electrical and sonic properties

5. Flow characteristics of the carrier gas.
   (a) Flow rate
   (b) Variations in flow rate
6. Specific properties of the contaminant
   (a) Composition
   (b) Contaminant phase
   (c) Solubility
   (d) Combustibility
   (e) Reactivity
   (f) Toxicity
   (g) Hygroscopicity
   (h) Agglomerating characteristics
   (i) Electrical and sonic properties
   (j) Catalyst poisoning
7. Allowable pressure drop
8. Contaminate disposal
9. Capital and operating costs of the equipment
10. Ease of maintenance and reliability.

It should be noted that the minimum removal efficiencies of one or more components may be dictated by applicable local, state, provincial, or national regulation.

Further, in some cases only one particular type of control equipment is suitable; in other cases different types are possible. The choice will be a compromise based on collecting efficiency, resistance to wear, space available, installation cost, operational cost, maintenance, etc.

13.13 Economical Aspects

Installation cost, power consumption and maintenance cost for different types of collectors must be considered before installing such a plant. In one case, it might pay to spend a large amount on the installation which can be recovered through low costs for operation and maintenance. In another case it might be more profitable to choose a cheap collector requiring more power and maintenance. It is not possible to frame definite rules in this respect. The economical aspects of various control equipment are briefly described below.

1. **Cyclones:** Cyclones are generally cheap to install. The power consumption is moderate. The maintenance cost is also normally not very high. However, it must be noted that special conditions such as corrosive gas or abrasive dust considerably influence the life and thereby affect maintenance cost.

2. **Filters:** A bag filter is comparatively expensive to install. Its power consumption is moderate. In most cases, the maintenance cost is high because the bags have to be repaired or replaced regularly. The nature of the gas and the dust decide the frequency of such maintenance work.

3. **Electrostatic precipitators:** They are generally the most expensive dust collectors as regards installation cost. However, they can be expected to have the longest life. Further, the power consumption is moderate to low as the pressure drop is very low and the required high voltage power is moderate. An ESP is not exposed to much mechanical wear if measures are taken to avoid condensation and corrosion. Therefore, the maintenance cost is moderate.

4. **Scrubbers:** The installation cost for a wet collector is moderate. Also, the maintenance cost is generally not very high. But the comparatively high pressure drop, especially in a high efficiency scrubber, can involve rather high power consumption.

QUESTIONS

1. What are the main principles of air pollution abatement?
2. What are the objectives of using control equipment?
3. List the various types of collection equipment for particulates.
4. How do you calculate the efficiency of a separating device?
5. What are the advantages of using collectors in series?
6. Explain with a neat sketch, the principle, construction. and working of a settling chamber. How can its efficiency be improved?
7. What are the advantages and disadvantages of settling chambers? Give their applications.
8. What is the principle of inertial separators? Mention the types of inertial separators.
9. Explain with neat sketches
   (a) Baffle type separator
   (b) Louvre type separator
   (c) Dust traps
10. Mention the advantages and disadvantages of the following inertial separators:
    (a) Baffle type separator
    (b) Louvre type separator
11. With a neat sketch explain the principle, construction, and working of a cyclone separator.
12. Draw a neat sketch of a typical cyclone separator and label the parts. Also, give the proportions of the various parts with respect to cyclone diameter.
13. Enumerate the factors on which the collection efficiency of a cyclone depends.
14. Distinguish between 'conventional' and 'high efficiency' cyclones.
15. Compare tangential inlet and involute inlet for cyclones.

16. Discuss the operating problems associated with cyclonic separators.
17. What are the merits and demerits of cyclones?
18. Write short notes on:
    (a) Multiple cyclones
    (b) Cyclones in series
19. Distinguish between fabric and fibrous filters.
20. Explain with a neat sketch, the principle and construction of fabric filters.
21. What are the factors affecting the efficiency of fabric filters?
22. Discuss the operating problems associated with fabric filters.
23. With sketches, explain the various methods of cleaning a fabric filter.
24. Explain the factors to be considered while selecting the filter medium for bag houses.
25. Give the properties of common bag fabrics.
26. What are the merits and demerits of fabric filters?
27. Explain with sketches
    (a) Reverse jet filter
    (b) Envelope-type fabric filter
28. Write short notes on
    (a) Multi-compartment type bag house
    (b) Fibrous filters
29. Distinguish between viscous and dry fibrous filters.
30. What are the advantages of fibrous filters?
31. With a neat sketch, explain the principle, construction and working of an electrostatic precipitator.
32. List the major components of an electrostatic precipitator.
33. Distinguish between
    (a) Single-stage and two-stage precipitators
    (b) Pipe-type and plate-type precipitators
    (c) Dry and wet precipitators
34. Explain with a neat sketch, the construction and working of a pipe-type electrostatic precipitator. When is it preferred?
35. Explain with a neat figure, the construction and working of plate-type electrostatic precipitator. When is it preferred?
36. Write a note on collection efficiency of an ESP.
37. With reference to electrostatic precipitators, explain the following:
    (a) Particle resistivity
    (b) Particle re-entrainment
38. Write a note on design parameters for ESP.
39. What are the advantages and disadvantages of electrostatic precipitators?
40. What are the general applications of the following air pollution control equipment?
    (a) Cyclones
    (b) Fabric filters
    (c) Electrostatic precipitators
41. What is the minimum size of the particulates removed through the following control equipments?
    (a) Settling chambers
    (b) Cyclones
    (c) Fabric filters
    (d) Electrostatic precipitators

42. Write a note on electrostatic precipitators for air cleaning.
43. Explain the mechanism of working of scrubbers for particulate removal.
44. List the types of scrubbers.
45. With neat sketches, explain the following:
    (a) Spray towers
    (b) Cyclone scrubber
    (c) Venturi scrubber
46. Draw neat sketches of
    (a) Counter-current packed tower
    (b) Cross-flow packed tower
    Explain the working of one of the above scrubbers.
47. When are scrubbers preferred for particulate control?
48. What are the merits and demerits of scrubbers?
49. Discuss the factors to be considered while selecting a suitable control equipment for particulate removal.
50. Discuss the economic aspects of various control equipment for particulates.

14

Smoke and Its Control

14.1 Introduction

Smoke particles are mainly urburnt carbon resulting from incomplete combustion. In other words, carbon particles, which make smoke are unburnt fuel. Therefore smoking chimneys are indications of inefficiency in the process of converting fuel into useful energy.

Generally, the sizes of fine particles in smoke will be less than 1 μ. Coal smoke particles range from 0.01–0.2 μ and oil-smoke particles from 0.03–1.0 μ.

Smoke as we all know is a nuisance. Hence, in the field of air pollution, smoke, particularly black smoke, is considered to be public enemy No. 1. Smoke affects public health and causes property damage.

Further, black smoke apart from being a public nuisance is an economic waste, as it is due to incomplete combustion of fuel.

But one interesting and well known fact is that all fuels can be burnt smokelessly and efficiently. Therefore, it is surprising that many chimneys emit black smoke.

14.2 Sources of Smoke

Important sources of smoke are automobiles, domestic heating plants, industrial power plants, railway engines, ships, and refuse incinerators. It is also due to oil burnt in diesel engines.

The amount of smoke given out from the above sources varies considerably. It depends on the type of fuel and the efficiency of the combustion device.

14.3 Measurement of Smoke

There are many methods for estimating the density of the smoke as it emerges from the stack. The most commonly used method is the Ringelmann Chart. Accurate measurement of the smoke density (smoke quality) is essential for the enforcement of ordinances regulating smoke emissions.

14.3.1 Ringelmann Chart

The Ringelmann Chart is a scheme where graduated shades of gray vary by five equal steps between white and black. The chart shown in Fig. 14.1, shows four grades of gray, as well as pure white and an all black section. The details about the width of black lines, width of white spaces, and percentage of black is given in Table 14.1.

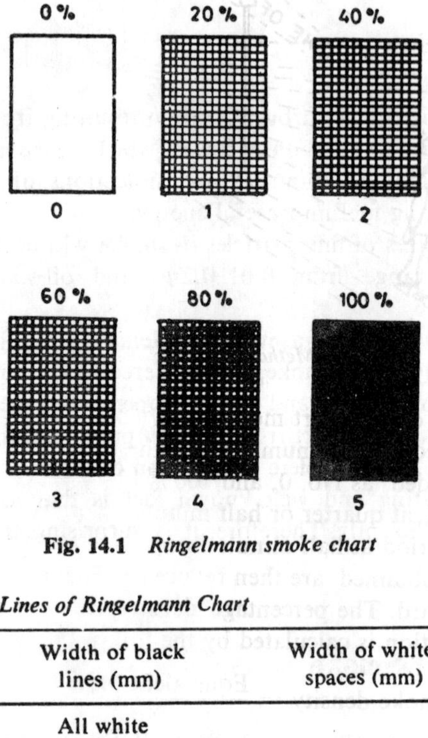

Fig. 14.1 *Ringelmann smoke chart*

Table 14.1 *Spacing of Lines of Ringelmann Chart*

Ringelmann chart No.	Width of black lines (mm)	Width of white spaces (mm)	Per cent black
0	All white		0
1	1	9	20
2	2.3	7.7	40
3	3.7	6.3	60
4	5.5	4.5	80
5	All black	—	100

The chart should be used under daylight conditions and held or fixed facing the observer in a vertical plane, as shown in Fig. 14.2. The chart should be preferably in line with the top of the chimney, and placed so that the chart and the smoke have a similar sky background. Also, the chart should be at a sufficient distance from the observer for the lines to appear to merge until each square forms a uniform shade. Generally this distance is about 15 m. Details about the method of use of the chart is given in IS: 9078 (1979).

The observer glances at the smoke as it comes out from the chimney, and

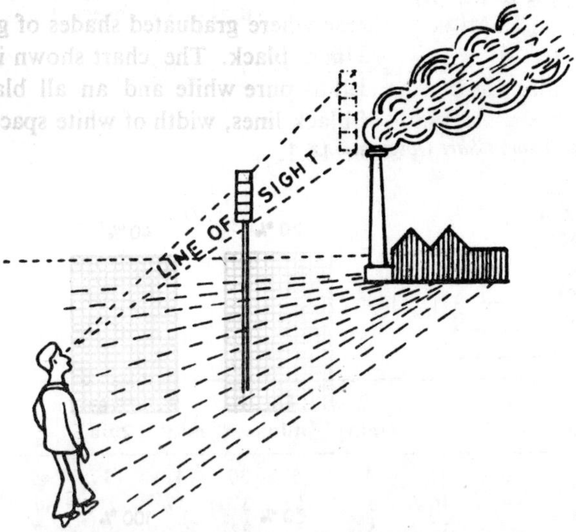

Fig. 14.2 *Method of using Ringelmann chart*

notes the number of the chart most closely corresponding with the shade of the smoke, and records this number with the time of observations. A clear chimney is recorded as No. 0, and 100% black smoke as No. 5. Observations are repeated at quarter or half minute intervals. A set of observations for a one hour period at half minute intervals is given in Table 14.2.

The readings obtained are then reduced to the total equivalent of No. 1 smoke as a standard. The percentage density of the smoke for the entire period of observation is calculated by the following formula:

$$\text{Percentage smoke density} = \frac{\text{Equivalent units of No. 1 smoke} \times 0.20}{\text{Number of observations}}$$

In the above formula, in the numerator, equivalent units of No. 1 smoke is multiplied by 0.20, as No. 1 smoke is 20% dense.

The method of using this formula is indicated in Table 14.2. In the example illustrated the period of observation is from 10.00 a.m. to 11 a.m. that is, one hour duration. Observations have been recorded at every half minute intervals, that is, 120 observations in one hour duration.

Calculation of Smoke Density		Equivalent No. 1 units
6 units of No. 5	=	30
12 units of No. 4	=	48
20 units of No. 3	=	60
23 units of No. 2	=	46
26 units of No. 1	=	26
33 units of No. 0	=	0
120 units		210

$$\text{Percentage smoke density} = \frac{\text{Equivalent units of No. 1 smoke} \times 0.20}{\text{Number of observations}}$$

$$= \frac{210}{120} \times 0.20 = 35\%$$

Table 14.2 *Ringelmann Chart Observations*

Location: ________
Time: 10.00-11.00 a.m. Date: ________
Direction of wind: ________
Velocity of wind ________
Direction of stack: ________
Distance to stack: ________
Diameter of stack: ________

10. a.m	0 min	½ min		0 min	½ min		0 min	½ min		0 min	½ min
0	—	—	15	4	5	30	4	3	45	1	1
1	—	—	16	5	5	31	2	2	46	1	—
2	—	—	17	5	5	32	2	1	47	—	—
3	—	—	18	5	4	33	1	1	48	—	—
4	—	—	19	4	3	34	1	1	49	—	—
5	1	1	20	2	1	35	1	1	50	—	—
6	1	1	21	2	2	36	1	1	51	1	2
7	1	1	22	2	2	37	—	—	52	2	2
8	2	2	23	2	2	38	—	—	53	2	3
9	2	2	24	2	2	39	—	—	54	3	3
10	2	3	25	3	3	40	—	—	55	3	4
11	3	3	26	3	3	41	1	1	56	4	3
12	3	3	27	3	3	42	—	—	57	2	2
13	3	4	28	3	4	43	—	—	58	1	1
14	4	4	29	4	4	44	1	1	59	—	—

INSPECTOR:

The Ringelmann Chart is a simple and effective method of estimating smoke density. But one main drawback of this method is the involvement of the human element in judging smoke shade. Other factors that may affect the results are cloud and wind conditions. However, when the main purpose is to lessen the smoke emission and create a clean environment, the practical usefulness of the Ringelmann Chart can be appreciated.

The Ringelmann Chart is commonly used by smoke inspectors. The chart cannot be used at night.

14.3.2 Miniature Smoke Chart

It is difficult to use a Ringelmann Chart conveniently in practice. Generally, the use of the standard chart requires the assistance of a helper to hold the chart for the observer. Hence, the necessity for the use of a miniature smoke chart.

As per IS: 9078 (1979), the miniature smoke chart shall have the same precision as the Ringelmann Chart. The miniature chart is not intended for use as a substitute for the Ringelmann Chart but the results obtained are expected to be similar to those from the Ringelmann Chart. A miniature smoke chart may be used for readings of a preliminary nature.

The gray shades marked on the chart correspond to similar shades of the Ringelmann Chart. The chart is held at arm's length or preferably at a distance of 1.5 m from the observer's eye, with the help of a holder while recording observations. For details IS: 9078 can be referred.

14.4 Other Methods of Measurement of Smoke

Nowadays many other devices like photoelectric and photographic devices are also being used for the measurement of smoke density. Even optical methods are being employed. A good example of an instrument used for estimating smoke density based on the principle of optics is the 'smoke-scope'.

A smoke-scope is an instrument developed by the Mine Safety Applicance Company, USA. It is an optical instrument used for estimating the density of the smoke coming from a stack. It has several features which help in maintaining accuracy and reproducibility of observations even though the stack is viewed from different angles or under different conditions of lighting and background.

While making observations, the stack is viewed through one barrel of the instrument, and the smoke is observed through an aperture in the centre of the screen. Light from the area adjacent to the stack comes through a second barrel and illuminates a standard density film which is used for comparison with the smoke. Consequently, an image of the reference film appears on the screen surrounding the aperture. This helps in easy comparison with the smoke. Refocussing of the eye is not required while making observations because, with the help of a lens, the image of the reference is made to appear to be at a distance equivalent to that of the stack. Also, as illuminations of the smoke and the reference are simultaneously influenced by the same factors, automatic compensation takes place under different conditions.

14.5 Smoke and Public Health

Smoke may affect public health through:

1. Irritation of the eye membranes
2. Irritation of the respiratory tract
3. Infection of the ear

4. Increased morbidity rate through diminished resistance to disease because of decrease of sunlight
5. Increase of nervous and psychological disorders because of SMOG formation

14.6 Smoke and Property Damage

1. Smoke causes property damage by disfigurement of exteriors of buildings and also their interiors.
2. Clothing is spoiled and consequently, cleaning and laundry costs are increased.
3. Smoke injures the growth of vegetation, by coating the leaves with soot to prevent the process of photosynthesis.
4. Daylight hours are shortened in a smoky city resulting in economic loss from extra use of electricity for illumination.

14.7 Smoke Prevention and Control

Following are some of the methods which can be adopted for smoke prevention and control:

1. Use of low volatile fuels such as bituminous coal and anthracite coal.
2. Use of smokeless fuels, e.g., coke.
3. Strict enforcement of 'smoke ordinances' i.e. regulations regarding smoke emissions.
4. Proper design and maintenance of stacks.
5. Use of proper combustion equipment (proper design of furnaces). This involves the fulfilment of the following conditions for smokeless combustion of a fuel:
   (a) Proper air-fuel ratio
   (b) Sufficient mixing of air and fuel
   (c) Sufficient ignition temperature
   (d) Sufficient space to permit time for proper burning
6. Use of particulate control equipment like electrostatic precipitators, scrubbers, etc.

QUESTIONS

1. What are the sources of smoke ?
2. Explain the principle, construction, and use of the Ringelmann Chart for smoke measurement.
3. State the effects of smoke on:
   (a) Public health
   (b) Property

4. Explain the characteristics and measurement of smoke.
5. What are the methods of smoke abatement?
6. Write short notes on:
   (a) Ringelmann Chart
   (b) Miniature smoke chart
   (c) Smoke-scope

15

Control of Gaseous Contaminants

15.1 Introduction

The control of gaseous pollutants from stack gases depends on their properties. The methods of control include

1. Combustion
2. Absorption
3. Adsorption
4. Closed collection and recovery systems
5. Masking and counter action (for odours)

15.1.1 Combustion

Combustion processes like flame combustion or catalytic combustion can be utilised to greatest advantage when the gases or vapours to be controlled are organic in nature. Equipment employing the principle of flame combustion include

(a) Fume and vapour incinerators
(b) After-burners
(c) Flares, either with steam injection or venturi flare

The use of after-burners on incinerators has met with varying success depending on the kind of after-burner used and the type of incinerator. Flare design should provide for smokeless combustion of gases of variable composition and a wide range of flow rates. Venturi flares mix air with the gas in the proper ratio prior to ignition to achieve smokeless burning. Steam injection flares mix steam with the stack gases as they reach the stack.

When the concentration of the combustible portion of gas stream is below flammable range and when lower operating temperatures are desired, catalytic combustion processes are used. Catalytic combustion process is used with success for the control of effluent gases, fumes and odours from refineries, burning waste, cracking gases, phenolic-resin curing ovens, paint and enamel ovens, coffee roasting processes, foundry core baking ovens and chemical plants discharging maleic and phthalic anhydrides. Gases and fumes containing excessive amounts of particulate matter reduce the effectiveness of catalytic combustion units due to a coating that forms on the catalyst.

Major Limitations

The method is too expensive when,

1. Fuel values of the gaseous discharge are low
2. Moisture content of the discharge is high
3. Exhaust volume is extremely large

However, combustion control equipment has been used with advantage, in petrochemical, paint and varnish and fertiliser industries.

Factors to be considered in the design of the Incinerator are

1. Sufficient air (O_2) for combustion reaction
2. Adequate temperature for continuous oxidation
3. Adequate retention time

Generally, incineration equipment consists of a single combustion chamber with provisions for maintaining continuous and intimate mixing of the contaminant-laden gas and the flame from the gas burner. The combustion chamber is so proportioned that the gas velocity and gas flow patterns established will produce adequate retention time in the combustion zone. Average retention time is 0.2–0.3 s at temperatures of 650°C and higher.

15.1.2 Absorption

In this process, effluent gases are passed through absorbers (scrubbers) which contain liquid absorbents that remove one or more of the pollutants in the gas stream. The efficiency of this process depends on

1. Amount of surface contact between gas and liquid
2. Contact time
3. Concentration of the absorbing medium
4. Speed of reaction between the absorbent and the gases

Absorbents are being used to remove SO_2, H_2S, SO_3, F and oxides of Nitrogen. The absorbents may be either reactive or non-reactive with the pollutants removed by them. Some of the reactive absorbents are regenerative (i.e. they may be treated and reused), while others are of non-regenerative type.

The equipments using the principle of absorption for the removal of gaseous pollutants include

(a) Packed tower
(b) Plate tower
(c) Bubble-cap plate tower
(d) Spray tower
(e) Liquid jet scrubber absorbers

Selective chromatographic absorption of gases on small pellets may offer much higher rates of absorption than those achieved in packed towers.

The absorbents commonly used for different gases are given in Table 15.1

Table 15.1 *The Common Absorbing Solutions Used for Removing Different Gaseous Pollutants from Gas Streams*

Gaseous pollutant	Common absorbents used in solution form
SO_2	Dimethylaniline, mixture of xylidine and water (1 : 1), ammonium sulphite, basic aluminum sulphate, ethanol amines (mono ethanol amine, diethanol amine, methyl diethanol amine or triethanol amine), sodium sulphite, ammonium sulphite and bisulphite, water, alkaline water, a suspension of $Ca(OH)_2$, calcium sulphite and calcium sulphate, barium thionates and sulphites.
H_2S	NaOH and phenol mix (mole ratio 3 : 2), tripotassium phosphate, sodium alamine or potassium dimethyl glycine, ethanolamines, soda ash solution containing suspended iron oxide or hydroxide, soda ash alone, sodium thioarsenate, ammoniacal liquor from coke ovens.
HF	Water, sodium hydroxide
Oxides of nitrogen	Water, aqueous nitric acid.

Packed Tower

A packed tower consists of a vertical shell filled with a suitable packing material and the liquid flows over the surface of the packing in thin films. The efficiencies of packing towers are being improved in recent years by use of new kinds of packing material.

Plate Tower

A plate tower consists of a vertical shell in which are mounted a large number of equally spaced circular perforated plates; gases and vapours bubble upward through the liquid seal above each plate.

Bubble-Cap Plate Tower

This consists of a vertical shell in which are mounted a large number of equally spaced circular bubble cap plates. In a spray tower, the absorbing liquid is sprayed through the gas. By this contact of gas and liquid is possible.

Efficiency of gas absorption depends upon the amount of surface contact between the gas and the liquid, time of contact between gas and liquid, concentration of the absorbing liquid, and the rate of reaction between the gas and the absorbent.

Sometimes, the carrier-gas may ontain particulate matter. In such a case the gas absorption equipment should also function as a wet dust collector.

15.1.3 Adsorption

In this process the effluent gases are passed through adsorbers which contain solids of porous structure. The commonly used adsorbers include activated carbon, silica gel, activated alumina, lithium chloride, activated bauxite, etc. Active carbon appears to be the adsorbent most suitable for recovering organic solvent vapours. The steps necessary for effective removal of gaseous pollutants by adsorbents are

1. Contact of the gaseous or vaporous pollutant with the solid adsorbent
2. Separation (desorption) of the adsorbed gaseous pollutant from the solid adsorbent by regeneration or replacement of the adsorbent
3. Recovery of the gases for final disposal

The efficiency of removal of gases by adsorbents depends on

(a) The physical and chemical characteristics of the adsorbent
(b) The concentration and nature of gas to be adsorbed

Desorption is accomplished by raising the temperature of the granular bed above the boiling temperature of the pollutant by superheated steam, submerged heating elements or combustion gases. Desorption may also be performed by reducing the pressure. The adsorbents commonly used for removal of different gases are given in Table 15.2.

Table 15.2 *Common Adsorbents Used for Removing Different Gaseous Pollutants*

Gaseous pollutant	Adsorbents used in solid form
SO_2	Pulverised limestone or dolomite, alkalised alumina (aluminium oxide plus sodium oxide)
H_2S	Iron oxide
HF	Lump limestone, porous sodium fluoride pellets
Oxides of nitrogen	Silica gel
Organic solvent vapours	Activated carbon

Adsorption is a surface phenomenon and requires very large solid surface areas to be effective. In almost all cases these areas are internal as with porous material and may be almost unbelievably great per unit of adsorber volume (e.g 1 cu in. of activated carbon has a surface area of 20,000 sq. yds). Adsorption is primarily a physical phenomenon.

Adsorption as a means of air pollution control, offers certain advantages for some classes of contaminant. It includes gases and vapours for which other means of collection are uneconomical, hazardous or impossible. For example, many flammable organic compounds may be recovered with relatively high efficiency by adsorption at concentrations safely below the lower explosive limits of the compound. Also, since adsorption gives very complete clean-up even when pollutants concentrations are very low, unusually toxic or odorous compounds can be sometimes be removed by adsorption on activated carbon, when all other methods fail.

Adsorption equipment is generally the packed bed of some adsorbing material. Installation costs for adsorbers are high but maintenance and operating costs are not excessive. Sometimes, the value of the recovered material enhances economic feasibility.

15.1.4 Closed Circuit and Recovery Systems

Gases like SO_2, Oxides of N, and hydrocarbons can be recovered from the waste gas streams if they are present in sufficient concentrations. For example, where the concentration is of the order 5–10% SO_2 as in smelter gases, the sulphur content may be recovered economically. The most usual method at smelters is to use the SO_2 stream as the raw material for the manufacture of H_2SO_4. Similarly, the vapour recovery methods used in refineries are useful when the concentration of hydrocarbons in the effluent stream is high and relatively uncontaminated.

Oxides of N from waste gas streams in a HNO_3 plant are recovered using commercial zeolite. Oxides of N absorbed in the bed are recovered as enriched oxides of N and HNO_3 by regenerating the bed at elevated temperatures with hot air or steam.

In the alkalised alumina sorption process, Oxides of sulphur in the stack gas are absorbed on spheres (1.6 mm) of alkalised alumina (a mixture of aluminium oxide and sodium oxide) in a bed suspended in the gas stream. The oxides are then removed from the spheres by reaction with a reducing gas containing hydrogen and CO, producing CO_2 and H_2S. The H_2S is converted to elemental S which can be sold and the regenerated alkalised alumina is recycled. The process would remove about 90% of the oxides of S in the stack gas. On a 800 MW power plant burning coal of 3% sulphur content, it would produce about 180 t of S per day.

In another process known as wet lime process for removing SO_2 from power plants, pulverised limestone is injected into the boiler furnace, where the heat drives off CO_2, converting the $CaCO_3$ to the reactive oxide form. The oxide then reacts with the SO_2 to form solid sulphites and sulphates. Some of the conversion takes place before the stack gas reaches the water scrubber, but most of it takes place in the scrubber after the reactants dissolve in the water. The resulting solids, as well as the fly ash removed in the scrubber, go to the settling pond, and water from the settling pond is recycled to the scrubber.

QUESTIONS

1. Mention the common methods of control of gaseous contaminants and describe any one of them in detail.
2. What are the factors to be considered in the design of incinerators? What are the limitations of this process?
3. When do you recommend absorption as a method of control of gaseous contaminants? List out the absorbents commonly used for removing SO_2, H_2S and NO_x.
4. Write explanatory notes on
   (a) Packed tower
   (b) Bubble-cap plate tower
5. What are the common adsorbents used in solid form to remove SO_2, H_2S, HF, NO_x and organic solvent vapours?
6. Write a note on the importance of activated carbon as an adsorbent
7. Write short notes on the recovery systems used for recovery of SO_2, NO_x and hydrocarbons from the waste gas streams.

16

Odours and Their Control

16.1 Introduction

Odour is undoubtedly the most complex of all the air pollution problems. Odour is caused in many cases by very minute quantities of substances. The only good measuring device for odour is the human nose, which as we all know is undependable. Further, people have mixed reactions with respect to the offensiveness of odours. Perfume liked by some, may be disliked by others!

Odours are emitted by many industries. Food processing, oil refining, paper and rubber industries and tanneries are among the major odour-emitting industries. Though foul smells may not cause direct damage, they are as much a nuisance as noise, dirt or corrosion. The primary effect of disagreeable odours on people is the nuisance effect. Secondary effects, in some cases, may be nausea, insomnia and discomfort. On the economic front, the loss of property values near poorly operated dumping yards and slaughter houses is partly a consequence of offensive odours.

Until recently, very little attention was given to the control of odourous air contaminants. Consequently, the field of odour control has remained unexplored to some extent.

16.2 Background Information About Odours

In tackling the problems of odour control, it is necessary to know something about the background of odours and the reaction of the individual to odours.

Odour can be defined as the 'perception of smell'. The experience of smell can be taken to mean any perception that results from nasal inspiration, or may refer to sensations received through the receptors of the olfactory epithilium, situated within the upper nasal cavity.

At present there is no instrument or means of chemical analysis that can replace the function of the nose in its ability to detect and to decide the intensity-rating of odours. Even otherwise, the development of such a unit or means of analysis would not necessarily solve the problem, since the perception of odour is an extremely subjective function.

Following are the well known facts about odours:

1. Some substances are odorous and some are not

2. Substances of similar or dissimilar chemical constitution may have similar odours
3. Nature and strength of the odour may change on dilution
4. Weak odours are not perceived in the presence of strong odours
5. Odours of the same strength blend to produce a combination in which one or both may be unrecognizable
6. Constant intensity of odours causes an individual to quickly loose awareness of the sensation and only notice it when it varies in intensity
7. Fatigue for one odour may not affect the perception of dissimilar odours but will interfere with the perception of similar odours
8. An unfamiliar odour is more likely to cause complaint than a familiar one
9. Two or more odorous substances may cancel the smell of each other
10. Odours travel downwind
11. Man can smell at a distance
12. Many animals have a keener sense of olfaction than man
13. Likes and dislikes often depends on the association of the scent with pleasant or unpleasant experiences
14. The number of different and distinct smells is great

16.3 Source of Odours

An odorant generally originates from a solid, a liquid, or a concentrated gas. Odour sources may be confined in space, like emission from ducts, or they may be unconfined, like drainage ditches, and settling lagoons.

Hydrogen sulphide, carbon disulphide, mercaptans, products of decomposition of proteins (especially those of animal origin), phenols, and some petroleum hydrocarbons are the malodours which are very common.

Odorous compounds are also generated due to various human activities. Garbage dumps, sewage works, and agricultural activities are typical examples. Bad smell is also produced by decaying vegetation. The exhaust from motor vehicles is also a common source of malodour.

The sources of odour are so many, that it is almost impossible to prepare a complete list of then.

Table 16.1 indicates the various odorous industrial operations.

16.4 Measurement of Odour

The measurement of odour falls into two categories:

1. Determination of the threshold concentration (minimum identification level) of odoriferous gases
2. Determination of the type and intensity of atmospheric odours

Table 16.1 *Odorous Industrial Operations*

No.	Industry	Odorous Material
1.	Pulp and Paper	Mercaptans
2.	Tanneries	Hides, flesh, hair
3.	Fertilizer	Ammonia, nitrogen compounds
4.	Petroleum	Sulphur compounds from crude oil, cresols
5.	Chemical	Ammonia, phenols, mercaptans, hydrogen sulphide, chlorine, organic products
6.	Foundries	Quenching oils
7.	Pharmaceutical	Biological extracts and wastes, spent fermentation liquors
8.	Food	Cannery waste, dairy waste, meat products, packing house wastes, fish, cooking odours, coffee roaster effluents
9.	Detergent	Animal fats
10.	General	Burning rubber, solvents, incinerator smoke.

16.4.1 Determination of the Threshold Concentration

For each odour there exists a concentration below which no perception is possible for most individuals. This concentration is called the threshold concentration or minimum identification level.

Most of the odour-measuring devices (osmometers) are based on the vapour-dilution method. In this method, the amount of odour producing substance added to purified air is generally increased until an observer or 'sniffer' can just barely detect the odour.

The odour threshold for acetone is reported to be 100 ppm, acetic acid 1 ppm and for pyridine 0.02 ppm.

Generally, odour intensity increases with the odorant concentration. The relationship between odour intensity and concentration can be expressed as

$$P = K \log S \tag{16.1}$$

where P = odour intensity
K = constant
S = odour concentration

Thus, we see that the odour intensity varies as the logarithm of the concentration.

16.4.2 Determination of the Type and Intensity of Odours

For odour intensity measurements generally a panel of 6 to 12 persons in normal health are employed. The panel members sniff the air at a given location at the same time and report individually the nature and intensity of the odour. Some experience may be required to identify an odour. The odour intensity is usually stated according to a predetermined rating system.

A widely used scale for odour intensity is the following: 0, no odour; 1, threshold level (barely perceptible); 2, definite odour; 3, strong odour; and

4, overpowering odour. Half scores may be used when the observer is undecided. By averaging the values recorded by members of the panel, a single value can be assigned to the odour intensity at a given location. If measurements are made by the same panel at various locations, the source of the particular odour may sometimes be found.

Several descriptive words are used to characterize the type or quality of an odour. For example fruity, flowery, burnt, aromatic, fragrant, sweet, garlic, foul, repulsive, nauseating, etc. Sometimes lay terms are used for odour identification. For example acetic acid has a vinegary odour, hydrogen sulphide has an odour of rotten eggs and sulphur dioxide has an irritating odour.

The expert panel method is fairly expensive and limited to short periods of use. Also, care must be taken because mixtures of two or more different odours can be very misleading. Thus, the identification and determination of odours in the open air is even less precise than the determination of threshold values in the laboratory.

Another method used by control officers, is to have residents in certain areas note the time of day when a particular odour is noticed and the number of times over a specific period. This data may be collected and reported graphically.

For determining the concentration of odoriferous compounds in ducts and stacks, a syringe technique is employed. Measured samples of odour-laden gas are taken in hypodermic syringes and brought to an odour-free room. A small quantity of this gas is then added to a syringe into which odour-free air is then drawn. The observer holds the syringe to his nostrils and drives the mixture into his nasal passages by means of plunger. Hence, the observer can determine the amount of dilution necessary to bring the odour to a barely perceptible level. Concentration is expressed in terms of *odour unit*, which is defined as the amount of odour necessary to contaminate 1 cu ft (0.027 m^3) of air to the threshold level.

Odorants in the atmosphere or gas streams may also be collected by passing known volumes of air or gas through a column of activated carbon or by condensation techniques. The collected odorants may then be taken to a laboratory and released by vacuum desorption and distillation, respectively.

16.5 Odour Control

An important problem of air pollution is the control of objectionable odours, which may be gases, mists, or solids discharged into the atmosphere from industrial, commercial, and municipal operations. Source control is the most effective means of abating odour. In many cases, this can be achieved by good sanitation practices, as the most persistent and offensive odours arise due to putrefaction.

Following are the methods which may be employed alone or in various combinations to eliminate or diminish odours.

1. Modification of the process
2. Dilution by ventilation or dispersal
3. Absorption
4. Adsorption
5. Combustion or oxidation
   (a) Direct incineration
   (b) Catalytic incineration
6. Odour masking
7. Odour counteraction or neutralization
8. Injection of a reactive substance
9. Irradiation

16.5.1 Modification of the Process

In some cases a change in the process, either by way of change in the composition of process materials or removal of impurities may help in odour control. Methods include substitution of low-odour solvents for highly odorous ones, adjustment of process variables like temperature, residence time, etc. If such alteration makes the resultant source less intense or more tolerable from the point of odour, and if such modification is technically and economically feasible, then it merits first consideration.

16.5.2 Dilution by Ventilation or Dispersal

Odour intensity being a function of the odorant concentration, proper well-designed ventilation is the most common method for removal of odours from enclosed spaces.

A method sometimes used for odour abatement outdoors is to release odorous gases from tall stacks. It results in normal dispersion in the atmosphere and consequent decrease in ground-level concentrations below the threshold value. Dispersal by stacks requires careful consideration of the location, meteorological parameters, etc.

16.5.3 Absorption

Absorption is applicable when the odorous gases are soluble or emulsifiable in a liquid or react chemically in solution. Liquid scrubbing of the gases in a suitable absorption unit is an important method of odour control.

16.5.4 Adsorption

Adsorption, particularly on activated carbon, has been widely used in odour control because activated carbon has a preferential attraction and high retentivity for organic vapours. Activated carbon is highly porous and has great adsorptive power due to very large surface area. The highly porous structure permits the carbon to remove and hold the organic vapours, hydrogen sulphide, and other odour-producing substances. Further more, the retained material may be desorbed comparatively easily, and the carbon reactivated and used again. But, if the concentration of the odorous material is high, the method may not be economical.

16.5.5 Combustion or Oxidation

Direct Incineration

In this process, the odour generating gases are made to pass through a combustion chamber at a temperature of the order of 650-815°C, in the presence of excess oxygen. The main demerit of this method is the cost of the fuel to produce the required temperature. Sometimes heat exchangers may be employed to recover the heat from the hot incinerated gases. The heat recovered may be used for other uses like preheating the effluent gases, generating steam, etc. The optimum residence time of the gases in the chamber should be found out by trial, since the various gas constituents to be burnt react at different rates. Normally this residence time should not be less than 0.3 s.

Catalytic Incineration

When the cost of heating the gas stream to 650°C is very high, catalytic combustion may be the choice. In a catalytic unit, oxidation takes place at a much lower temperature than necessary for direct incineration. In this method, the process gases, pass through specially designed units containing catalyst elements, on the surface of which oxidation occurs. During the catalytic oxidation, the constituents in the gas stream such as hydrocarbons and other organic malodours, react with oxygen to form carbon dioxide and water vapour. In general, complete oxidation should take place to solve the odour problem. Catalytic combustion of most of the organic constituents that occur in the effluent gas stream is initiated and self-sustaining at about 260°C. In case the gas stream is colder than this before coming into contact with the catalyst, it must be pre-heated. However, the fuel cost here for pre-heating will be very much less, when compared to direct incineration.

Usually certain metals and metal-oxides are used as catalysts in oxidation reactions. It has been reported that platinum and platinum-rhodium alloy distributed on a suitable support are highly effective as catalysts.

The important properties for a commercial catalyst unit are high specific surface, low resistance to the flow of gases, and arrangement in a manner as to provide turbulent mixing of process gases. Gases entering the unit should be free from particulate matter such as dust particles that might cause catalyst loss due to abrasion action. The gas stream should also be free from fumes of mercury, lead and zinc, that might impair catalyst activity by coating the catalyst or deactivating its active centres. All necessary precautions must be taken to prevent the catalyst from becoming ineffective.

Commercial catalyst units are being used for the odour control of effluent gases from chemical plants, varnish manufacture, paint and enamel-baking ovens, refuse incinerators, refineries burning waste cracking gases, and coffee roasters.

16.5.6 Odour Masking

Odour masking is based on the principle that, when two odours are mixed, the stronger one will predominate. In other words, strong odours tend to mask weaker ones. Thus a strong, pleasant odour can be used to 'mask' or 'cover' a weak unpleasant odour. However, care must be taken to see that the odorant used for superimposing the pleasant odour is not flammable, corrosive, or allergic.

Following are the methods used for odour masking

1. Spraying, vaporizing, or atomizing the odorant chosen, into air-gas streams in stacks
2. Adding directly to a process wherever possible
3. Adding to scrubbing liquors

In cases where the masking odorants are directly added to the process, care should be taken to see that they will not affect the quality of the products.

Odour masking can also be used to control odours in outdoor places like refuse dumps or waste lagoons. Here, the masking compound must vaporize rapidly enough to overcome the unpleasant odour and slowly enough to last for a reasonable length of time.

Due to various practical problems involved in odour masking, the selection of the suitable odorant and its quantity should be left to specialists in the field.

16.5.7 Odour Counteraction

Certain pairs of odours in appropriate relative concentrations are antagonistic. Therefore, when they are mixed together, the intensity of each odour is diminished. This effect is known as counteraction or neutralisation. Examples of such odour pairs are musk and bitter almond; rubber and cedar wood.

Selection of the proper counteractant for a given odour is even more difficult than the selection of masking compounds, and therefore the selection should be left to experts.

16.5.8 Injection of a Reactive Substance

Odours can be controlled by injecting controlled quantities of chlorine or ozone into the process-gas stream.

Chlorine either as a gas or in solution form, hypochlorites and chlorine dioxide are used for odour control. Here, both bactericidal and chemical actions are involved. Hypochlorites are effective for partial oxidation of highly odorous mercaptans to less odorant disulphides. Chlorine is used in the oxidation of effluents from paper manufacturing plants.

Ozone converts organic matter by oxidative degradation usually to aldehydes, acetones, and acids. Such conversions can be carried out in stacks carrying odorous exhaust gases, where ozone is produced by corona discharge. Ozone also destroys or prevents the growth of odour-producing micro-organisms such as fungi and bacteria.

16.5.9 Irradiation

Ultra-voilet irradiation can be used for control of odours. Here, the action is probably due to ozone formation or bactericidal effect.

16.6 Nuisance Regulations

Generally, most municipal authorities resort to nuisance prevention laws to abate odours, but it is being followed more in breach than in practice.

Perhaps, the most effective method of overcoming the problem is to have regulations which can be directed towards specific emissions from specific industries. For example, regulation of pulp mill odours which limits emission of mercaptans.

QUESTIONS

1. What are the sources of odours?
2. List the various odorous industrial operations. Also, mention the odorous materials produced during these operations.
3. Explain how the threshold concentration of odoriferous gases is determined.
4. Explain how the odour intensity is measured.
5. List the various methods of odour control.
6. Write short notes on:
   (a) Odour masking
   (b) Odour counteraction
7. Explain briefly the various methods of odour control.
8. "Odour is undoubtedly the most complex of all the air pollution problems." Do you agree with this statement? Give reasons.

17

Control of Air Pollution by Process Changes

17.1 Introduction

In controlling air pollution by process changes, the method to be employed naturally depends upon the particular process involved. As a result, no fixed set of rules can be applied on a universal basis.

Generally, four methods are available to control the pollutants by process changes. They are,

1. Substitution of raw materials or fuels
2. Modification of the process itself
3. Modification or replacement of the process equipment
4. Changes in operational practices

17.1.1 Substitution of Raw Materials or Fuels

This method has been used successfully in many cases for controlling atmospheric pollution. For example use of low-volatile coals in place of high-volatile coals has proved quite effective in eliminating smoke and soot in many industrial and commercial heating applications. Similarly, substitution of low-sulphur fuels for high-sulphur fuels has reduced considerably the sulphur dioxide discharge into the atmosphere. Another method of decreasing emission of air pollutants is substituting bauxite flux for fluorspar in an open hearth furnace.

17.1.2 Modification of the Process

While modifying a given process, a unit operation may be eliminated or altered, or other unit operations may be substituted or added. For example, in disposing combustible refuse, the practice of incineration may be discontinued in favour of sanitary land fill. Another example is in Brass Foundry Practice, where an additional operational step has been used to reduce air pollution. Here, a fluxing material is applied to the surface of the molten brass which serves as an evaporation barrier and consequently reduces the emission of brass fumes. This additional step has been included strictly as an air pollution control measure. A third example of process modification is, substitution of oxygen for air in gas manufacture and in blast furnaces. The vent gases produced as a result are lower in volume and contaminant content.

17.1.3 Modification or Replacement of the Process Equipment

This may include (a) modification of one or more items of the process equipment, (b) replacement or repair of faulty or malfunctioning equipment, or (c) substitution of one type of equipment for another type.
For example

1. Use of vapour recovery systems to control vapour losses in handling volatile materials—absorbers, condensers, and compressors may be used
2. In cast iron foundries, substitution of reverberatory furnaces for cupolas has resulted in reducing atmospheric pollution from this type of operation

17.1.4 Changes in Operational Practices

A good example for this is a thermal power plant. Here by using a low-sulphur fuel in place of high-sulphur fuel, the rate of release of sulphur dioxide from coal burning operations can be reduced, especially during the periods of adverse meteorological conditions, and consequently reduce the air pollution problem.

Investigations have shown that fuel additives are effective in reducing smoke emissions from gas turbines. Hence, probably they may be used to reduce emissions of smoke, soot, carbon monoxide, and hydrocarbons associated with incomplete combustion, and thus reducing the air pollution problem.

17.2 Case Studies

Two case studies of process changes bringing reduction in air pollution are discussed in the following paragraphs.

17.2.1 Chemical Industry

The development of the Double Catalysis Double Absorption (DCDA) process was found to reduce SO_2 emission from 2000–5000 ppm to less than 500 ppm in the stack gases.

17.2.2 Description of Conventional Contact Process

Sulphur is burnt with preheated air to produce SO_2 which after cooling is converted to SO_3 by passing over vanadium pentoxide catalyst in four stages. From the point of view of equilibrium the maximum conversion that can be achieved is 97–98% depending on the type and depth of the catalyst.

The SO_3 gases leaving the converter are then passed through a tower where 98% sulphuric acid is circulated. The SO_3 is absorbed and unconverted SO_2 leaves the absorption tower. The strong acid leaving the tower is then diluted with water and cooled before transfer to the product storage. The overall conversion efficiency for various designs supplied by Process Designers of international repute, namely, Monsanto, Chemico, Lurgi, Simon-Carves etc., can be guaranteed for 95–96.5%. This would mean raw material consumption of 0.34 tons of sulphur per ton of 100% sulphuric acid.

17.2.3 Description of the DCDA Process

In this process the sulphur is burnt with air in a horizontal spray-type combustion chamber. Before the air is admitted to the combustion chamber, it is dried by scrubbing with 98–99% sulphuric acid.

The temperature of the SO_2 gas from the sulphur burner is higher than is required for inlet to the conversion system. Therefore, the gas is cooled in a waste heat boiler, which recovers the surplus heat as the by-product steam.

From the waste heat boiler the gas flows to the first pass of the converter system where it is partially converted to sulphur trioxide catalyst. The conversion reaction produces heat. Gases leaving the first converter pass flow to the No. 2 waste heat boiler where they are cooled. Temperature of the gas stream downstream of the No. 2 waste heat boiler is controlled in the proper range by-passing a portion of the gas flow around the No. 2 waste heat boiler. The cool gas stream flows from the No. 2 waste heat boiler to the second converter pass where additional conversion of sulphur dioxide to sulphur trioxide takes place accompanied by the generation of additional heat. Hot gases leaving the No. 2 converter pass are cooled by sending them through the tube side of the hot inter-pass heat-exchanger. Temperature downstream of the heat-exchanger is controlled in the proper range by by-passing a portion of gas around the shell side of the heat exchanger. The gas then flows to the third converter pass where additional conversion of sulphur dioxide to sulphur trioxide takes place. Hot gases leaving the third converter pass are cooled by sending them through the tube side of two gas heat exchangers, called cold interpass heat exchangers, connected in series.

Cool gas leaving the interpass absorbing tower, containing unreacted SO_2 flows to the cold interpass gas heat exchangers where it is heated by gases leaving the No. 3 converter pass.

From the cold interpass heat exchangers the gas stream flows to the hot gas heat exchanger where it is further heated by gases flowing from the second converter pass. From the hot interpass heat exchanger, the gas stream flows to the fourth converter pass where final conversion of SO_2 in the gas stream to SO_3 is accomplished.

The gas stream from the fourth converter pass flows to the super-heater economizer where it is cooled by the incoming boiler feed-water, and steam from the boiler system and then flows to the final absorbing tower.

In the final absorbing tower, SO_3 in the gas stream reacts with water in the 98–99% circulating acid. The temperature of the strong acid circulated over the final absorbing tower increases due to the heat of formation and the sensible heat of the gas stream entering the tower. Acid from the bottom of the final absorbing tower is circulated through coolers and returned to top of the tower. Sufficient water is admitted to the final acid tower between 98–99%. The acid produced in the final absorbing tower underflows to the drying interpass acid pump tank.

Gas leaving the final absorbing tower flows to the atmosphere through a stack.

The basic additions for a sulphuric acid plant using the DCDA process as against conventional process are:

(a) An intermediate absorption tower, (b) counter-current heat-exchangers to preheat the gases, (c) additional catalyst loading, (d) blower to take care of increased pressure drop in the system, and (e) additional acid circulation pump and drive. An interesting feature of the DCDA process is that it can take up a higher strength of SO_2 in the converter since the product of reaction will be absorbed and sent for conversion again into the catalyst bed. Also, because of the pre-heating requirement of the return gases from the intermediate absorber, it is not necessary to pre-heat the air for burning the sulphur. A conventional plant presently operated on the old process when converted to DCDA process will, therefore, have an additional capacity.

17.2.4 Present Method of Aluminium Production

The metal 'aluminium' is produced by the electrolytic reduction of alumina, which is generally carried out in the electrolytic cells having a cathode and the anode. At present there appears to be no other commercial method available to replace the same.

Commercial electrolytic reduction cells are mainly classified with the types of anodes as follows:

1. Horizontal stud design (Soderberg type)
2. Vertical stud design (Soderberg type)
3. Prebacked design

In the first two types of design the anode is continuously baking during operation while in the third, it is prebaked. The chemical reaction usually takes place in the molten alumina which is adequately fluxed with cryolite. The reaction is as follows:

$$2Al_2O_3 + 4C \rightarrow 4Al + 2CO + 2CO_2 \quad (17.1)$$

The liquid aluminium is deposited at the bottom of the cathode and sucked out of the cells. It is then sent to the foundry for casting into various forms of ingots and pigs.

During regular operation, dissolved alumina in the cryolite is depleted continuously. Gases like CO, CO_2, HF, CF_4, SO_2 are liberated continuously along with dust, alumina, tar fumes, etc. These are taken out from the gas burners provided at the side of the electrolytic cell where CO is burnt into CO_2. These gases pass through the central duct line provided for this purpose and could be let off into the atmosphere with or without treatment. The gases liberated at the anode consist primarily of CO and CO_2 as already mentioned, and carbon monoxide burns when reaching the surface of the electrolyte. Thus the hazard of CO is reduced to minimum, but the gases still contain undesirable impurities primarily of fluorine compounds resulting from the breakdown of fluorides.

In modern aluminium plants, therefore, the gas is cleaned before it is released into the atmosphere. This serves three purposes viz. (a) it allows the recovery of alumina, cryolite and aluminium fluoride (b) it also ensures better working conditions for the workmen, and (c) it protects the vegetation around.

The gases when let off into the atmosphere without any treatment constitute the main source of air pollution from the aluminium industry. In general, these gases have been found to have the following composition.

Table 17.1 *Air Pollutants from Aluminium Industry*

Type of gas	Concentration gm/NM³
HF	0.126
SO_2	0.217
Dust	0.800
Tar	0.010

At present there are no standards (i.e., maximum allowable discharge to the atmosphere) available in India for these pollutants.

The waste gases containing fluorine compounds are the only source of recovery of the raw materials. The recovery of important materials like aluminium fluoride and cryolite becomes essential because these items are scarce. At present, the indigenous production of cryolite and aluminium fluoride cannot meet the demand of increased aluminium production. Since imports are becoming difficult, recovery of the above valuable materials through chemical treatment of these gases will result in considerable saving of valuable foreign exchange.

There are two important methods available for reducing the concentration of harmful gases generated during production of aluminium before they are let out into the atmosphere. The methods are: (a) wet srubbing, and (b) dry scrubbing described in Chapter 13.

17.2.5 New Process for Aluminium Production

It may be possible in future to replace the available commercial method by a direct reduction process. In this process, in short, bauxite is reduced with carbon in an electric furnace to an alloy of aluminium and the other reduction products of other constituents of bauxite such as iron, silicon and titanium. This alloy is further treated at high temperature and pressure with aluminium trichloride gas, whereby the Al of the alloy is volatilised as aluminium subchloride. The gaseous aluminium subchloride is further decomposed by changing temperature and pressure to aluminium and gaseous aluminium trichloride. This gaseous aluminium trichloride is further recycled into the system.

The distinct advantage of this process is that it is a close process. It not only replaces the use of fluxing media (mainly cryolite, aluminium fluoride etc.), but also eliminates the conventional use of alumina in the extraction of aluminium from bauxite. This process, therefore, eliminates the air pollution problem posed by the existing commercial method. Research, however, is still continuing and there may be a breakthrough in the near future.

QUESTIONS

1. What are the methods available to control air pollution by process changes? Illustrate with examples.
2. List the various methods of control of air pollutants by process changes.
3. Explain how 'process changes' can be adopted in a thermal power plant to reduce the problem of air pollution.

18

Air Pollution from Major Industrial Operations

18.1 Introduction

Even though there are many different sources which contribute to air pollution, industries contribute a major share. There are a number of industries which are responsible for air pollution, but it is not possible to cover them all in this book. As such, only major air polluting industries are dealt with in this chapter. Table 18.1 shows the specific pollutants found in the emissions of some typical industrial and other sources.

Table 18.1 *Air Pollution Problems from Some Typical Industrial and other Sources*

Source	Besides smoke, sulphur dioxide, oxides of nitrogen and fly-ash, the following specific pollutants may also be found.
Fertilizer industry and aluminium manufacturing plants	Hydrogen fluoride, ammonia, fluorides, fertilizer dust and sulphuric acid mist
Lead casting and melting, pigments, etc.	Tin, lead etc., fumes and oxides, solvents and thinners
Heavy chemical industry like acid plants, synthetic fibre, etc.	Acid fumes
Tanneries and leather industry	Mercaptans and sulphides
Cement industry	Cement and lime dust
Carbon black manufacture	Polynuclear hydrocarbons, carbon soot and hydrogen sulphide
Paints and pigments	Nitrobenzene and aniline, thinners, solvents and base material
Coal tar industry	Polynuclear hydrocarbons and aerosols of tar
Paper and paper products	Hydrogen sulphide and mercaptans
Refinery and petrochemical industry	Hydrogen sulphide, hydrocarbons, odours of mercaptans
Metallurgical industry	Metallic fumes, dust
Electrolytic manufacture of chlorine	Chlorine
Coal burning (power plants)	Soot
Vehicle emission	
(a) Petrol	Hydrocarbons, HCHO
(b) Diesel	Hydrocarbons, HCHO

18.2 Mineral Product Industries

The conversion of naturally occuring minerals into saleable products involves various operations and processes. Operations are primarily concerned with chemical alteration. Certain unit operations are common to practically all mineral production procedures.

General Operations

Mining

Air pollution problems may be created by the discharge of deep mine ventilation air. Such problems are infrequent and easily controlled, most often by relocating the discharge vent.

Open Pit Mining

Particulate matter produced during open pit mining operations is usually discharged directly to the atmosphere, rather than being captured by a local exhaust ventilation system, and is thus difficult to control.

Transportation and Storage

Scientific materials handling techniques and automation have reduced air pollution from these operations. Loading and unloading, conveyor belt discharge and transfer, and stock filing operations can produce particulate matter emissions. Lack of attention to wind erosion of stock piles has often created an 'air pollution control paradox'. Gaseous emissions can be produced by evaporation of materials from storage piles.

Size Reduction

Operations are commonly classified as either crushing or grinding according to the size range of feeds and the size reduction ratio achieved. Various types of crushers and grinders are usually operated in series to obtain desired size reduction. Dust is discharged from crusher and grinder inlet and outlet ports. For most effective dust control, crushers and grinders should be enclosed and provided with exhaust ventilation, discharging to a suitable collector. Although generally not as effective as local exhaust ventilation, wet sprays can be used in controlling dust emissions from crushing, grinding and other mineral production operations.

Classification, Concentration and Mixing

Following size reduction operations, ores are generally size classified, concentrated and/or mixed to obtain a desired size or composition.

Drying

Wet mixes, slurries and filter cakes are dried in either direct or indirect dryers. Direct drying usually creates more air pollution than indirect drying since the hot gases remove dust from the material being dried.

18.2.2 Specific Processes

Crushed Stone

With the exception of fuel production, the crushed stone industry is the largest non-metallic mineral industry in developed countries like the United States with respect to total value of production.

1. Process description and emissions

After the stone is removed from the quarry, it is processed in a crushing plant for size reduction and classification. Tertiary grinding and air classification may be employed if fine aggregate is required. Particulate matter is the only significant emission produced by these operations.

2. Dust control and compliance with regulatory standards

Most control standards are based on opacity. Process sources should be enclosed and vented by mechanical or fabric collectors. Fugitive dust sources should be controlled by (a) elimination of sources (b) dust suppression systems. Effective control of particulate matter emissions, as indicated by compliance with governmental regulations, should result in no 'significant' visible emissions beyond the plant property line.

Lime Production

In terms of total tonnages processed, the rotary kiln is the most commonly employed unit process for the conversion of non-fuel minerals into saleable products. Most of the lime manufactured in the United States is produced in rotary kilns.

Dust control

Kiln emissions are controlled by fabric collectors, water scrubbers and electrical precipitators. The choice of the collector depends upon factors such as

1. The type of fuel used for the kiln
2. The use of water scrubber in processes associated with the lime producing operation
3. The expected life of the plant

Regulatory emission standards have been based upon performance tests of fabric collectors. Regulatory agencies prefer fabric collectors, over scrubbers, because there are less interferences with compliance inspections i.e., opacity readings.

Coal Preparation

The major air pollution problems associated with coal preparation procedures are gaseous emissions from ignited coal refuse disposal areas, dust from refuse and coal storage piles, and particulate matter from coal dryers and dedusting operations.

1. Cleaning and drying

Operations of coal preparation involve the processing of run-off mine coal to produce a saleable product. Run-off mine coal contains impurities in varying amounts and ranges in size from fine to sizes having equivalent diameters of over 3 metres. The primary purpose of a preparation plant is to crush the coal, remove the impurities and classify the product into standard sizes. There are four principal coal cleaning techniques.

(a) Cleaning at the mine face
(b) Picking by hand or by mechanical means
(c) Froth flotation
(d) Gravity concentration

After wet process cleaning operatives, the coal is usually subjected to thermal dripping. The size of the coal being dried and the velocity of the gases through the bed are the major factors in determining the air pollution potential of the plant. Emissions include products of combustion and entrained coal fines.

2. Emission control

The use of inertial collectors and wet scrubbers, or packed towers, in series, can provide efficient control of emissions. Fabric collectors are not usually used because of fire hazard. Controlling operating procedures and providing for proper maintenance of equipment are the most important factors in preventing air pollution from thermal driers. Even with the best air pollution control equipment, consideration must be given to

(a) Not overloading the plant
(b) The prevention of clogging of ducts, cyclones and other equipment by caked coal fines

3. Coal Refuse Disposal

With the potential growth of the coal industry, especially in areas where refuse content is high, control efforts must be increased. About 20–50% of the raw coal processed in a cleaning plant is rejected as refuse. Coal refuse piles ignite either through spontaneous combustion, carelessness, or deliberate action. Significant concentrations of H_2S and SO_2 have been measured in communities adjacent to burning refuse piles. The products of combustion react with each other and with materials in the pile to form a number of noxious gases including carbon monoxide, ammonia, hydrogen-sulphide etc. Proper construction including compaction and sealing to prevent the circulation of air within refuse piles will prevent ignition.

Asphaltic Concrete

The aggregate is proportioned and charged on a rotary dryer and heated to about 150°–175°C. The dried aggregate is screened, transferred to storage bins and introduced into a mill, where it is mixed with asphaltic oil.

Emissions and compliance with regulatory standards
Air pollution sources include the rotary dryers, transfer points, bucket elevators, screens, weighing and mixing operations, storage piles, and sometimes the boiler facility. The dryer is the major source of dust emissions. Because of the many variables that effect emissions and control equipment costs, the establishment of a United States federal 'New source performance' standard for asphalt concrete plants has been a long and controversial process.

Glass Manufacture

The glass industry makes products from common minerals, silicates, and alkalies employing a high temperature process. This industry has usually not been considered as a major source of air pollutants. Recent studies have confirmed that reduction in melter temperatures significantly affects particulate generation. Uncontrolled particulate matter concentrations in glass furnace stacks are normally less than 0.23 gm/s cm and therefore would comply with most grain loading regulations.

1. Melting

Glass furnaces are of various sizes and configuration depending upon the type and quantity of glass desired. Pot furnaces for production of special glasses are batch operations. Large direct fired continuous melting furnaces are used in the production of container and flat glass. Regenerative firing systems using brickwork chambers are employed to recover heat from furnace gases.

2. Emissions

During charging of the furnace, dust may be released at the changing port or part of the batch may be swept through the melting area by the furnace draft. Such emissions are usually of minor significance. Fabric collectors have been used to capture dust released from automatic batch feeding equipment. Adjusting furnace operating conditions can reduce 'swept through' dust. Pelletizing the charge eliminates this source. Furnace particulate matter emissions consist primarily of Na_2SO_4 that is formed and condenses in the regenerator checker work. Therefore, modifying batch composition and furnace operating conditions can reduce particulate emissions.

3. Particulate control equipment

On a cost effectiveness basis, incremental reduction in particulate emissions through process modifications is the most desirable approach. Fabric filters have been employed to collect particulate from both batch charging and furnace emissions. Electrical precipitators have been installed on a number of container glass furnaces. High energy scrubbers have also been used in limited applications.

Concrete Batching

Concrete batching plants are generally simple arrangements of steel hoppers, elevators and batching scales for proportioning rock, gravel and sand aggregates with cement for delivery, usually in transit mixer trucks. By careful use of sprays, felt or other filter material over breathers in the cement silos, and canvas curtains drawn around the cement dump trucks while dumping, dust losses can be controlled. Aggregate stocks in bunkers should be wet down with sprays to prevent dusting.

Asbestos Processing

Asbestos is unique as an air pollutant. Regulatory specification type standards have been promulgated and are being proposed to control asbestos emissions. Operating and design characteristics of the collectors are specified in these unique standards. Recent regulatory standards are addressed primarily towards the control of asbestos emissions from milling and manufacturing operations. Milling operations usually are crushing and drying the ore, fibre separation and grading. Emissions are usually controlled by a wet type collector. Emissions from all collectors are required to meet zero opacity standards. Large quantities of dust are generated during dry asbestos milling. Central air handling and filtering systems usually provide for combined process and dust control air.

Asbestos fibre is used to manufacture textiles, cement products, plastics and many other products. Because of the awareness of the industry to the occupational health aspects of asbestos, processing and handling equipment is routinely equipped with local exhaust ventilation systems.

18.2.3 Emissions Reduction During Air Pollution Episodes

Reduction in emission is required through curtaining, postponing or deferring production. Considerations should also be given to preventing excessive emissions while increasing production to normal levels, following an episode.

18.3 Cement Industry

Portland cement is manufactured from a suitable mixture of limestone and clay, or from marls which are first crushed and ground, either in the dry state or with water. The raw mixture is thereafter burnt at a sintering temperature and the clinker thus obtained is ground to a fine powder with the addition of gypsum, to give cement. Cement is packed in jute bags and despatched in this form in railway wagons or in road trucks. Alternatively, it is also despatched in bulk as loose cement.

Thus, by the very nature of above processes, there is considerable generation of dust which needs to be contained at all stages.

The size of these dust particles is very small ranging from as low as 1 μ to as high as 100 μ and above. While considerable effort has gone into the aspect of controlling dust generation and collecting the dust so that the gases let out in the atmosphere from the various processes are free from dust, the results achieved are not entirely satisfactory.

The prevailing environmental conditions in the cement plants in the country have been studied by the Central Mining Research Station and the concentration of air borne dust at different operations in ten cement factories reported. The results of these studies are reproduced in Table 18.2. It was found that the concentration of airborne dust at limestone crushing plants, cement packing machines, around wagons during loading, and at the coal grinding and coal drier areas was rather high, whereas the same around the kiln firing end, clinker cooler and cement mills was within the permissible limits.

Table 18.2 *Air-borne Dust Concentrations at Different Locations*

No.	Operation or locations	Concentration of Air borne dust in (ppcc*)		
		Maximum	Minimum	Average
1.	Limestone crushing	6905	957	2367
2.	At the kiln firing end	1596	110	580
3.	At clinker cooler area	6430	430	1394
4.	Around cement mill	3267	146	1214
5.	Packing of cement	8480	1024	3330
6.	Loading of cement into wagon	18020	3670	6723
7.	Around a coal crushing plant	4180	771	1843
8.	Around a coal drier	3385	1920	1609
9.	Around a coal mill	4000	325	1769
10.	General atmosphere within factory area	950	145	567
11.	In front of general office	918	35	181

*Particles per cubic centimetre.

There were wide variations in the dust concentrations recorded at any one location on different occasions, and these were due to several factors such as rate of material input in the process at the time of measurement, the standards of plant maintenance, limitations in accuracy of measurement, etc.

18.3.1 Case for Dust Control in Cement Plants

The benefits to be achieved from efficient dust control in a plant are enumerated below:

1. It improves conditions of industrial hygiene for greater worker comfort and efficiency
2. It eliminates or substantially minimises air pollution by preventing the travel of dust to surrounding areas
3. It reduces equipment maintenance and wear by preventing the dust from re-entering the working atmosphere of a machine or process.
4. Valuable products are recovered which otherwise might escape into the atmosphere
5. It achieves better housekeeping in the plant

18.3.2 Methods of Dust Control Being Adopted

The methods at present adopted by cement plants for control of dust nuisance at the various locations of operations are as follows:

Limestone Crushing Plant

Generally suppression of dust nuisance at crushing plants is carried out by means of spraying water at the areas of dust generation viz., damping hopper, crusher, etc., and at the various material transfer points, especially in the dry months. The use of an atomised spray with sufficient water to settle the dust will give good results for small crushing plants up to a capacity of 200 TPH provided the same is judiciously employed and transfer points are carefully engineered. However, for larger crushing plants, there is a need for providing suction at the crusher discharge and the areas of dust generation and transfer points should be suitably enveloped and dust collected by means of a cyclones or bag type dust collectors.

Grinding Mills

In dry grinding of raw materials and cement clinker, the dust carried by the air stream is conveniently collected by means of conventional bag type dust collectors using canvas, wool or terylene filter bags. However, difficulties have been encountered in coal grinding where a bag type dust collector cannot be employed on account of the hazard of the bags catching fire. At the same time, cyclone type dust collectors have so far been found to be ineffective because of the very fine particle size of 50 μ and below, encountered. At some cement works, water scrubbers have been employed for cleaning coal laden air and have proved quite effective. However, in this case the coal dust mixed with water forms a coal slurry which cannot be easily disposed off. One of the methods of preventing venting of coal dust laden air into the atmosphere is to utilise the same as primary air for firing pulverised coal in the kiln; this may however result in reduction in thermal efficiency of the process.

Kilns

The older cement plants in India are provided with wet process rotary kilns with chains at the slurry feed end. These chains act not only as heat exchangers but also help in arresting the dust picked up by the exhaust gas stream. The kilns were provided only with dust settling chambers to further settle the dust, although some of them have now been provided with multicyclones for further arresting the dust. Later cement plants, however, were of the semi-dry process (Lepol Grate) type and subsequently, around 1966, kilns with Air Suspension Preheaters were installed. The Lepol Grate kilns had an inherent feature that the exhaust gases were passed through the cyclones to arrest the dust further. The Air Suspension Preheater kilns are equipped with electrostatic precipitators which for various reasons explained have not given the desired efficiency of dust collection. Table 18.3 gives an idea of the quantity of dust encountered from the three types of kilns.

Table 18.3 *Quantity of Dust Emitted from Different Processes*

Description	Kiln type		
	Wet	Lepol	Dry (preheater)
Quantity of kiln exhaust gas m³/kg clinker	3.5	2.6	1.7
m³/kg clinker (EFF) temp. °C	6.1	3.75	3.85
(Gas exit)	200	120	340
Dust loading in exit Gases g/nm³	40	5	70
g/m³ (EFF)	23	3.5	31
90% collecting efficiency Dust in flue gases with designed efficiency g/m³	2.3	0.35	3.1
95% collecting efficiency Dust in flue gases with designed efficiency g/m³	1.15	0.175	1.55
98% collecting efficiency Dust in flue gases with designed efficiency g/m³	0.23	0.035	0.31
99% collecting efficiency Dust in flue gases with designed efficiency g/m³	0.46	0.07	0.62

The common difficulties encountered in the satisfactory operation of the electrostatic precipitators are

(a) CO analysers installed are difficult to maintain due to lack of imported parts. The CO analysers detect presence of CO in kiln exit gases so that when it exceeds a set value, the H.T. (high tension) supply to the ESP is tripped off to eliminate the possibility of an explosion.

(b) The control of coal feed is inadequate, so that at times there is an excess supply of coal resulting in higher CO content in the flue gases and even passage of unburnt coal to the ESP This unburnt coal, when supplied with enough oxygen provided by leakage or raw mill venting air, may cause an explosion even when the ESP is not charged with the high tension voltage. Thus a pulverised coal feeding device which can be accurately controlled and which can operate without coal flushing is one of the essential requirements for efficient ESP operation. Sophisticated feeders of the gravimetric type which accurately control the weight of coal supplied to the burning process within the desired limits are available abroad, but are yet to be developed in our country.

(c) The ESP achieves optimum functioning when gas temperatures are in the range of 90–180°C with moisture content of 15–30%; whereas in dry process kilns with preheater, the temperature is around 330°C with negligible moisture and under these conditions the efficiency of the ESP is unsatisfactory. To overcome this problem a water spray is injected into the exit gas stream before the ESP. However, this system has not worked satisfactorily as it was not possible to have the necessary retention time for evaporation of sufficient quantity of water so as to lower the gas temperature to the desired value. This problem can be overcome by the installation of a gas conditioning tower before the ESP, in which the quantity of water is automatically adjusted to give the required temperature. The tower should be of sufficient size to provide the necessary time for water to evaporate.

Dust is generated in the packing plant initially at the packing machine. The cement is poured into jute bags through a filling spout on which the bag is inserted. Here, there is heavy dust emission as the cement escapes through the weave of the jute bags and causes dust nuisance, affecting not only the packing machinery but also the packer operators and other workers in the vicinity. Normally, the packer machine spouts are covered by a hood or a canopy, which is put under suction so that air-borne dust is exhausted through a bag type dust collector by means of a fan. The jute bags are conveyed from the packing machine to wagons or trucks by means of belt conveyors and undergo several drops, where again, due to porosity of the jute bags, air-borne dust is created. This means that the bags need to be deposited gently from the conveyor. However, this is not always possible, the slope of the chute is very critical—if it is held up along the chutes. At changes in the direction of conveying, appropriate spiral chutes or if need be, turn tables are installed, in order to ensure smooth transfer of bags. Further dust nuisance is generated in case the bags burst during transport from the packing machine to the wagons or trucks. Thus, while it is extremely difficult to prevent dust nuisance in this area, it can be to some extent ameliorated by good engineering of the conveying system. This problem is particularly encountered in India as almost the entire quantity of cement manufactured is despatched in jute bags, whereas in other countries a major part of cement despatches are made as bulk cement and the balance in paper bags only.

Dryers

When materials such as limestone, slag, and coal contain excessive moisture, they need to be dried in rotary dryers. These dryers are normally provided with cyclones and it is our experience that even with small diameter multicyclones, the efficiency achieved is only of the order of 80% because most of the dust escaping is of a size below 20 μ, and this does escape through the stack. While in case of limestone and slag the dust in the exhaust gases is not so perceptible, the black coal dust shows appreciably. Also in case of coal dryers, bag type dust collectors cannot be installed on account of the fire hazard.

18.3.3 Factors Affecting the Efficiency of Dust Collecting Equipments

The following types of dust collecting equipment are generally employed for the various operations in a cement plant:

1. Dust settling chambers
2. Cyclones and multicyclones
3. Bag type dust collectors
4. Water scrubbers
5. Electrostatic precipitators

While in the case of bag type fabric filters a good efficiency of the order of 99% is achieved under all operating conditions provided the bags are maintained properly and replaced in time, optimum efficiency from other types of dust collectors can be realised only if gas-flow rates and dust loadings are kept within the specified limits. In the case of cyclones and multicyclones, the efficiency increases up to a certain limiting inlet velocity beyond which the efficiency of the cyclone decreases. Similarly, the efficiency of ESP is also dependent on the gas velocity through it. It is therefore, necessary that means are available to accurately determine the volumes of gases, the concentration of the dust in the same and particle size of the dust. Whilst means of carrying out such measurements are available in the country, the tests to be carried out are not only time consuming and cumbersome but also require great skill to give reliable results.

18.3.4 Dust Emission Standards

The question of fixing dust emission standards for cement kilns has been engaging the attention of various countries. Table 18.4 gives the standards adopted in some of the developed countries in the world.

Table 18.4 *Dust Emission Limitations Specified for Cement Plants in Other Countries*

Country	Actual standards for kilns	Converted to mg/m³ for different kilns		
		Wet	Lepol	Preheater
Switzerland	100 mg/m³ effective	100	100	100
Germany	150 mg/Nm³	55	88	88
USA Michigan	0.15 kg/mt kiln feed	25	44	48
Mexico	72 kgs/hour-wet 66 kg/hour-dry	289	476	518
Great Britain	0.2 grain/cu. ft Wet gas at 60°F and 30″ Hg	253	309	309
Australia	0.2 grain/cu. ft Dry gas at 0°C and 760 mm/Hg	169	268	270
Spain	800 mg/m³	800	800	800
South Africa	Proposed 460 mg/Nm³	170	270	272

As regards our country, a recommendation was made early that dust emission should be restricted as follows:

Wet process plants	200–300 mg/m^3
Dry and semi-dry process plants	300–340 mg/m^3

18.3.5 Economic Aspects

In any endeavour to minimise air pollution in the surrounding areas, the main approach is to clean the gases coming out of the rotary kilns as these gases distribute the dust to the vicinity. The initial capital investment for installing ESP along with a conditioning tower (for dry process kilns) and other auxiliaries such as a carbon monoxide analyser and gravimetric weigh-feeder for coal, forms a substantial part of the total outlay on a cement plant. In addition, there will be some recurring cost for the operation of the same with the result that it will appreciably increase the cost of cement. Thus the installation of ESPs would need to be done primarily from ecological considerations, in which case it would be a social service, and it needs to be decided whether it would be in the larger interests of the country to incur this additional expenditure as against diverting our limited resources to improving the productivity in the cement industry and/or its expansion with a view to boosting the national economy.

18.3.6 Some Points for Consideration

The following points made by an expert body earlier need careful consideration:

1. Considering all the factors, the ESP designed for 99% collection efficiency for both wet and dry process kilns is quite appropriate and necessary.
2. The ESP installation is more for reducing pollution hazards than for economic considerations. However, two marginal benefits result from this installation, e.g., conservation of valuable raw materials and some additional clinker production. These benefits alone do not, of course, justify the high capital cost and low return on the ESP installation particularly for the existing smaller plants.
3. It must be compulsory for all the cement industries to install the necessary auxiliaries along with the precipitator to ensure the safe and satisfactory operation of these units. The gas conditioning tower (for dry process kilns only), coal weigh-feeder and CO analyser must compulsorily be installed along with the kiln ESP installation.

18.4 Petroleum Refining

Because of the technological intricacy of petroleum refining processes, an understanding of air pollution control techniques is necessarily dependent on a general knowledge of oil industry manufacturing operations. Although each modern refinery is unique in design and contains a composite of many processes employing a magnitude of towers, vessels, piping, valves, tubes, exchangers and storage tanks, the operations can be classified into four basic procedures—separation, conversion, treating and blending. Individual refineries differ widely, not only as to crude oil capacity, but also as to the degree of processing sophistication employed.

18.4.1 Oil Refining Technology

Separation

Crude oil is a mixture of many different hydrocarbons combined with small quantities of S, O_2, N_2, and traces of other elements. Distillation is employed to separate crude oil into its various components with different boiling ranges such as wet gas, gasoline, kerosene, fuel oil and heavy bottoms. The crude oil is distilled in furnaces followed by vaporization and separation in fractionating towers.

Conversion

Conversion processes are employed to increase the yield of the more desirable products and to provide the refiner with the necessary flexibility to meet seasonal changes in product demand. The most familiar technique is cracking. Cracking by catalytic action is still the most significant conversion process employed. Earlier catalytic processes employed fixed-bed or once through catalysts. Moving-bed and fluid bed systems are now commonly used. Hydrocracking, cracking in the presence of a high H_2 partial pressure, is experiencing rapid growth. This process is characterised by high liquid yields of saturated isomerized products. Virtually supplanted by catalytic cracking and hydrocracking, some thermal cracking units are still operating. Catalytic reforming is a rearrangement of molecular structure to produce higher quality gasoline and large quantities of H_2. These processes utilise catalysts such as phosphoric, sulphuric, and hydrofluoric acids, platinum and aluminium chloride.

Treating

Crude oil may contain small quantities of impurities, such as sulphur and certain trace amounts of metals. While sulphur removal from basic crude is neither technically nor economically feasible at this time, desulphurization of products and intermediate stocks has come into wide use. Desulphurization of products is practiced because of the effect on product quality, catalyst sensitivity, odour and corrosivity. Both physical and chemical procedures are available for treating products and feed stocks. Solvent extraction for removal of aromatics is also widely practiced. The use of H_2 for desulphurization of petroleum products has become widespread. Hydrogen required for the process is frequently supplemented by generated H_2 from processe such as steam methane reforming.

Blending

The relatively few base and intermediate stocks are blended in innumerable combination to produce over 2000 finished products including liquefied gases, motor and aviation fuels, lubricating oils, greases, waxes and heating fuels.

Desulphurization

Hydrogen treating at high pressure in the presence of a catalyst is by far the most feasible means of effecting a significant degree of sulphur reduction for most refinery stocks. Hydro desulphurization process for treating gasolines, middle distillates and lubricating oils are in extensive use. Extension of this technology towards residual oils has proved costly and has found only limited application. Factors influencing residual fuel sulphur content include the sulphur content of the crude oil prior to processing, availability of low sulphur crude and blending stocks, and whether all residual fuel produced at a particular refinery has to meet a single sulphur specification or whether two or more sulphur levels are produced.

Lead Antiknock Additives

The primary cost to the petroleum industry will be for revision of the gasoline distribution system to handle the unleaded fuel and to keep it segregated from leaded grades. If unleaded fuel of higher than 91 octane is required, additional refinery processing will be necessary.

18.4.2 Types of Emission

Individual refineries vary greatly as to the character and quantity of emissions. Controlling factors include crude oil capacity, type of crude processed, type and complexity of the processing employed, air pollution control measures in use, and the degree of maintenance and good housekeeping procedures in force. Refinery emissions may be classified as smoke and particulate matter, hydrocarbons and other gaseous compounds, principally oxides of sulphur and N_2 but also including malodorous vapours. While the quantitative aspects of refinery emissions are generally considered by type of equipment, the major refining operations offer a guide as to the type of releases that are encountered. The molecular rearrangement processes—alkylation, reforming, polymerization and isomerization—are significant in that they handle volatile hydrocarbons; andtight equipment such as valves and pumps becomes of greater emission control importance than in the heavier oil processes yielding fuel oils, lubes or asphalts. Pollution aspects of treating operations are concentrated on methods of disposing of the spent chemicals and extracted impurities such as spent acids, spent caustics, and H_2S. Air agitation and blowing, vesselvents, drains, valves and pump seals are possible sources of loss or leakage of sulphur oxides and hydrocarbons in any of the above processes or handling procedures. Before considering individual emission sources, an over all estimation of refinery releases will be of interest.

The extent and need for control depend on local meteorological and topographical conditions, size and type of plant, quantity and height of release, and proximity to community activity. Emissions of gaseous contaminants other than hydrocarbons, principally oxides of sulphur and N_2 originate mainly from combustion sources. Along with hydrocarbons and sulphur oxides, refineries discharge relatively small quantities of carbon monooxide, nitrogen oxides, aldehydes, organic acids, ammonia and particulate matter. The other miscellaneous discharges originate from combustion reactions in boilers, process heaters and catalyst regeneration units.

18.4.3 Source and Control

Refinery atmospheric emissions and control procedures are customarily considered by types of equipment employed rather than by refinery process operation.

Storage Tanks

Hydrocarbon vapours may be released through a number of mechanisms in storage tanks, including tank breathing due to temperature changes, direct evaporation and displacement during filling. The principal source of potential loss is from crude oil and light distillate products. Vapour conservation storage may involve tanks with floating roof covers, pressurised tanks and connections to vapour recovery systems.

Catalyst Regeneration Units

Coke formed on the surface during catalytic cracking, reforming and hydrogenation is burnt off in regenerating vessels by controlled combustion.

Loading Facilities

Careful operation to minimise spillage and vapour collection and recovery equipment will control vapour loss from this operation.

Pumps and Compressors

Hydrocarbons can leak at the contact between the moving shaft and stationary casing in pumps and compressors. Inspection and maintenance, sealing glands under pressure and use of mechanical seals in light hydrocarbon service are useful control measures.

Blow Down Systems, Flares and Shut Downs

Refinery process units and equipment are periodically shutdown for maintenance and repair. For aesthetic purposes ground flares are becoming more popular.

Boilers and Process Heaters

Refineries depend on boilers and heaters to supply high pressure steam at elevated temperatures. Stacks on boilers and heaters are elevated to improve atmospheric dispersion and further diminish resulting ground level concentrations of gases such as SO_2 and nitrogen oxides.

Incinerators

There are a number of incinerator processes available for the incineration of refinery sludges, solids and spent caustics. These fluid bed incinerators operate at temperatures of about 700°C. As the sludge is burned, the solids from the sludge remain in the bed while gaseous products of combustion, water vapour and fine particulate matter pass overhead through a cyclone separator and a water scrubber before venting to the atmosphere.

Miscellaneous

Steam-driven vacuum jets, employed to induce negative pressure in process equipment, may discharge light hydrocarbons with the exhaust system. These gases may be vented and burnt in an adjacent boiler or heater fire box. Fumes from air blowing operations may be consumed by incineration. Refinery process changes may have the net effect of reducing over all emissions to the atmosphere.

18.4.4 Estimation of Quantities

When an accurate estimate of refinery atmospheric emissions is required, there is no substitute for inplant field testing, preferably by plant personnel familiar with the equipment and operations involved.

18.4.5 Economics of Control

The economics of air pollution control are a function of specific sources and methods of control in individual refineries. Partial economic return may be realised when pollution control facility recovers such components as hydrocarbons, sulphur and catalyst. In general, rates of return on pollution control expenditures do not approach those for normal processing installations. In some cases processing revisions have been combined with plant modernization steps resulting in modest economic return. Odour and particulate matter control are particularly unattractive categories for economic justification.

Impact of Pollution Control Costs on Overall Refinery Expenditures

The refining segment of the oil industry spends from 5–10% of its capital investment in new refining facilities towards the control of air pollution.

Examples of Cost-effectiveness to Achieve Incremental Levels of Control

A useful tool in determining a reasonable level of pollution control is the method of cost-effectiveness analysis.

For example,

Hydrocarbon Emissions Hydrocarbon vapours are emitted to the atmosphere during storage and transfer of crude oil and refined petroleum products. The three main emission control techniques are (a) floating roof tankage, (b) submerged or bottom loading during shipping and transfer of products, (c) vapour recovery using refrigeration, condensation or absorption. Recovery of valuable gasoline vapour more than pays for the first two steps of control, but the added cost for vapour recovery equipment appears expensive.

Particulate Emissions

The major sources of particulate emissions in petroleum refining are fluid catalytic cracking units. These emissions are normally controlled by the use of internal cyclones. The cyclone is cheaper than either high or low efficiency electrostatic precipitators.

18.5 Ferrous Metallurgical Operations

Iron and steel industry production facilities can be grouped generally into steel mills and ferrous foundries. Steel mills are concerned with the production of steel shapes for subsequent use in other manufacturing industries, whereas ferrous foundries produce iron and steel castings.

18.5.1 Coke Production

General Process Description

Carbon has been found to be the reducing agent best suited for the reduction of iron ore to metallic iron. Coke is made by heating blended coals to a temperature of 900–1100°C over a period of 10–20 hrs to drive off the volatile matter, while retaining certain physical and chemical properties. The manufacture of metallurgical coke is accomplished by the non-recovery beehive process or the byproduct process, the latter accounting for the vast majority of world coke production.

18.5.2 Emissions and their Control

The generation of air pollutants in the coke making process is associated with (a) coal and coke handling, (b) coke oven charging, (c) coke oven discharging, (d) coke quenching, (e) leaking oven doors, (f) byproduct processing

Coke Oven Charging

As coal is charged into the hot oven, the gases in the oven are displaced, and the coal immediately begins to volatilize. Pollutants are emitted through the charging ports at this time. These pollutants consist of vaporized hydrocarbons, graphite, tars, char and coal dust. Efforts to control them can be grouped into five general categories.

(i) Aspiration systems
(ii) Larry-mounted scrubbers
(iii) Fixed duct secondary collectors
(iv) Sequential charging practices
(v) Closed charging systems

Coke Oven Discharging

When the hot coke is discharged or pushed from the oven into an open coke car, thermal drafts in the vicinity of the operation give rise to the emission of coke dust or in the event of incomplete coking, dust and hydrocarbons from the partial combustion of uncoked coal. Methods of controlling emissions are (a) bench mounted self contained hoods, (b) coke car mounted hoods, (c) fixed duct hoods, (d) spray systems, (e) coke side enclosures.

Coke Quenching

Uncontrolled particulate emissions from quenching have been reported in a range of 0.06–0.24 kg/ton of coal. However, these emissions are routinely reduced by approximately, 80% by simple baffles mounted in the quench tower. Dry quenching is widely used in the USSR. The primary incentive for dry quenching is energy conservation through heat transfer.

Leaking Oven Doors

Shortly after the coke ovens are charged, sufficient internal positive pressure is developed to result in leakage at all available ports in the ovens, including the end doors. With the exception of improved maintenance and operating practice, the principal effort made to control door leaking emissions is in the use of ventilated coke side sheds.

Coke Oven Gas Desulphurization

Sulphur occurs in raw coke oven gas primarily as H_2S from the reaction of ferrous sulphide and H_2. Worldwide there are eight commercial processes for coke oven gas desulphurization. They can be subdivided into three categories—dry oxidative, wet oxidative and liquid absorption. The presence of cyanides in coke oven gas has been a major deterrent in the development of suitable desulphurization processes and has prevented transfer of much of the technology of the petroleum coke industry.

18.5.3 Sintering

General Process Description

Sintering was developed to improve low grade ores and to effectively utilise ore fines which, unless agglomerated are detrimental to efficient blast furnace operation. Sintering machines process a wide variety of materials and produce substantially different sinter depending upon blast furnace requirements.

18.5.4 Emissions and their Control

The discharge of pollutants from sinter plants can occur principally at four locations within the plant (a) windbox waste gases, (b) the discharge end of the sinter machine, (c) material handling points throughout the plant, (d) sinter cooler.

Windbox Waste Gases

Particulate matter is the predominant contaminant in windbox waste gases. The dust is generally relatively coarse. Most sinter plants are equipped with a series of gravity separators or dropout chambers followed by mechanical collectors such as cyclones, to remove the coarse, abrasive dust to protect the induced draft fan from severe wear.

Electrostatic precipitators are widely used. A fabric filter has been installed on the windbox discharge of one United States plant with success. Wet scrubbers have been installed on sinter plants in the USA and the USSR. Water scrubbers have a substantial advantage in that, gaseous sulphur oxides, fluorides or chlorides can be removed if they are present in excessive quantities.

Sinter Machine Discharges

Most modern sinter plants are equipped with crushers or breakers at the discharge end of the machine. The breaker and the hot screen area are the source of dust and they are normally hooded and exhausted to the atmosphere. Mechanical collectors, fabric filters and low energy wet scrubbers have been used. Another technique used is to direct the exhaust discharge to the hood located over the sinter machine and use it as a portion of the sinter machine air.

Materials Handling

The vast quantity and variety of materials handled in sinter plants invariably require that certain materials handling ports be hooded and exhausted and the gases cleaned prior to discharge. Fabric filters are the common means of controlling such dust problems, but mechanical collectors may be adequate. Water sprays and chemical dust suppressants have been used successfully.

Sinter Cooler

After hot screening, the sinter is cooled prior to final sizing. Cooler exhausts can contain entrained sinter dust, but most cooler discharge stacks contain no visible emissions if the sinter is properly screened ahead of the cooler

18.5.5 Iron Making

General Process Description
Most iron is produced in blast furnaces which are large refractory lined structures producing up to 10,000 tons per day of molten iron.

18.5.6 Emissions and their Control

The operations associated with potential blast furnace emissions consist of

1. Charging
2. Blast furnace gas handling
3. Casting
4. Slag handling

Blast Furnace Gas Handling

Dust generated and discharged with raw blast furnace gas can range from 14–150 kg/ton of molten metal depending upon the type and amount of burden preparation and screening. Multiple stage gas cleaning is generally capable of reducing dust concentrations.

Casting

Graphite called 'Kist' as well as other minor metallic oxide emissions, escape during the casting operation. These emissions can be minimised by using shorter runs to the cars.

Slag Handling

Most sulphur emissions are associated with the subsequent slag handling and result from quenching with water in various ways. Except for changes in slag practice, no successful control efforts have been widely used.

18.5.7 Direct Reduction Processes

These direct reduction processes hold promise for ore-rich areas seeking a steel industry but lacking capital or product demand to justify a blast furnace/basic O_2 steel making complex. A direct reduction or electric furnace steel making complex can be economical for 3.5% of the size of a conventional complex. Emission levels would be expected to be much less than in a blast furnace operation due to elimination of casting and slag handling requirements.

18.5.8 Open-Hearth Steel Making

The open-hearth furnace consists of a shallow rectangular hearth with a brick arch roof and refractory ducts at each end of the furnace for removal of hot gases to regenerative checker chambers prior to their discharge into the atmosphere in order to pre-heat combustion air. The flame for heating is swept across the length of the open bath. Fuel may consist of oil, tar, coke, oven gas or natural gas.

Emissions and their Control

Minor emissions of dust are associated with charging and tapping the open hearth and teeming the molten steel into ingots. Control of these emissions in open-hearth shops has not been defined or controlled to the extent of those in electric furnaces or basic O_2 shops. The control of open-hearth emissions is accomplished with electrostatic precipitators and high energy scrubbers.

18.5.9 Basic Oxygen Steel Making

General Process Description

Basic O_2 steel making was developed in Linz-donawitz, Austria in the 1950's and is a variation of the older Bessemer process. The basic O_2 process now accounts for the majority of steel making capacity world wide.

18.5.10 Emissions and their Control

The potential sources of emissions in a conventional basic O_2 shop are associated with the following activities.

1. Hot metal transfer (slag skimming and desulphurization)
2. Charging and tapping
3. Furnace waste gases
4. Minor emissions may occur during teeming or casting, but these have been neither defined nor controlled

Hot Metal Transfer

Early control efforts consisted of the installation of multiple type cyclones which removed most of the kist particles but passed the finer, more visible iron oxide particles. Fabric filters are now widely used to reduce particulate discharges from hot metal transfer operations.

Charging and Tapping

The control of charging emissions can be accomplished by changes in practice or with control systems. The most common practice is to install auxiliary hoods on the charging side above the vessel and direct the exhaust fumes to a dust collector. Fabric filters and wet scrubbers have been used. Auxiliary canopy hoods may also be installed in the roof trusses of the building to extract fumes that the local hoods do not capture. Tapping emissions are generally of less consequence than charging emissions, but can be substantial depending upon ladle additions. Control efforts have been few but have been similar to charging emission controls to the extent that the same system often serves both purposes.

Basic Oxygen Furnace Waste Gases

Control of basic oxygen emissions is accomplished effectively with both electrostatic precipitators and high energy scrubbers. Moisture conditioning and gas distribution to and through the collector have been reported as the primary factors for high efficiency. Precipitators are often preceded by gravity separators to remove the larger particles for improved performance.

18.5.11 Electric Furnace Steel Making

The electric process as it is commonly called today, employing a cylindrical vessel and three phase power, was first installed in 1909. Electric arc furnace steel making is a batch process with heats generally ranging in duration from 1½–5 hrs for carbon steel production and 5–10 hrs for alloy steel production. The cyclic operation consists of one or more charges of scrap steel from the scrap charging bucket, followed by melting, refining and tapping through the tap spout into the steel ladle.

Emissions and their Control

Pollutants generated during operation of electric furnaces are primarily limited to particulate matter and carbon monoxide. The control of emissions from electric furnaces has revolved around variations in the method of fume capture and in the method of gas cleaning. There are basically three methods of fume capture which are employed for electric furnace fume control systems:

(a) Direct shell evacuation
(b) Roof mounted hoods
(c) Canopy hoods.

The vast majority of electric furnace control systems have employed fabric filters as the gas cleaning device of choice. Bag houses offer advantage in overall cost, simplicity, reliability, performance, and ease of expansion. Electrostatic precipitators have also been used primarily in Europe.

18.5.12 Rolling, Processing and Finishing

General Process Description

After steel is tapped in molten form from an open-hearth, basic O_2 or electric furnace, it is either poured into ingots or continuously cast into slabs or billets. Whether the steel is made into ingots or continuously cast, it must be further heated to proper temperature before being further reduced to the desired shape and gauge in a rolling mill. Slabs or blooms made from ingots often require surface preparation to remove defects which would otherwise be rolled into the steel. Numerous other metallurgical or chemical coating processes for finishing steel may be conducted at any given steel mill. These steps may consist of normalising, annealing, heat treating, galvanizing, tin plating, aluminizing or painting.

Emissions and their Control

The potential sources of emissions are reheat furnaces, scarfing, pickling and galvanizing units. Particulate emissions are generated from high speed hot rolling mills. Oil mist can be generated from the intense heat and pressure on rolling oils used on high speed cold reduction mills, but simple exhaust hooding and wetted baffles serve to minimise these emissions.

18.5.13 Ferrous Foundry Operations

General process description

Ferrous foundries produce castings of grey iron, ductile iron and steel. Molten steel used to produce castings in steel foundries is produced in electric melting furnaces or open-hearth furnaces. The predominant means of melting iron, is the cupola, a refractory lined cylindrical vessel equipped with combustion air tuyeres at the base and a charging door or opening near the top, through which coke, limestone and scrap metal are charged.

18.5.14 Emissions and their Control

Potential emissions from ferrous foundry operations are primarily associated with:

(a) The melting facility
(b) The shake out and sand conditioning system
(c) Coremaking

Cupolas

Particulate emissions emitted from a cupola consist of metallic oxides, coke ash and volatilised materials driven from the scrap. Gaseous emissions from cupolas may include carbon monoxide and sulphur dioxide. Control of cupola emissions has been primarily confined to the use of fabric filters and wet scrubbers although several installations have included electrostatic precipitators.

Sand Handling

Control of emissions in foundries from sand conditioning, shake out, and mould preparation areas is well developed due to the health hazard from occupational exposure to silica and dust. Local hooding and ventilation systems are installed and dust laden gases are directed to wet scrubbers, fabric filters or in some cases, mechanical collectors.

Coremaking

Gases and odorous compounds derived from binders in the coremaking process are emitted from core baking ovens. Thermal or catalytic oxidation has been employed to burn these gases.

18.6 Non-Ferrous Metallurgical Operations

Non-ferrous metals production has been very important in the development of the science of air pollution control. Smelting processes are carried on at high temperatures and large quantities of dust and metal oxide fumes are generated to contaminate the huge volumes of gas which flow through process equipment. In addition to the potential for substantial emissions of particulate matter some of which could be toxic, there is a concomitant high production and emission of sulphur dioxide. Sulphur is a mixed blessing. Its oxidation provides valuable process heat and conserves energy that would have to be supplied from other sources. Also SO_2 can be useful raw material for sulphuric acid production in areas where acid can be used and where smelter acid can compete in the market with acid derived from burning elemental sulphur. But, in the main sulphur dioxide has been an onerous problem. Damage from particulate smelter emissions has also been the subject of litigation.

18.6.1 Copper

Mining, Milling and Concentrating

Both open pit and underground mining are practiced. However, the principal method employed is open pit. Open pit operations are conducted successfully with ores containing as little as 0.5% copper. A major development in open pit mining has been the leaching of waste dumps. Most of the primary domestic copper is recovered from low grade sulphide ores using pyrometallurgical procedures. The four hydrometallurgical methods practiced are in-place leaching, heap leaching, dump leaching and vat leaching. Copper-containing solids are leached with a dilute solution of sulphuric acid in all these methods. Separation of the traces of copper bearing minerals from the mass of waste is essential to economic recovery of the metal.

Smelting

A typical copper smelter in the United States uses roasters, reverberatory furnaces and converters.

Refining

A small percentage of virgin copper is sufficiently refined by special furnace treatment at the smelter to be used directly in certain applications. Very small amounts of certain impurities greatly reduce the electrical conductivity of copper and adversely affects its annealability. Anodes of blister copper are arranged alternately, face to face with thin sheets of pure copper in large tanks. The sheets are the cathodes of multiple electrolytic cells to be formed in each tank. The tank is filled with a dilute solution of copper sulphate and sulphuric acid. A low voltage current passes between anodes and cathodes, causing dissolution of copper at anodes and deposition at the cathodes. Trace metals present at the impure anode dissolve in the electrolyte or settle to the bottom of the tanks as a black sludge, called anode mud or slimes. The slimes may contain selenium, tellurium, gold, silver, platinum and palladium. All these elements are frequently associated with copper in its ores.

Emissions and Controls

(*a*) *Mining operations* Dust can be a problem in underground mines in spite of wet collaring and drilling. Ventilation must be provided to keep concentrations of dusts down to safe levels. In most instances power ventilation is also necessary to remove nitrogen oxides and carbon monoxide from blasting and to control heat and humidity. Blasting gases are absent except during the short period of blasting and then are highly diluted by the large volume of ventilating air constantly being moved through the mine. Open pit mining may create localised dustiness near operating drills, power shovels and other equipment. But, it is a standard practice to use water while drilling and to wet down ore and waste piles, when necessary. Roadways are sprinkled to reduce dusting. Blasting when properly done, disperses surprisingly little dust in open pit mining but may produce nitrogen dioxide. Improved formulations and practices have virtually eliminated production of nitrogen oxides from exploding ANFO (ammonium nitrate-fuel oil). Ore crushing generates considerable dust. Wetting must usually be supplemented by appropriate exhaust ventilation for good dust control. In recent years, cohesion compounds have been used to suppress dust emissions from tailing dams. Planting and cultivation of ground cover and shrubs to act as wind breaks are also successful. They also improve appearance of the landscape.

(*b*) *Smelters* The high temperatures necessary in roasting, smelting and converting cause volatilization of a number of trace elements such as As, Pb, Cd, etc., which may be present in copper ores and concentrates. Balloon flues in which gases move at low velocities serve as gravity collectors of the large particles and provide low resistance ducting for the large volumes of combustion gases and ventilating air that must be removed. Cyclones also may be used. For collection of the finer particulates, electrostatic precipitators, in which collection efficiencies up to 99.7% for copper dust and fume are attained by careful conditioning of the flue gases are most often used. Cleaned hot flue gases are vented to the atmosphere via tall stacks for maximum dispersion and dilution of contained sulphur dioxide. To make sulphuric acid efficiently, SO_2 concentrations in the gas must be between 4-8%. Utilisation of SO_2 reduces the potential air pollution problem associated with copper smelting; hence, acid production would seem to be an obvious and simple solution.

SO_2 monitoring stations at selected points around a smelter are helpful in carrying out a 'Sea Captain' method of controlling SO_2 emissions, [illegible]wn as 'closed loop control' or 'intermittent control'. This technique ma[illegible] use of measurement of meteorological conditions followed by their expert evaluation. If conditions are unfavourable for adequate dispersion of gas or are predicted to be unfavourable, SO_2 emitting operations are curtailed. However, forecasting is not perfect and detection of significant SO_2 at ground level by an automatic instrument may be the first indication that dispersion and dilution are not adequate. Telemetering of detector information to the smelter control centre computer system permits immediate curtailment action to be taken.

(*c*) *Refinery Emissions and Control* Electrolytic copper refining operations by their nature do not create significant air pollution. In one part of the copper refining process called electrolytic purification, there are possibilities of excessive acid mist generation and arsine evolution.

Impact of Regulations on Emission Control Practices

The impact of the 1970 United States Clean Air Act regulations on copper operations has been severe. Alternative methods to reduce sulphur oxide emissions include process changes such as the use of electric smelting and flash smelting furnaces. In addition, some of the contemplated SO_2 removal processes have the potential of causing severe adverse effects on water and land. There are also problems of an increased supply of controlled SO_2 by-products on markets traditionally supplied by elemental sulphur and of high energy consumption by SO_2 recovery facilities.

18.6.2 Lead

Mining, Melting and Concentrating

Most lead ores are mined underground. The lead mineral of greatest importance is galena (PbS). Traces of silver, iron, zinc and other metals are usually present in lead ores and they accompany the lead in the iron ore concentrate.

Smelting

The sulphur content of the lead concentrates is reduced by sintering them on Dwight-Lloyd sintering machines, most commonly on updraft machines.

Refining

Lead bullion is purified by a number of different processes like Parkes process, Harris process, the Betts process, etc. An important part of lead refining is the recovery of silver and gold.

Emissions and Controls

(a) Mining operations: Dust problems in the mining, milling and concentrating are the same as those for copper.

(b) Smelter operations: Hot gases from the lead concentrate sintering process carry SO_2, dust and the oxide fumes of volatile metals such as antinony lead and zinc. Also, some lead is volatilised and oxidised. Blast furnaces emit similar particulates plus low concentrations of SO_2 and carbon monoxide. Dust and fume are recovered from the gas stream by settling in large flues and by precipitation in electrostatic precipitators or filtration in large bag houses, more commonly the latter. SO_2 derived from sintering may not be concentrated enough to be used directly by available commercial process for sulphuric acid production. It is possible, however by recirculation of gases, to build up the SO_2 concentration sufficiently to permit production of H_2SO_4 or liquid sulphur dioxide from a portion of the gases. Even though the use of an acid plant appears superficially to be a feasible solution for SO_2 control at a lead smelter, acid production, because of technical and disposal problems, may be a less satisfactory and reliable control method. Fugitive emissions from sinter plant operations are in many instances collected by appropriate exhaust hoods and are passed into bag houses or high efficiency scrubbers. Wetting with water spray nozzles is effective in abating fugitive dust in the sinter plant.

(c) Refinery emissions and controls: Lead refineries have bag houses for recovery of fumes from softening furnaces and cupeling furnaces.

Impact of Regulations on Emission Control Practices

The impact of air pollution regulations on lead operations is similar to the impact on copper operations.

18.6.3 Zinc

Mining, Milling and Concentrating

The bulk of zinc ore contains zinc as sphalerite, ZnS, which is separated as a concentrate from accompanying minerals by selective flotation. Concentrates contain about 60% zinc. Another important source of raw material for zinc metal is zinc oxide from fuming furnaces.

Roasting and Retorting

For efficient recovery of zinc, sulphur must be removed from concentrates to less than 2%. This is done by roasting. Multiple hearth, flash or fluid bed roasting may be followed by sintering, or double pass sintering may be used alone.

Leaching and Electrolysis

Zinc of high purity may be obtained from roasted concentrates, from densified zinc oxide from fuming furnaces or from impure metallic zinc by solution in sulphuric acid, removal of impurities from the solution by appropriate chemical treatment, and finally electrolysis of the purified electrolyte. Electrolysis is done in tanks containing alternating anodes of lead and cathodes of aluminium.

Emissions and Controls

Dust, fume and SO_2 are evolved from zinc concentrate roasting or sintering. Particulates are caught in conventional bag houses or electrostatic precipitators. Sulphur dioxide in the more concentrated gas streams is utilised for direct production of sulphuric acid by the contact process, while more dilute gas streams are scrubbed with aqueous ammonia solution. Leaching and electrolysis do not emit significant amounts of particulates or gases. Tanks in which leaching and electrolyte purifications are done are covered and ventilated to prevent worker exposure to possible toxic gases or mists. Stack sampling surveys and continuous ambient monitoring have shown that the outplant emissions from zinc leaching and electrolysis operation are insignificant.

Impact of Regulations on Emission Control Practices

In recent years the United States zinc smelting industry has undergone tremendous change due to imposition of environmental control measures, sharply rising costs and the lack of an attractive domestic climate for the investment of capital in production facilities.

18.6.4 Aluminium

Mining and Ore Treatment

Bauxite is the base ore for aluminium production. Most bauxite ore is purified by the Bayer process.

Electrolysis

Commercial recovery of aluminium from the oxide is accomplished by a unique electrolytic process discovered simultaneously in 1886 by Hall in United States and Heroult in France.

Emissions and Controls

Calcination of aluminium hydroxides for the production of alumina entails mechanical dust dispersion. The valuable dust is recovered from kiln effluents by a combination of cyclone collectors and electrostatic precipitators. The type and character of emissions depend on the type of electrolytic cell used to electrolyse the alumina feed material. The major pollutants for all types are, gaseous and particulate fluorides. Minor gaseous emissions are CO_2, CO, N_2O_2, H_2S, carbon disulphide, sulphur hexafluoride and gaseous fluorocarbons.

Non-fluoride dust emissions are principally mechanically generated carbon dust from the electrodes and mechanically dispersed alumina dust from the charge materials.

The anode plant for a pre-baked electrode operation can be the source of significant SO_2 and hydrocabon emissions particularly tarry and distillate hydrocarbons from the pitch in the anodes. Given the same operating conditions and metal production rates, the vertical and horizontal stud Sodenberg cells generate about the same quantity of total emissions, the total being higher than emissions from pre-baked electrode cells. Volatile constituents of pitch binder used in the electrodes are removed during baking, which is done in facilities separate from the port rooms. In Sodenberg cells baking occurs in the reduction process, using the heat of the process to bake or coke the carbon electrodes in place.

Some dust collection occurs in all of the systems designed for gaseous fluoride control. Multiclones, venturi scrubbers, and wet and dry electrostatic precipitators are used.

Aluminium chloride may also be released by aluminium smelters because either it, or chlorine gas, is used to treat aluminium in holding furnaces to flux and de-gas the molten metal.

Impact of Regulations on Emission Control Practices

Historically emission regulations for aluminium production facilities have been designed to prevent damage to vegetation and livestock in the surrounding areas. These effects have been largely avoided by dependence on a combination of emission controls and environmental dispersion of emissions through tall stacks. Regulations were written around acceptable environmental fluoride concentrations which would not lead to damage to vegetation and animals. The pressure for low fluoride emissions and the associated cost of control will most likely lead to a clear preference for pre-baked electrode facilities when new primary aluminium plants are being developed. However, economic and operational factors may force the improvement of controls for Sodenberg cells and/or may lead to entirely new primary production processes.

18.6.5 Beryllium

Production and Fabrication

Beryllium is widely distributed in the crust of earth, constituting approximately 0.001%. The only beryllium containing ores presently mined for this beryllium content are beryl and bertrandite. The hydroxide is subsequently converted into the desired product-metal, oxide or alloy.

Emissions and Controls

The emissions generated by beryllium extraction processes include beryllium salts, acids, beryllium oxides and other beryllium compounds in the form of dust, fume or mist.

Emissions generated by beryllium operations are successfully controlled by:

1. Dry mechanical collectors
2. Wet collectors

3. Fabric filters
4. High efficiency particulate air filters.

Impact of Regulations on Emission Control Practices

In 1973, the USEPA. classified beryllium as a hazardous air pollutant. Beryllium emissions to the atmosphere are limited to a maximum of 10 g over a 24 hour period, or as an alternative ambient concentration in the vicinity of the source shall not exceed 0.01 $\mu g/m^3$, averaged over a 30 day period. The greatest impact has been the added emphasis on stack testing and ambient monitoring.

18.6.6 Mercury

Mining

Mercury is produced commercially by processing mercury sulphide ore, cinnabar.

Production

Mercury sulphide is thermally decomposed by a variety of means such as retort or a wedge or Herreshoff roaster, to produce elemental mercury and sulphur dioxide.

Emissions and Controls

There are three major sources of mercury vapour emissions from a mercury ore processing facility.

1. Mercury vapour not captured by the condensers
2. Leaks from broken retorts, roasters or condensers
3. Poor house keeping practices

Mercury capture by the condensers is at least 95%. Stack losses of the mercury from ore processing facilities range approximately from 2-3% of total production capacity, but higher losses have been recorded. A major concern with mercury emissions lies with emissions from other sources where mercury is a trace contaminant. For example, coal burning electric power plants and chloro-alkali plants.

Impact of Regulations on Emission Control Practices

Strict regulations have been promulgated by the U.S. Environmental Protection Agency (USEPA) for all operations emitting mercury. No more than 2300 g of Hg may be emitted per 24 hours.

18.6.7 Non-ferrous Metals of Minor Significance

Arsenic

Although arsenic occurs naturally in some minerals, it is not commercially mined. It is produced only as a by-product of various primary smelting operations. Arsenic is produced commercially by heating arsenic-containing fumes and dust in Godfrey roasters or in reverberatory furnaces to volatilise arsenic trioxide. It is used commercially in herbicides, insecticides, rodenticides and in paints as a fungicide. It is also used in glass manufacture and for therapeutic use. There is small but increasing demand for high purity arsenic for use in semi-conductors. Emissions from arsenic production facilities are primarily uncondensed and unsettled arsenic trioxide in the gases leaving the kitchens. These releases are controlled by passing gases through electrostatic precipitators and/or bag houses. Recovered dusts are, reprocessed. It is anticipated that there will be little direct impact of present pollution control regulations on arsenic production because it is considered that best available control technology is being used.

Cadmium

Most cadmium production is the result of processing primary smelter by-products which are rich in cadmium, particularly those resulting from zinc smelting. Most of the cadmium metal produced in the United States is sold for use in electro-plating and alloying applications. Bag houses are considered to be the emission control devices of choice and the best available technology for preventing cadmium releases into the atmosphere.

18.6.8 Secondary Copper, Lead, Zinc and Aluminium

Sources

Scrap provides an important source of metals for the market.

Recovery Process

Scrap may be converted directly to usable metal by simple melting and casting into saleable ingots.

Emissions and Controls

Little or no sulphur dioxide is evolved, and in general, smaller quantities of metal oxide dusts and fumes are produced. Bag houses, more commonly, and electrostatic precipitators are successfully used for collection. Properly designed scrubbers or condensers are effective control devices.

Impact of Regulations on Emission Control Practices

The impact on the larger manufacturer is minimal. In some instances it has been necessary to hood sources of fugitive emissions and vent them through bag houses in order to comply with opacity regulations.

18.6.9 Non-ferrous Foundries

Alloys and Operations

(a) Copper based alloys: There are 2 groups of these alloys; the brasses and the bronzes. There are four basic types of furnaces used to melt and prepare these alloys—pit, tilting crucible, induction and cylindrical rotary or stationary reverberatory furnace.
(b) Zinc based alloys: Zinc die casting metal accounts for most of the zinc used in alloys, excluding brass. Pit and tilting crucible, pot or induction furnaces are all used, usually without fluxes, to melt and prepare zinc alloys.
(c) Emissions and controls: Metal fume is evolved during the melting and casting of alloys. Zinc is the major fume in brass and bronze foundry operations. Fuming does not occur for die casting metals because they are not heated to sufficiently high temperatures. Under such conditions, high lead alloys will produce lead fumes. Another source of emissions is smoke production. The type and degree of emission controls required depends on the furnace, and the alloys. Operating procedures can be as important as the type of furnace, in determining emissions. Shake out of castings from sand-moulds is a dusty operation and may be conducted under hood or over down draft gratings.

Impact of Regulations on Emission Control Practices

Emission control technology is well developed for non-ferrous foundries.

18.7 Thermal Power Plants

Thermal Power Plants utilise fuel to produce steam for power generation. Combustion of fuels produces significant amounts of air pollutants. The types of pollutants depend on the nature of fuel used. If coal is used, fly ash, sulphur dioxide and oxides of nitrogen are the major pollutants. Combustion of coal is a more important source of particulate air pollution than that of fuel oil. In the case of fuel oil, sulphur dioxide and oxides of nitrogen are the major pollutants emitted to the atmosphere. The amount of fly ash and SO_2 released depend on sulphur and ash content of the fuel used. Data on particulate emission from coal fired boilers without air pollution control is given in Table 18.5.

The three major air pollutants from power stations are thus particulate matter (fly ash and soot), sulphur oxides (SO_2 and SO_3) and oxides of nitrogen (NO and NO_2). Besides these, there is a possibility of emission of carbon monoxide and unburnt carbon. However in the modern thermal power stations with automatic combustion control systems, the formation of these products is eliminated. Another pollutant in coal fired power stations is coal dust emitted from the coal handling plant.

Table 18.5 *Particulate Emission from Coal Fired Boilers*
(Without rigorous pollution control)

Types of furnaces	Particulates in kg/ton of coal burnt
Pulverised	
General	7.3 A
Dry bottom	7.8 A
Wet bottom without fly ash re-injection	6.0 A
Wet bottom with fly ash re-injection	10.8 A
Cyclone	0.9 A
Spreader Stoker	
Without fly ash re-injection	6.0 A
With fly ash re-injection	9.1 A
All Other Stokers	2.3 A

Note: A is a multiplication factor representing % ash in coal. Values represent mass of particulates reaching control equipment used on this type of furnace; they are not emissions.

Studies made on the fallout rates of fly ash from a thermal power plant located at Korba (Madhya Pradesh, India) have shown interesting results. The power station with a production capacity of 540 MW and a coal consumption of 1500–1700 tonnes per day has five stacks and the dust catching equipment consists of mechanical dust collectors. The fallout rates at a distance of about one kilometre around the power plant ranged from 291.0–848.2 tonnes/km^2/month; those at an average distance of about 2.5 kilometres ranged from 134.1–548.2 tonnes/km^2/month and those at a distance of 4 kilometres ranged from 64.3–478.2 tonnes/km^2/month. With the onset of the dry season an increase of the fallout rates was observed.

18.7.1 Pollution Due to Oil Fired Boilers

Many types of boilers are under operation in different industries. Depending on the requirement of steam, boilers can be operated either on full or partial load continuously or even on variable load. The black smoke that is emitted during the furnace oil burning consists of unburnt carbon particulates and its intensity depends upon various factors such as type of boiler, air-fuel ratio, burner condition, control of draft and type of fuel. The particulate emission rate for a package type automatic fuel fired boiler used in a textile mill was found to be 2.22 kg/kilo litre of furnace oil burnt under normal working conditions.

When the boiler is stopped for sometime, its surface cools down and impingement of oil flame on a cool surface can partially quench the flame and cause incomplete combustion with a release of the particulates. When the boiler is overloaded, the flame envelope will be larger than that for which the combustion chamber is designed and the particulate emission will be high. Under such disturbed conditions the particulate emissions from the boilers is reported to be 50–60% more than that during normal conditions.

QUESTIONS

1. What are the main air pollution problems encountered in
   (a) Fertiliser industry
   (b) Glass manufacture
   (c) Asbestos processing
2. (a) What are the sources of air pollution in the cement industry?
   (b) What methods of control are adopted to control dust in cement plants?
3. What are the sources of emissions in a petroleum refinery and how are they controlled?
4. Describe briefly the sources of air pollution in the following industrial operations and suggest suitable methods of control:
   (a) Ferrous metallurgical operations
   (b) Copper mining and smelting
   (c) Lead refining
5. Discuss the air pollution problems associated with the following industrial operations:
   (a) Aluminium production
   (b) Mercury production
   Suggest suitable methods of control for them.
6. Explain sources of air pollution and methods of abatement in thermal power plants.

19

Air Quality and Emission Standards

19.1 Introduction

The preferred sequence of development of air quality standards is given below:

(a) Prepare air quality criteria which are analyses of the relationship between pollutant concentrations in the air and the adverse effects associated therewith. The WHO calls these as 'guides'.

(b) From the air quality criteria develop air quality goals, which are the concentration of pollutants with which we believe we can live without adverse effects on health and welfare.

(c) From the air quality criteria, develop air quality standards, which are the concentrations of pollutants that we intend to achieve in the immediate future but that may fall short of our air quality goals because the standards must give consideration to feasibility of achievement within the immediate foreseeable future.

(d) In order to develop the above standards, there must be standards for measurement and testing of the ambient air and air pollution effects.

19.2 Air Quality Standards

19.2.1 Definition

Air Quality Standards are legal limits placed on levels of air pollutants in the ambient air during a given period of time. As such, they characterize the allowable level of a pollutant or a class of pollutants in the atmosphere and thus define the amount of exposure permitted to the population and/or to ecological systems.

Air Quality Standards are expressions of public policy and thereby requirements for action. Thus, they are not based solely on air quality criteria but are also based on a broad range of economic, social, technical and political considerations. Air quality standards have evolved differently in different countries depending on exposure conditions, the socio-economic situation, and the importance of other health related problems.

19.2.2 Single or Multiple Standards

A major decision that must be taken in the adoption of air quality standards is whether there should be one standard for an entire jurisdiction or different standards for different areas within the jurisdiction. All political jurisdictions within an air quality region would have the same air quality standards. In general, air quality regions would not be contiguous but would be separated by non-urban hinterland. If two separated regions were to expand in area in the future until they become contiguous, they would presumably then be merged into one larger air quality region. Since each air quality region would adopt its own air quality standards, they could differ among the several regions. However, there should be a datum below which no region's air quality standards should drop.

19.3 Kinds of Air Quality Standards

19.3.1 Ambient Air Quality Standards

These are the legal limits placed on the concentration of air pollutants in a community where people and things are exposed. Enforcement of ambient air quality standards is difficult. To know if the actual concentrations are in compliance requires an air monitoring programme. Ambient air quality standards need to be distinguished from threshold limit values (TLV) for work-room atmospheres, e.g., those published by the American Conference of Government Industrial Hygienists. These latter values are the permissible exposure levels for healthy adult worker for eight hours per day, five days per week. In contrast, air quality standards are permissible exposures of all living and non-living things for 24 hours per day, 7 days per week.

19.3.2 Other Air Quality Standards

(a) Quasi-emission standards called 'point of impingement standards' which are the limit on specific pollutants of the ambient air at ground level required by national, state, or local regulations to be used in diffusion computations to determine limits of emission from specific sources.

(b) Soiling index, which is the measurement of transmitted or reflected light through or from a spot of particulate matter collected on a filter for a prescribed period of time. (The darkness of stains produced by drawing polluted air through filter paper has been used to estimate suspended particulate matter concentration.)

(c) Odour standards

(d) Visibility standards

(e) Standards for particulate matter deposited, etc.

19.4 Emission Standards

Definition Emission standards establish permitted emission levels for specific groups of emitters and require that all members of these groups emit no more than these permitted emission levels. These standards can be applicable to any selected group of emitters and can be national, regional or local in application. They can be based on some air quality standards (thus making them a tactical subsidiary to an air quality management strategy) or can be entirely independent of any such air quality standards, serving as an entirely separate and different type of strategy from air quality management.

19.4.1 Emission Standards for Mobile Sources

Generally, mobile sources include aircraft, ships, motor vehicles and railroad locomotives. Motor vehicles are sub-divided into automobiles, trucks, buses and motor cycles. Emission standards for motor vehicles have been one of the most significant developments in the field of air pollution control.

Emission standards prescribe limits of contaminants discharged into the atmosphere, so that when the standards are met, adverse effects from air pollution will be minimized or eliminated. Once air quality standards are established, such standards can be used as the basis for formulating emission standards. The emission standards, therefore, represent emission levels not to be exceeded if air quality goals are to be achieved.

Table 19.1 *Pollutants from the Crankcase, Fuel System, and Exhaust of an Uncontrolled Vehicle*

Emission source	Percentage emission		
	CO	NO_x	Hydrocarbons
Crank case	1-2	1-2	25
Fuel system	0	0	10
Exhaust	98-99	98-99	65

19.4.2 Emission Standards for Stationary Sources

Stationary source emission standards are standards relating to a stationary site, process, stack, chimney, or vent intended to help achieve the desired air quality. They include standards for buffer zones, stack height, equipment design, and fuel composition, and those that directly limit the amount or concentration of a pollutant emitted from a source. Emission standards may be derived from process and equipment considerations, air quality considerations or both. In some cases, we have the technological ability to control certain pollutants but are not doing so for economic, sociological or political reasons.

19.5 Air Quality Standards: A Comparison

The cost to a community is the paramount consideration in deciding which of the devices or techniques available to control air pollutants will provide an acceptable atmosphere. To eliminate air pollution entirely would be excessively expensive. Then, the economics involved demands that some level of pollutant be permitted in the atmosphere. This permissible level as mentioned earlier is called 'air quality standards'.

Thus air quality standards imply acceptable pollutant concentration in the atmosphere enveloping a community.

Several countries in the world which are advanced industrially and which have an air pollution problem have developed standards for air quality which they strive to attain by adopting suitable air pollution control measures. Table 19.2 gives the air quality standards adopted by the Environmental Protection Agency, USA. Table 19.3 gives the standards for suspended particulate matter for various countries. Table 19.4 gives the standards for Sulphur dioxide for various countries. Table 19.5 gives the standards for dust fall for various countries. Table 19.6 gives the standards for sulphation rate in Canada.

Original standards expessed in different units in original data have been converted to suitable units for easy understanding. Due to this change, the original value may match to the nearest expressed values as mentioned in the various tables.

Table 19.2 *Air Quality Standards Adopted by EPA*

No.	Parameter	Standard concentration		Remarks
		$\mu g/m^3$	ppm	
1.	Suspended particulate matter	75		Annual geometric mean
		260		24 h
2.	Sulphur dioxide	80	0.83	Annual mean
		365	0.14	24 h
3.	Carbon monoxide	10000	9	8 h. Not more than once per year
		40000	35	1 h. Not more than once per year
4.	Nitrogen oxides	100	0.05	Annual mean
5.	Ozone	235	0.12	1 h daily maximum. Not more than once per year
6.	Non-methane hydrocarbons	160	0.24	6–9 h. Annual mean Not more than once
7.	Lead	1.5		3-month average

In India, on the basis of land use and other factors, the various areas of a state may be classified into three categories by concerned State Pollution Control Boards:

A. Industrial and mixed—use areas,
B. Residential and rural areas,
C. Sensitive areas.

Table 19.3 *Standards for Suspended Particulate Matter*

Country	μg/m³	Averaging Time
Bulgaria	150	24 hours
Czechoslovakia	150	,,
East Germany	150	,,
Finland	150	,,
Poland	200	,,
Israel	150	,,
USA		
Delaware—rural	60	,,
Delaware—residential	75	,,
Delaware—commercial	95	,,
Delaware—industrial	125	,,
South Carolina	25	,,
Texas—residential	125	,,
Texas—commercial	150	,,
Texas—industrial	175	,,
Texas—others	200	,,
USSR	150	,,
Canada		
Ontario—residential—rural	60	1 Year
Ontario—industrial-cum-commercial	110	,,
France	60	,,
USA		
Missouri	75	,,
Montana	75	,,

Category (A) becomes self evident by the intensity of industrial activity in an area and is bound to have somewhat inferior quality of air compared to other categories. The category (C) covers hill stations, tourist resorts, sanctuaries, national parks, national monuments, health resorts and other such areas where the nation conserves its clean environment even if that implies some curbs on economic activity. All areas not specifically declared by the concerned State Pollution Control Board to be classified in category (A) or category (C) will be automatically deemed to fall in category (B). Table 19.7 gives the Indian Ambient Air Quality Standards.

When monitored uniformly over the 12 months of an year with a frequency of not less than once in a week with a sampling time of eight hours for any sample and analysed according to procedures specified by the Central Board, the concentrations for the pollutants shall be 95% of the time within the limits prescribed.

Whenever and wherever three consecutive measurements spaced at least one week apart, or any three out of ten consecutive measurements spaced by at least one week apart are found to exceed limits specified in Table 19.7 for the respective category it would be considered adequate reason to institute regular weekly monitoring and further investigation.

Table 19.4 *Standards for Sulphur Dioxide*

Country	μg/m³	Averaging Time
Belgium	150	24 hours
Bulgaria	50	,,
Canada		
Alberta	150	,,
Newfoundland	290	,,
Ontario	290	,,
Czechoslovakia	150	,,
East Germany	150	,,
Finland	250	,,
France		
Paris	75	,,
Italy	380	,,
Netherlands	125	,,
Poland	75	,,
Rumania	250	,,
Sweden	250	,,
Switzerland	75	,,
Turkey	150	,,
USA		
Colorado	260	,,
Delaware—rural	50	,,
Delaware—residential	80	,,
Delaware—commercial	100	,,
Delaware—industrial	160	,,
Missouri	120	,,
USSR	150	,,
Canada		
Alberta	30	1 Year
Manitoba	60	,,
Ontario—Industrial—commercial	130	,,
Ontario—Residential—rural	50	,,
Spain	150	,,

Table 19.5 *Standards for Dust Fall*

Country	Dust fall rate MT/km²/Month
Canada	
Ontario—residential—rural	8
Ontario—industrial—commercial	15
Federal Republic of Germany—industrial	13
Poland	3
Canada	
Alberta—residential	6
Alberta—industrial—commercial	17
USA	
Missouri—non-residential	4
Missouri—heavy industrial	10
Montana—residential	6
Montana—heavy industrial	12
Oregon—residential—commercial	6
Oregon—industrial	12

Table 19.6 *Standards for Sulphation Rate*

Country	Sulphation rate mg SO_3''/100 cm^2/day
Canada	
Manitoba	1.00
Ontario	0.40
Saskatchewan	0.80

Table 19.7 *Indian Air Quality Standards*

Category	Area	Concentrations in $\mu g/m^3$			
		SPM	SO_2	NO_x	CO
A.	Industrial and mixed use	500	120	120	5000
B.	Residential and rural	200	80	80	2000
C.	Sensitive	100	30	30	1000

The World Health Organisation (WHO) Technical Report discusses the effects of sulphur dioxide and particulate matter, carbon monoxide, photochemical oxidants and nitrogen dioxide. It was the opinion of the committee that exposure to these pollutants should be kept as low as possible. Taking into consideration all the evidence available to it, the committee reached a consensus on long-term goals for these pollutants (Table 19.8).

Table 19.8 *Recommended Long-Term Goals*

Pollutant	Measurement methods	Limiting value
Sulphur dioxide*	British standard procedure**	Annual mean 60 $\mu g/m^3$ (0.02 ppm) 98% observations below 200 $\mu g/m^3$ (0.060 ppm)
Suspended particles*	British standard procedure**	Annual mean 40 $\mu g/m^3$ (0.016 ppm) 98% observations† below 120 $\mu g/m^3$ (0.048 ppm)
Carbon monoxide	Non-dispersive infrared	8 h average 10 mg/m^3 (9 ppm) 1 h maximum 40 mg/m^3 (30 ppm)
Photochemical oxidant	Measured by neutral buffered KI method expressed as Ozone	8 h average 60 $\mu g/m^3$ (0.03 ppm) 1 h maximum 120 $\mu g/m^3$ (0.06 ppm)

* Values for sulphur oxides and suspended particulates apply only in conjunction with one another.

** Methods are not necessarily those recommended but indicate those on which these units have been based. Where other methods are used, an appropriate adjustment may be necessary.

† The permissible 2% of observations over this limit may not fall on consecutive days.

19.5.1 Emission Standards

The emission standards for cement, thermal power, iron and steel, fertiliser (urea), nitric and sulphuric acid plants, as framed by the Central Board for Prevention and Control of Water Pollution (Emission Regulations Part I, July 1984) are given below in abstract. The emission standards for calcium carbide, copper, lead and zinc smelting, carbon black, fertiliser (phosphatic), oil refineries and aluminium extraction, framed by the Central Pollution Control Board (Emission Regulations Part II, Jully 1985) are also given below in abstract.

The State Boards may adopt standards that are more stringent than those given depending on the location of the industry and specially if it is in a protected area. They shall not, however, relax the standards.

A protected area is one that is already polluted being in a metropolitan/industrial location. A protected area also includes a sensitive area because of its proximity to national parks, forests, historical monuments and health resorts.

19.6 How to Express the Emission?

While measuring the emission in a stack, the temperature, pressure, and humidity are to be recorded. The emission limits prescribed are expressed as concentration of pollutants per unit volume of air under *standard or normal conditions, written as* Nm^3. The standard conditions for air in India, herein after termed as Standard Air, are 25°C temperature, 760 mm Hg pressure, and zero per cent moisture. Emissions measured at other conditions of air are to be computed for Standard Air for reporting compliance to the prescribed limits.

For example, if the volume of the gas drawn through the sampling train, as measured by the dry gas meter, is V_m cubic metres under a negative pressure of P_m mm Hg and a temperature of T_m°C; then the standard volume, V_{std} is given by

$$V_{std} = V_m Y \left(\frac{298}{273 + T_m}\right)\left(\frac{P_{bar} - P_m}{760}\right)$$

where P_{bar} is the barometric pressure, at the sampling site, in mm Hg and Y is the calibration factor of the dry gas meter. Thus, if the total particulates collected is Mn, mg, then the emission concentration is Mn/V_{std} mg/Nm^3.

PART-I

1. Cement

Standard for particulate matter emission

Capacity	Protected Area	Other area
200 tpd and less	250 mg/Nm^3	400 mg/Nm^3
Greater than 200 tpd	150 mg/Nm^3	250 mg/Nm^3

2. Thermal power

(a) Standard for particulate matter emission

Boiler size	Protected area	Other area	
		Old (before 1979)	New (after 1979)
Less than 200 MW	150 mg/Nm³	600 mg/Nm³	350 mg/Nm³
200 MW and above	150 mg/Nm³	—	150 mg/Nm³

(b) Standard for sulphur dioxide control (through stack height)

Boiler size	Stack height
Less than 200 MW	$H = 14(Q)^{0.3}$
200 MW to less than 500 MW	220 m
500 MW and more	275 m

Q = Sulphur dioxide emission in kg/h
H = Stack height in metres

3. Iron and steel
Standard for particulate matter

Process	Emission limits
Sintering plant	150 mg/Nm³
Coke oven	—
Blast furnace	—
Steel making—during normal operation	150 mg/Nm³
During oxygen lancing	400 mg/Nm³

4. Fertiliser (Urea)
Standard for particulate matter emission

Process	Emission limits
Prilling Tower	50 mg/Nm³

5. Nitric acid
Standard for oxides of nitrogen

3 kg of NO_x per tonne of weak acid (before concentration) produced.

6. Sulphuric acid
Standard for sulphur dioxide and acid mist emission

Process	Sulphur dioxide emission	Acid mist emission
Single conversion Single absorption	10 kg/tonne of concentrated (100%) acid produced	50 mg/Nm³
Double conversion Double absorption	4 kg/tonne of concentrated (100%) acid produced	50 mg/Nm³

Guidelines for Minimum Stack Height

1. Plant Type	Stack Height
For all plants except thermal power plant	30 m

2. For Plants where the sulphur dioxide emission is estimated as Q (kg/h) the stack height, H in metres is given by

$$H = 14\,(Q)^{0.3}$$

3. For plants where the particulate matter emission is estimated as Q (tonnes/hr) the stack height, H in metres is given by

$$H = 74\,(Q)^{0.27}$$

4. If by using the formula given in 2 or 3 above, the stack height arrived at, is more than 30 m, then this higher stack should be used.
   In no case should the height of the stack be less than 30 m.

PART-II

1. Calcium carbide
   Standard for particulate matter emission

Source	Emission limit
Kiln	250 mg/Nm^3
Arc furnace	150 mg/Nm^3

2. Copper, lead and zinc smelting
   Standard for particulate matter and oxides of sulphur

Concentrator	150 mg/Nm^3 for particulate matter
Smelter and converter	Off-gases must go for H_2SO_4 manufacture. No release of SO_2/SO_3 shall be permitted from the smelter or converter.

3. Carbon Black
   Standard for particulate matter emission

Year of Commissioning	Emission limit
New plants (Built and commissioned after January 1, 1985)	150 mg/Nm^3
Existing Plants (Built and commissioned before December 31, 1984)	250 mg/Nm^3 (till December 31, 1986) 150 mg/Nm^3 (from January 1, 1987)

4. Fertiliser (Phosphatic)
   Standard for fluoride and particulate matter

Process	Emission limit
Acidification of rock phosphate	25 mg/Nm^3 as total fluoride (F^-)
Granulation, mixing, rock grinding	150 mg/Nm^3 of particulate matter from each process

5. Oil Refineries
Standard for sulphur dioxide

Process	Emission limit
Distillation (atmospheric plus vacuum)	0.25 kg/Te of feed*
Catalytic cracker	2.5 kg/Te of feed
Sulphur recovery unit	120 kg/Te of sulphur in the feed

*Feed indicates the feed for that part of the process under consideration only

6. Aluminium
Standard for fluoride and particulate matter

Process	Standard
Calcination	250 mg/Nm³ of particulate
Aluminium smelting	1 kg (F^-)/Te of aluminium produced and 150 mg/Nm³ of particulate matter

ANNEXURE I

National Ambient Air Quality Standards for the United States

Contaminant	Primary*	Secondary**
Sulphur oxides	1. 80 μg/m³ (0.03 ppm) annual arithmetic mean 2. 365 μg/m³ (0.14 ppm) 24 h concentration†	1. 60 μg/m³ (0.02 ppm) annual arithmetic mean 2. 260 μg/m³ (0.1 ppm) 24 h concentration† 3. 1300 μg/m³ (0.5 ppm) 3 h concentration†
Particulate matter	1. 75 μg/m³ (0.03 ppm) annual geometric mean 2. 260 μg/m³ (0.1 ppm) 24 h concentration†	1. 60 μg/m³ (0.024 ppm) annual geometric mean 2. 150 μg/m³ (0.06 ppm) 24 h concentration†
Carbon monoxide	1. 10 mg/m³ (9 ppm) max 24 h concentration 2. 40 mg/m³ (35 ppm) max 1 h concentration	1. Same as primary 2. Same as primary
Oxidants	160 μg/m³ (0.08 ppm) max 1 h concentration†	150 μg/m³ (0.075 ppm) 24 h concentration
Hydrocarbons	160 μg/m³ (0.24 ppm) max 3 h (6–9 a.m.) concentration†	Same as primary
Nitrogen dioxide	100 μg/m³ (0.05 ppm) annual arithmetic mean	Same as primary

*Primary air quality standards define levels of air quality judged necessary, with an adequate margin of safety, to protect public health.

**Secondary air quality standards define levels of air quality judged necessary to protect public welfare from any known or anticipated adverse effects of a pollutant.

†Maximum values not to be exceeded more than once a year.

ANNEXURE II

*Ambient Air Quality Standards and Workroom Air Threshold Limit Values (TLV) for Some Gases (ACGIH)**

Country or state	Ambient air		Workroom TLV (8 h/day, 5 days/week), ml/m³
	Average concentration, ml/m³	Time h	
Carbon monoxide			
California (serious level)	30	8	100
,, ,,	120	1	100
Los Angeles (first alert)	100	—	—
,, (second alert)	200	—	—
,, (third alert)	300	—	—
USSR (max 4.8)	0.8	24	18
Chlorine			
USA	—	—	1.0
USSR (max 0.033)	0.01	24	0.33
West Germany (VDI)(max 1.0)	0.1	0.5	1.0
Hydrogen chloride			
USA	—	—	5.0
USSR (max 0.03)	0.009	24	—
West Germany (max 1.0)	0.5	0.5	5.0
Hydrogen Sulphide			
California (adverse level)	0.1	1.0	20.0
USSR (max 0.005)	0.005	24	6.7
West Germany (max 0.2)	0.1	0.5	20.0
Ethylene			
California (adverse level)	0.5	1.0	—
,, ,,	0.1	8.0	—
Ozone or oxidant			
California (adverse level)	0.15	1.0	0.1
Los Angeles (first alert)	0.5	—	—
,, (second alert)	1.0	—	—
,, (third alert)	1.5	—	—
Oxides of nitrogen			
Los Angeles (first alert)	3.0	—	5.0
,, (second alert)	5.0	—	—
,, (third alert)	10.0	—	—
USSR ($NO + NO_2$, max 0.18)	0.06	24	2.8
Czechoslovakia ($NO + NO_2$, max 0.24)	0.08	24	
West Germany (NO_2, max 1.0)	0.50	0.5	5.0
Sulphur dioxide			
California (adverse level)	0.3	8.0	5.0
,, (adverse level)	1.0	1.0	—
,, (serious level)	5.0	1.0	—
,, (emergency level)	10.0	1.0	—

(Contd.)

Country or state	Ambient air		Workroom TLV (8 h/day, 5 days/week) ml/m^3
	Average concentration, ml/m^3	Time h	
Los Angeles (first alert)	3.0	—	—
,, (second alert)	5.0	—	—
,, (third alert)	10.0	—	—
USSR (max 0.18)	0.05	24	4.1
West Germany (max 0.3)	0.20	0.5	5.0

*ACGIH: American Conference of Governmental Industrial Hygienists
1 ml/m^3 = 10^{-6} m^3/m^3 = 1 ppm

QUESTIONS

1. Explain what you understand by air quality standards.
2. Give the Indian air quality standards for residential, industrial, and sensitive areas.
3. Write a note on: Emission standards.
4. What are the Indian standards for particulate matter emission in the case of cement industries in
   (a) Protected areas
   (b) Other areas?
5. What are the guidelines prescribed by the Central Board for minimum stack height in polluting industries?
6. Write a brief note on maximum allowable concentration of air contaminants. Illustrate your answer with examples.
7. Discuss which should come first—the technology to clean up air or the setting up of standards which will require a clean up? Is it fair to expect industry to clean up a process if the technology is not yet available? On the other hand, will industry ever find a way to clean up if it does not have to meet emission standards?

20

Air Pollution Indices

20.1 Introduction

Today the problem of air pollution has become so important that there is a need for timely information about changes in the air pollution levels in cities. This should be done in a simple way as the layman does not understand the technical language of air pollution. Air pollution indices fulfil this important need by communicating with the public in a non-technical way.

There are many circumstances where it is desirable to have a means of describing 'air pollution' as one overall entity, rather than in terms of a series of concentrations of several individual pollutants. This should be done in terms more familiar to the public or in terms that incorporate the significance of the concentration in terms of effects. These indirect methods of expressing pollutant concentrations or levels are known as 'Air Pollution Indices'.

20.2 Definition of Air Pollution Index

An 'Air Pollution Index' can be defined as a scheme that transforms the (weighted) values of individual air pollution related parameters (for example, sulphur dioxide concentration or suspended particulate matter) into single number or set of numbers.

As a result, we get an equation which transforms the parameter values—by means of a numerical manipulation—into a more simple and precise form. This is indicated in Fig. 20.1.

However, it should be noted that the complexity of the problem of air

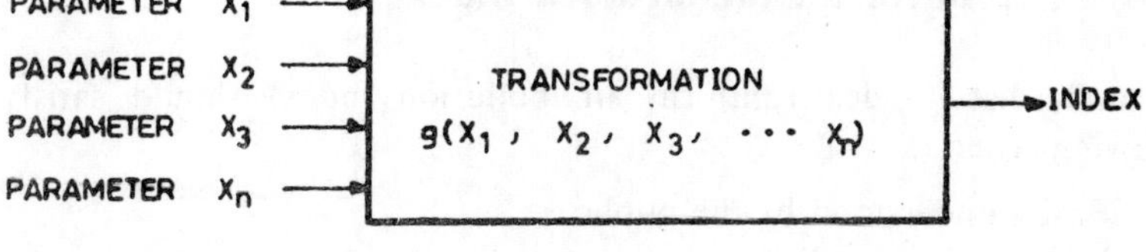

Fig. 20.1 *Index calculation*

pollution makes it difficult, if not impossible, to develop an index that is really meaningful.

20.3 Uses of Air Pollution Indices

1. It helps in informing the public about air pollution levels in a particular area
2. It helps in comparing air pollution levels in different cities
3. It may be used for development of a system for avoidance or management of a severe episode of air pollution

20.4 Types of Indices

Air quality indices can be broadly classified into two groups:

1. Short-term indices
2. Long-term indices

Short-term indices are usually intended to inform the public about daily changes in the air pollution levels. These indices are commonly used by the local and state air pollution control agencies.

Long-term indices are intended to evaluate changes in air quality over periods of several years or more. These indices are very helpful for the purpose of assessing the effectiveness of enforcement policies with regard to pollution control measures, in improving air quality.

20.5 Air Pollution Parameters

Generally the parameters used for calculating the air pollution index are suspended particulate matter (SPM), sulphur dioxide (SO_2), carbon monoxide (CO), nitrogen dioxide (NO_2), photochemical oxidant (principally ozone O_3) and hydrocarbons. 'Visibility' can also be taken into consideration while determining the air pollution index.

20.6 Criteria for a Standardised Index

A standardised index (uniform air pollution index) should satisfy the following criteria:

1. Easily understood by the public
2. Include major pollutants
3. Calculated in a simple manner
4. Rest on a reasonable scientific basis

5. Consistent with perceived air pollution levels
6. Spatially meaningful
7. Exhibit day-to-day variation
8. Relate to ambient air quality standards and goals
9. Relate to episode criteria
10. Can be forecast a day in advance (if possible)

Further, in addition to issuing a common index, it may be useful to also issue a running annual average of the index, as well as the number of days the standards have been exceeded.

It is better to have a single uniform air pollution index for the entire nation. Otherwise it may lead to lot of confusion, when an individual travels to different cities, with different air pollution index standards.

20.7 Determination of Index

There are several methods and equations used for determining the Air Pollution Index. A few of these methods and equations are given here.

In the first method, we try to relate the existing pollution levels of various pollutants to their ambient air quality standards, with the standard being assumed as the reference base line for each pollutant and then converting the concentration of pollutants into a percentage of the standard. The Air Pollution Index is then obtained by adding the percentages for the several pollutants considered.

For example, if there are six pollutants, then the equation for calculating the Index will be:

$$I = \frac{1}{6}\sum_{i=1}^{6} A_i; \qquad A_i = \frac{C_i}{S_i} \cdot 100 \tag{20.1}$$

where C_i = Concentration of pollutant i

S_i = Air quality standard for pollutant i

I = Air Pollution Index

In the second method, the average of the sum of the ratios of three major pollutant concentrations to their respective air quality standards is obtained. The average is then multiplied by 100 to get the index.

For example, if the three major pollutants in a city atmosphere are particulate matter, sulphur dioxide and carbon monoxide, then

$$\text{API} = \frac{1}{3}\left[\frac{(\text{PM})}{(S_{\text{PM}})} + \frac{(\text{SO}_2)}{(S_{\text{SO}_2})} + \frac{(\text{CO})}{(S_{\text{CO}})}\right] \times 100 \tag{20.2}$$

where S_{PM}, S_{SO_2}, and S_{CO} represent the ambient air quality standards for particulate matter, sulphur dioxide, and carbon monoxide respectively.

In the third method, air pollution index is calculated from five sub-indices. Each sub-index is obtained by assigning sub-index values for particular ranges of pollutant concentrations.

For example:

CO (ppm)	NO_2 (ppm)	Oxidant (ppm)	COH (units)	Visibility (kms)	Sub-index A_i
0-1	0-.005	0.0-.005	0.0-0.5	12-24	2
1-2	.005-.01	.005-.01	0.5-1.0	8-12	4
2-4	.01-.02	.01-.04	1-2	6-8	8
4-6	.02-.06	.04-.06	2-3	4-6	12
6-8	.06-.10	.06-.09	3-4	2-4	16
8-35	.10-.20	.09-.10	4-5	0-2	20

Note: COH refers to coefficient of haze

The index is calculated using the equation:

$$I = \sum_{i=1}^{5} A_i \tag{20.3}$$

where A_i = sub-index determined as given above

In the fourth method, known as the maximum type, only one pollutant index (the highest one) among the several pollutants is reported.

In the fifth method, only two common pollutants i.e., particulate matter and sulphur dioxide are taken into account while calculating the air pollution index.

In the sixth method, only one major pollutant, e.g., ozone concentration is taken as the basis for reporting the Air Pollution Index.

All the above methods have their own advantages and disadvantages. Research work is now being carried out to develop a scientifically defensible index which takes into consideration the effects of air pollutants on humans, on plants and animals, and on materials, to overcome the defects of indices which are now being arbitrarily designed.

20.8 Rating Scale for Indices

A typical rating scale can be as follows:

Index value	*Remarks*
0-25	Clean air
26-50	Light air pollution
51-75	Moderate air pollution
76-100	Heavy air pollution
> 100	Severe air pollution

The other terminology that are commonly used in rating scales are,

1. Good, acceptable, unsatisfactory, unhealthy
2. Good, satisfactory, unsatisfactory, unhealthful
3. Light, moderate, heavy, alert
4. Good, normal, moderate, heavy, severe

20.9 Display and Dissemination Techniques

The methods most frequently used in the USA to report air pollution indices to the public are newspapers, radio and television. Among these, newspapers are the most common dissemination method, and they give air pollution indices the maximum coverage. Some newspapers give only a narrative description, while others include a bar graph or table with a brief explanatory note of each day's index.

Following are some typical illustrations:

Illustration 1:

Place . . . .
Date . . . .
Air Pollution Index = 30
Rainfall = 00
Mean temperature = 35°C

Illustration 2:

Place . . . .
Date . . . .

The air quality index from 8 a.m. to 9 a.m. today was 15 for the pollutant sulphur dioxide. Index values between 0 and 25 indicate good air quality.

Illustration 3:

Place . . . .
Date . . . .

Air quality at 2 p.m. yesterday was good according to the Department of Health's Computerised air monitoring system.

Illustration 4:

Place . . . .
Date . . . .

The air quality index for yesterday between 2 and 3 p.m. was 20 for the pollutant, photochemical oxidants. Index values between 0 and 25 indicate good air quality. When the index exceeds 100, the air becomes hazardous and persons with lung, heart, and eye problems should restrict their activity.

Illustration 5:

Place . . . .
Date . . . .

Air Pollution Index: 36

The Air Pollution Index is an average degree of air pollution for a 24-hour period ending at 9 a.m. The index compiled at the Public Health Centre, combines measurements and estimates of the levels of three major air pollutants here: carbon monoxide, sulphur dioxide, and particulates (dust). A reading between 0 and 25 is 'good'; between 26 and 100 is 'satisfactory' between 101 and 150 is 'unsatisfactory', and between 151 and 200 is 'severe'.

Illustration 6:

Place . . . .
Date . . . .
Tomorrow's pollution forecast: 60 to 70 (moderate)

Television is another popular medium of reporting air pollution indices daily. Television reports may give only the index value or include visual displays with a short verbal explanation. Of course, radio is limited to verbal reports. Index reporting formats may vary from city to city.

20.10 Conclusions

Air pollution indices are very useful as they indicate day-to-day changes in air quality. Further, air quality levels observed in urban areas are also of interest to the Nation as a whole because they provide an indication of the Nation's progress towards cleaner air. Also, it facilitates comparison of air pollution problems in different cities. Such information is of vital importance to national decision makers with respect to prevention and control of air pollution.

As already mentioned, it is better to have a single, uniform air pollution index for the whole country, which will facilitate comparison of air pollution levels in different cities. In order to do this, the data on which it is based should be of the same high quality in all the cities. This necessitates proper measurement methods and laboratory practices in all the cities. To achieve this, standard procedures must be established for determination of the Air Pollution Index. Hence, the concerned authority (e.g., Pollution Control Boards) should publish a document containing 'Index Monitoring Guidelines'. The document should specify factors such as:

1. Pollutants used in the index
2. Method of index calculation
3. Descriptor categories
4. Averaging time for the pollutants
5. Period of the day in which averaging time should begin
6. Measurement methods to be adopted for each pollutant
7. Height of sampling probe
8. Sampling probe specification (e.g., inlet diameter and flow rate)
9. Procedure for selecting monitoring station sites.

Also, further research should be undertaken on important related topics like public attitudes towards index, index reporting systems, monitoring siting, etc.

In general, indices have some value. At the same time there is a possibility that they may give false impressions. Therefore, it is of utmost importance to note that those who prepare and use indices are aware of its basis for preparation, and the significance and limitations that should be attached to them.

QUESTIONS

1. Define Air Pollution Index. Explain its significance.
2. What are the uses of air pollution indices?
3. Distinguish between short-term indices and long-term indices.
4. What are the parameters generally used for calculating Air Pollution Index?
5. State the criteria for a standardized index.
6. Explain any three methods for determining air pollution index.
7. Write short notes on:
   (a) Air pollution indices
   (b) Rating scale for indices
   (c) Dissemination techniques for air pollution indices
   (d) Index Monitoring Guidelines

21

Air Pollution Survey

21.1 Scope

An air pollution survey is a critical study of a particular geographical area like Bangalore, for the purpose of determining the sources, nature, extent and effects of the air pollution that exists within the area concerned. Air pollution surveys may help in several respects. Among them, are urban and land and use planning, highly localized problems (for example: diseases like asthma and other health problems) and development or improvement of a community's air pollution control programme and research.

An air pollution survey in a town or city must provide answers to the following questions:

1. *Where* do the pollutants come from?
2. *What* are the pollution levels with respect to time and place?
3. *Why* do these levels vary?
4. *Which* adverse environmental effects are produced?
5. *How* to overcome the problem?

21.2 Objectives of Sampling the Atmosphere

1. To identify specific industrial and other sources of pollution
2. To determine the degree of air pollution control required for existing industries
3. To identify and control pollution from vehicular emission
4. To assess health hazards and potential damage to property
5. To determine the background pollution levels for application in industrial zoning, town planning or location of sites for certain types of industries requiring stringent air quality criteria
6. To collect data for formulating and testing air pollution models
7. For purely scientific investigations

21.3 Classification of Sampling Methods

Sampling methods can be broadly classified into two categories. They are:

1. In situ sampling
2. Remote sensing

The in situ sampling may further be sub-classified as: intermittent sampling and continuous sampling.

Intermittent sampling is less expensive and easy to operate. But it requires frequent personal attention and therefore more manpower and ancillary facilities. On the other hand, continuous methods are more expensive and are used only when long-term investigations are undertaken.

The remote sensing method of sampling is a recent development. It is possible to detect and measure the concentrations of several atmospheric pollutants by this technique. The main advantage of this method is that physical access to the location where the measurement is to be made is not needed unlike in the case of the in situ method. Another advantage of remote sensing is the instantaneous space averages of the constituent concentrations may be obtained. However, the remote sensing method is costly and requires sophisticated equipment.

21.4 Preliminary Information Required

Before planning an air pollution survey in a town or city, it is desirable to collect the following data:

1. Sources of pollution
2. Topography
3. Climatology
4. Vegetation
5. Population distribution
6. Exotic pollution, if any, i.e., pollution caused by distant sources
7. Area map

21.5 Location of Sampling Stations

When in situ sampling has to be carried out, sampling stations have to be located at suitable places depending on the main objective of sampling. For example, if it is a general air pollution survey of a city, sampling stations can be located in prominent residential, commercial and industrial areas.

When the objective is to study the effects of air pollutants on human health, the sampling locations should be as close as possible to specific locations where the effects are being studied. For example, population centres, hospitals and schools where a more vulnerable section of the public is present.

When the objective is to study pollution from vehicular exhausts the sampling points should be near traffic signals and traffic lanes.

Similarly, when the objective is to study material damage, the sampling stations should be located near the places of damage. For example, for assessment of damage to vegetation, sampling should be done at foliage level and in case of damage to overhead electrical wires, sampling should be done at the level of the wires.

21.6 Air Pollution Survey Programme for a City

When a general air pollution survey is to be undertaken in a city, the following factors should be taken into consideration:

21.6.1 Duration of Survey Programme

In order to get useful data from which proper conclusions can be drawn, the duration of survey should be spread over a minimum period of two years, preferably three years.

Generally, air pollution will be maximum during the winter months (November-March) and hence observations should be made particularly during this period.

21.6.2 Selection of Air Sampling Stations

To begin with, it is better to limit the survey for collection of samples from six sampling stations. While selecting the stations, various factors like population density, concentration of industries including specific industries if any, and intensity of traffic should be taken into account. After studying such factors, five or six stations should be set up, one or two for each distinctive area of a particular type of activity and shown on the city map. Also the setting up of an additional station a few kilometres outside the city limits in semi-rural surroundings, is of value as a 'control station', which will help in comparing with the pollution levels at the stations within the city. For example, a sampling station in Kengeri satellite town can act as a 'control station' when an air pollution survey has to be done in Bangalore city.

The air sampling stations should not be located near buildings liable to obstruct free flow of air movements *or* just close to the source of emission of pollutants to be estimated.

Another prime consideration is the safety of the equipment to be installed for monitoring the air quality. From this point of view, the sampling station should be located in such a way that it will not be easily accessible to outsiders and entry to which can be controlled and limited only to the workers engaged in the survey. Therefore two or three storeyed buildings with open terraces are convenient for installing the equipment.

21.6.3 Selection of Air Pollution Parameters

In order to select the major pollutants for the purpose of the survey, a preliminary inspection of all the major industrial plants in the area and other human activities constituting the sources of pollution, is desirable. Based on such an inspection survey, it is easy to identify the major air pollutants which are of special significance for the area in which the air pollution survey is intended to be taken up. Generally, in urban areas, the pollutants which are of major importance are sulphur dioxide, nitrogen dioxide, hydrogen sulphide, oxidants and dust. Depending on the type of industries in operation in the specific areas, other pollutants like hydrocarbons, mercaptans, mercury, lead, chlorine, etc., may be included.

21.6.4 Sampling Instruments Required

After sampling stations have been selected as per the guidelines given above, the following sampling instruments should be installed:

1. Dust fall jar (Fig. 21.1).
2. Sulphation rate measuring unit
3. High volume air sampler
4. Multi-gas sampling kit (Fig. 21.2).

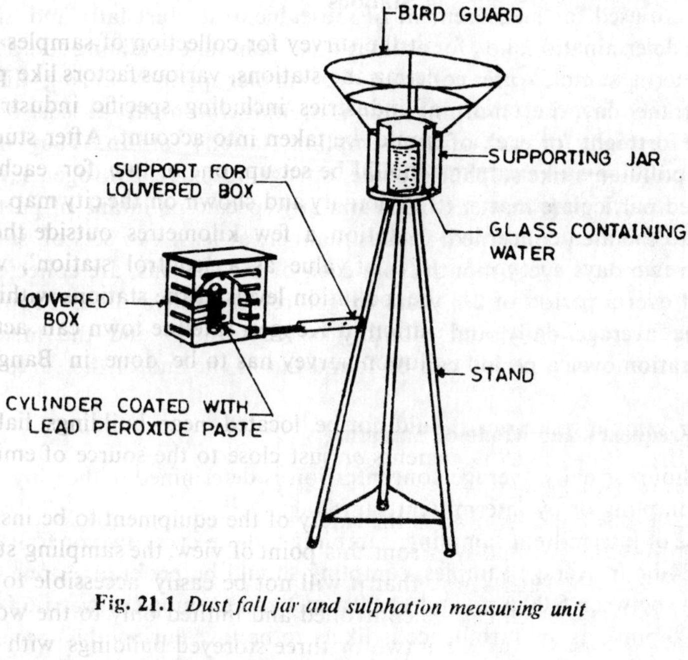

Fig. 21.1 *Dust fall jar and sulphation measuring unit*

The following meteorological instruments should also be installed for measurement of meteorological variables:

1. Wind speed recorder
2. Wind direction recorder
3. Whirling psychrometer.

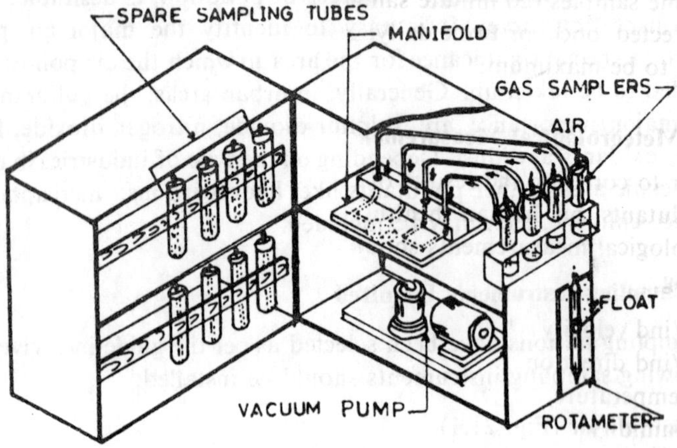

Fig. 21.2 *Internal arrangement of the sampling kit*

21.6.5 Cumulative and Short-term Samples

The cumulative samples are collected over a period of one month. These samples are used for measurement of settleable dust (dust fall) and sulphation rate determination, i.e., for estimation of sulphur containing compounds.

Short-term samples are collected for smaller periods during different times of the day, i.e, morning, noon, evening and night, at least one day during a fortnight for each of the sampling stations set up. They include gaseous pollutants like sulphur dioxide, nitrogen dioxide, hydrogen sulphide, suspended particulate matter (SPM) and any specific pollutant of particular interest to the area concerned. In this manner, at least two sets of samples, taken on two days every month from each of the stations, are collected and analysed over a period of 2-3 years to get representative data. From such data, the average daily and monthly concentrations and the maximum concentration over a period can be found out at the end of the survey.

21.6.6 Frequency and Mode of Sampling

The 24-hour or daily average concentration is determined either by continuous sampling or by intermittent sampling.

In case of intermittent sampling, sampling is done every two or four hours. To get 24-hour average samples, sampling should be preferably done in the morning (between 5.00 a.m. and 11.00 a.m.) and evening (between 6.00 p.m. and 10.00 p.m), as air turbulence is likely to be low during this period. In case maximum discharge occurs during other hours of the day, the sampling time should be suitably changed. In order to get reliable data, it is better to collect 24-hour average samples about 10-12 times per season.

If the object of the survey is to determine the pollutants which irritate the mucous membranes of the eye or the upper respiratory tract, then it is of utmost importance to note the maximum concentration. For this purpose, short-time samples (20 minute samples) are of value. The 20-minute samples are collected once or twice daily, during the period when the air pollution is likely to be maximum.

21.6.7 Meteorological Measurements

In order to correlate the results of the analysis obtained with the effects of air pollutants on human health, materials damage, etc., the following meteorological measurements should be recorded regularly at the time of sampling.

1. Wind velocity
2. Wind direction
3. Temperature
4. Humidity
5. Barometric pressure
6. Subjective sensations of the observer (eye irritation, throat irritation, irritation of the respiratory tract, offensive odour, etc.)

A specimen data sheet for an air pollution study programme is shown in Tables 21.1 and 21.2.

21.7 Continuance of Air Quality Studies

Air pollution studies during the first phase of the programme as described earlier, is more or less a routine air monitoring programme, which will be of help in getting an idea about the air quality of the city concerned. It will also help to pin-point the major contributors to pollution. Now a detailed study can be undertaken with respect to these particular contributors to pollution. This will form the second phase of the programme. This will help in assessing the effectiveness of the air pollution control measures which might have been adopted. Of course, the regular air monitoring programme similar to the first phase, should be maintained during the second phase also. This will help in determining any future increase in industrial-emission load added to the city atmosphere during subsequent years.

During the second phase of the programme, air pollution studies can also be undertaken to study the typical sources of pollution in the city, e.g., auto-exhaust emissions. In addition, medical investigations can also be undertaken in different parts of the city, to study the physiological effects of air pollution on the local population. Such studies can be carried out with the cooperation of the local health departments, hospitals, research institutes and voluntary organisations. It should be a systematic study spread over a considerably long period.

Table 21.1 *Air Pollution Study Programme*

DATA SHEET

Location of the sampling station	Zone: Residential/commercial/Industrial			Surveyor:		Analyst:		Month:	
1. Dust fall jar	Exposure period of the jar			Open surface area of the jar in cm^2	Water soluble (mg)	Water insoluble (mg)	Total dust fall (mg)	$Tons/km^2$, 30 days	Remarks
	From	To	Total number of days						
2. Sulphation rate	Exposure period of the cylinder			Weight of $BaSO_4$ (mgms)		Sulphation rate mgms SO_3/100 cm^2/day			Remarks
	From	To	Total number of days						
3. Suspended particulates	Date of sampling			Weight of deposit (mg)	Flow rate (m^3/min)	Total volume of air sampled (m^3)		$\mu g/m^3$ of particulates	Remarks
Filter No.	From	To	Total hours run						
1									
2									
3									
4									
5									
6									

Table 21.2 *Air Pollution Study Programme*

DATA SHEET

Date of sampling:

Gaseous pollutant	Period of the day	Time			Volume of air sampled (litres)	Total μg in sample	Concentration of pollutant		Wind		Temperature			Weather conditions (cloudy, rainfall, etc.)	Remarks
		From	To	No. of hours			μg/lit	PPM	Direction	Velocity	Dry	Wet	% humidity		
1. Sulphur dioxide	Morning														
	Afternoon														
	Evening														
	Night														
2. Nitrogen dioxide	Morning														
	Afternoon														
	Evening														
	Night														
3. Hydrogen sulphide	Morning														
	Afternoon														
	Evening														
	Night														
4. Any other	Morning														
	Afternoon														
	Evening														
	Night														

21.8 Conclusion

Continuous air monitoring is an essential part of any air pollution control programme. The data collected indicates the ambient air quality and gives an idea regarding the degree of success of the air pollution control programme initiated in a city.

In our country, a network of air quality monitoring stations have been established by the NEERI (National Environmental Engineering Research Institute) in cities where its zonal laboratories are based. The ten cities covered under this programme are Ahmedabad, Bombay, Calcutta, Cochin, Delhi, Hyderabad, Jaipur, Kanpur, Madras and Nagpur. The air quality data in these cities are being collected on a long-term basis. These data are expected to provide valuable insights into the environmental impact of industrialisation and urbanisation, and the degree of abatment and the control measures required.

QUESTIONS

1. What are the objectives of sampling the atmosphere?
2. Explain the following sampling methods for air pollutants
   (a) In situ sampling
   (b) Remote sensing
3. How are sampling stations located for an air pollution survey?
4. Explain how a short-term air pollution survey is conducted in a city like Bangalore.
5. Write short notes on:
   (a) Selection of air sampling stations
   (b) Selection of air pollution parameters
   (c) Instrumentation in air sampling
6. What are the meteorological measurements required to correlate the results obtained in an air pollution survey?
7. Explain the importance of air quality studies.
8. Explain the importance of establishing a 'control station' during an air pollution survey.

22

Air Pollution Legislation and Regulations

22.1 Introduction

The Air (Prevention and Control of Pollution) Act, 1981 of the Government of India came into force from May 16, 1981. The Act is applicable throughout the country unlike the earlier Central Water Pollution Control Act, which is applicable only in states, which have adopted it. The Act provides for an integrated approach for tackling environmental problems relating to the pollution.

For administering the Act, the State Governments have been authorised to declare, after consultation with the State Boards, any area or areas within the state as air pollution control area or areas. These areas may be extended, reduced or even merged together in course of time. Pollution by ships and aircraft is not covered by the provisions of the Act.

The Act includes the following:

1. Definitions of terms used
2. Constitution of the Central Board and State Boards
3. Functions of Boards
4. Powers of Board
   - (a) Power to declare Air Pollution Control Areas
   - (b) Power to give instructions for ensuring standards for emission from automobiles
   - (c) Restriction on use of certain industrial plants
   - (d) Power of entry and inspection
   - (e) Power to take samples of air or emission
   - (f) Establishment of State Air Laboratory
5. Penalties and procedure

In States where Water Pollution Control Boards have been constituted, they will also look after the administration of the Air Pollution Control Act. In case of States which have not adopted the Water Pollution Control Act, only Air Pollution Control Boards will be constituted. The boards will have a Chairman, Member-Secretary and some official and non-official members.

Two important functions of the Central and State Boards are laying down standards for emission of air pollutants into the atmosphere from industrial plants and the establishment of laboratories.

State Board's officers have been authorised to enter and inspect any place to ensure implementation of the Act. They can collect samples of air or emission for analysis from any chimney, flue, duct or any other outlet in the prescribed manner. Violation of the various provisions of the Act has been made punishable with imprisonment and a fine.

22.2 Constitution of the Board

A State Board constituted under the Act consists of the following 17 members—all nominated by the State Government.

1. A full time or part time Chairman
2. Five official members to represent the Government
3. Five persons representing the local authority
4. Three non-official members
5. Two persons representing companies/corporations owned or managed by the State Government
6. A full time Member-Secretary

22.3 Functions of the Central Board

Main functions of the Central Board are as follows:

1. To improve the quality of air and to prevent, control or abate air pollution in the country
2. To advise the Central Government on any matter concerning the improvement of the quality of air and the prevention, control or abatement of air pollution
3. To plan and execute a nationwide programme for the prevention, control or abatement of air pollution and arrange training programmes
4. To coordinate the activities of the State Boards, provide technical assistance, carry out and sponsor investigations and research relating to the prevention and control of air pollution
5. Organise through mass media, a comprehensive programme regarding prevention and control of air pollution
6. Collect, compile and publish technical and statistical data on air pollution and the measures devised for its prevention and control and prepare manuals, codes or guides
7. Lay down standards for the quality of air
8. Establish or recognise laboratories

22.4 Functions of the State Boards

Following are the main functions of the State Boards:

1. To plan a comprehensive programme for prevention and control of air pollution and implement the same
2. Advise the State Government on any matter concerning air pollution including siting of any industry which is likely to cause air pollution.
3. Collect and disseminate information, to collaborate with the Central Board in organising training programmes and mass education programmes
4. Inspect, any control equipment, industrial plants etc., and give by order such directions to such persons as may be necessary to take steps for prevention and control of air pollution
5. Inspect air pollution control areas, assess the quality of air therein and take appropriate steps for prevention and control of air pollution in such areas
6. Lay down, in consultation, with the Central Board, standards for emissions of air pollutants into the atmosphere from industrial plants and automobiles
7. To establish or recognise laboratories

22.5 Classification of Pollution Sources

Section 19–26 of the Act can be broadly stated to be operative sections of the Act. On perusal of these sections, the sources of air pollution can be divided into three categories.

1. Major air polluting industries (20 industries listed in the Schedule attached to the Act)
2. Automobiles excluding ships or aircraft
3. Other sources including domestic sources

It is necessary to bear in mind that the approach to pollution control, problems in implementation and the success of enforcement will vary significantly for these three sources. Keeping this in view, different approaches have been proposed for these three classes of sources as detailed below:

22.5.1 Scheduled Industries

These are supposed to be the major air polluting industries which require comprehensive and individual pollution control measures and these measures are practicable. Pollution control in such industries is mainly through the grant of Consent and enforcing the requirements and conditions laid down in that consent. As specified under Sec. 21 of the Air Act, such a consent would clearly specify any equipment or appliances to be installed and operated or equipment or appliances prohibited from being operated, and thus pollution control is through a proper choice of process technology together with pollution control equipment, if necessary. Section 21(5)(iii) also ensures that the industry will have to maintain such equipment in a proper running condition. Section 21(5)(iv) provides for requirement in respect of stack heights, dimensions, etc., while Sec. 21(5)(v) permits the Board to lay down 'any other conditions'. Since Section 21, constraint with grant of consent does not specifically talk of permissible limits on emission and yet any consent would essentially be incomplete without such limits being prescribed, it is obvious that such limits will have to be covered under 'any other conditions'.

22.5.2 Automobiles

Pollution due to automobiles is primarily intended to be controlled through changes in the design and installation of suitable appliances right at the time of manufacture of the vehicle. Under Sec. 20 of the Air Act, the State Government, in consultation with the State Board, has been empowered to give necessary instructions to the Regional Transport Authority. Such instructions are binding on the State Registration Authority, thus ensuring that no vehicle which does not meet such requirements shall be on the road.

In relatively more polluted and high traffice intensity regions, pollution due to automobiles can be controlled by restricting the use of certain types of fuels as empowered under Sec. 19(3). Petrol mixed with lead could be banned or in extreme cases, a ban on petrol-driven or diesel-driven vehicles can be imposed within certain areas.

The Act does not envisage compliance of emission standards by an individual automobile or granting of a consent to them or a prosecution under this Act in case a particular automobile is not meeting the standards except it's being generally covered, under Sec. 22. where essentially all types of emissions are covered.

22.5.3 Domestic and other Sources

Control of air pollution from domestic, commercial and other sources including very small industries, could possibly be by either prohibiting the burning of certain type of fuel as provided under Sec. 19(3) or by prohibiting the use of unapproved appliances provided under Sec. 19(4). Insisting on consent of emission standards from such sources is bound to be impractical. At best, the Board could ban the use of hard coke in residential areas or could insist on the use of particular type of chulas or ovens to ensure control of smoke from domestic fires.

22.6 Significance of Air Pollution Control Areas

Keeping to the philosophy of controlling Air Pollution from three major sources as indicated in the earlier paragraphs, Sec. 19 of the Act provides for the notification of air pollution control areas by the State Government in consultation with the respective State Boards.

Air pollution control areas obviously means an area where air pollution is considered to be critical requiring immediate control to be effected by the State Boards. In the air pollution control area, all types of sources including domestic, small industries and others will be subject to control by State Boards through control on the fuel used, control on the appliances used, control on processes employed for industrial production through issue of consent, and where necessary, installation of pollution control or emission treatment devices and also where necessary monitoring of the emissions concerned and prosecution where the emissions do not meet the standards laid down by the Board under Sec. 17 and thus are in contravention of Sec. 22 of the Act. The scheduled industries may be put to more severe constraints under Sec. 21(5) of the Act if they are located in the air pollution control area, than what would have been required by Sec. 17 which provides standards applicable over the entire State and not just in air pollution control areas.

Against this stringent control in the air pollution control areas on all types of sources, in the remaining area not declared as an air poullution control area the Act envisages control over only two major sources of air pollution namely scheduled industries under Sec. 17(1)(g) and Sec. 22 and automobiles under Sec. 20. In these cases, there would be no consent, no special conditions and probably not too much of monitoring.

Even though the Act specifies only scheduled industries and is silent in indicating the size or capacity, it is obvious that some limiting capacity for these highly polluting industries needs to be considered so that the pollution control measures do not become counter productive irrespective of their location. Thus, small boilers, mini cement plants, or small chemical industries cannot be treated on par with big thermal plants or major cement plants. The Policy Planning Committee set up by the Central Board has recommended an approach and classified the capacities of industries for the consideration of the Government and the various State Boards. This takes into account following factors:

1. Polluting potentiality of the industry
2. Age of the industry
3. Location of the industry (with respect to the classification of area for ambient air quality stipulations)

22.7 Ambient Air Quality Standards

Section 16(h) empowers the Central Board to lay down standards for the quality of air in ambient air for the entire country. In exercise of these powers, the Central Board has notified the ambient air quality in its notification dated 16.12.82. For this purpose, the entire country is classified into three main categories as indicated below:

A. Industrial and mixed use areas
B. Residential and rural areas
C. Sensitive areas

To begin with, standards for four major parameters like SPM, SO_2, CO, NO_x only are prescribed. As can be seen from the standards laid down, the air quality is stringent as we go from category A to C. The standards laid down envisage a monitoring period of eight hours.

22.8 Emission Standards

Section 17(1)(g) empowers the State Board to lay down, in consultation with the Central Board and having regard to the standards for the quality of air laid down by the Central Board, standards for emission of air pollutants into the atmosphere from industries, plants and automobiles or for the discharge of any air pollutant into the atmosphere from any other source, whatso-ever not being a ship or an aircraft. This section also provides for laying down different standards for emission for different industrial plants considering the quantity and composition of emissions of air pollutant into the atmosphere from such industrial plants.

Thus, it is observed that there are two types of limits on the emissions from major polluting scheduled industries:

i Standards laid down under Sec. 17(1)(g) and applicable over the entire state which the industry concerned will have to comply with wherever it might be located.
ii Specific emission limits that may be prescribed by the pollution control Boards in the consent conditions under Sec. 21(5). These limits should generally be more stringent than the standards set under Sec. 17(1)(g).

The Central and State Boards are evolving industry specific emission standards duly considering the prevailing ISI standards, and the guidelines given by the Policy Planning Committee set up by the Central Board. Emission standards stipulated should not only be easily enforceable but also be achievable with the present stage of technology and the equipment available in the country. The standards could be either based on total emission per unit of production or per unit volume of air emitted. As air pollution control is being adopted for the first time, the emission limits that are being laid down need to be revised progressively once in 3-5 years. Some of the State Boards have already laid down emission standards in respect of few specific industries.

22.9 Penal Provisions of the Act

Failure to comply with the specific provisions of Sec. 21(5), Sec. 22, Sec. 23(1) and any other provisions of the Act and directions issued under the Air Act, attract the penal provisions enumerated under Secs. 37–40 of the Act. The Act provides for stringent punishment including penalty. A Metropolitan Magistrate's Court or a First Class Judicial Magistrate's Court and higher courts are the courts where offences under this Act can be tried, that too on a written complaint made by the Pollution Control Board. Implementation of the Air Act calls for greater cooperation between the polluter and the implementing authority.

22.10 Industries Specified in the Schedule

A list of industries specified in the schedule under Air Act, 1981, is given below:

1. Asbestos and asbestos products industries
2. Cement and cement products industries
3. Ceramic and ceramic products industries
4. Chemical and allied industries
5. Coal and lignite-based chemical industries
6. Engineering industries
7. Ferrous metallurgical industries
8. Fertiliser industries
9. Foundries
10. Food and agricultural products industries
11. Mining industry
12. Non-ferrous metallurgical industries
13. Ores/mineral processing industries including beneficiation, pelletization, etc.
14. Power (coal, petroleum and their products) generating plants and boiler plants
15. Paper and pulp (including paper products) industries
16. Textile processing industry (made wholly or in part of cotton)
17. Petroleum refineries
18. Petroleum products and petrochemical industries
19. Plants for recovery from and disposal of wastes
20. Incinerators

22.11 Air Pollution Legislation in other Countries

In many parts of the world like the United States, United Kingdom, Soviet

Union and France, air pollution control is effected through legislation. All these legislations contain clauses relating to

(a) Control of sources of pollution by specifying the types of industrial and other processes which should operate under supervision of control authorities
(b) Air emission and air quality standards
(c) Regulation regarding the types of fuels to be used in a particular area
(d) Penalty clauses for violation of the Act
(e) Proper town planning and zoning of industrial sites for the purposes of reducing air pollution

22.11.1 United Kingdom

In relation to air pollution by discharges from chemical and other manufacturing processes, there was a public outcry in the middle of the 19th century. Consequently, the First Alkali Act was passed in 1863. Later, the Act was extend on several occasions to cover a number of processes, including many that are not related to the manufacture of alkalis. The Public Health Act was passed in 1936. After the disastrous smog in December 1952, which caused the death of about 4000 people in the London area, British Parliament passed the Clean Air Act in 1956.

The Public Health Act, 1936, laid down regulations regarding the type of fuel to be used and smoke emission limits for the industry. In the Clean Air Act (1956) also, specifications regarding smoke emission standards have been given. Further, according to the Act, certain areas can be declared as 'smoke control areas' in which only approved appliances and fuels can be used. The Act also regulates emission of dust from industries and the height of chimneys.

The Motor Vehicles (Construction and Use) Regulations, 1973, require every motor vehicle to be constructed, maintained and operated in such a way as to prevent avoidable emission of smoke, or visible vapour. Failure to comply with these regulations is an offence under the Road Traffic Acts.

22.11.2 United States of America

In the USA, unlike the United Kingdom, there is no nation-wide or general legislation directed towards the control of air pollution. Air pollution control is exercised under State or local powers relating to the control of nuisances and conditions affecting public health and welfare. State laws usually give authority to counties or cities. For example California has an Act providing for creation of 'air pollution control districts' within its various cities. Almost every city in USA now has a smoke or air pollution regulation of some kind. In some cases, there are interstate compacts to improve controls, and there is an international compact between the USA and Canada to cover adjacent areas in the two countries.

American Ordinances are often more severe than the general law for the control of smoke pollution in the United Kingdom, in that, they restrict the type and quality of coals that may be used in hand-fired industrial furnaces and in domestic heating appliances. For example, according to the St. Louis Ordinance, all plans for new fuel-burning plants and plans for the repair or reconstruction of existing plants are subject to the approval of the Smoke Commissioner. In the event of contravention of the regulations, the Smoke Commissioner has the power to shut down the plant.

The Federal air pollution control acts of the USA are listed in Table 22.1.

Table 22.1 *The Federal Air Pollution Control Acts (USA)*

1955	Air Pollution Control Act	First federal legislation
1963	Clean Air Act	Federal financial assistance but problem left to the States
1965	Amended Clean Air Act	First federal emission standards for cars
1967	Air Quality Act	First requirement for ambient standards, but these are to be set by the States
1970	Amended Air Quality Act	First national ambient standards and these are to be set up by federal action; first national emission standards for certain pollutants; continued state implementation.

Interest and action, backed by strong public opinion, have developed rapidly within the last twenty years in the USA, with obvious reduction in air pollution in many areas, especially pollution by smoke.

22.11.3 USSR

In the USSR, an order issued in 1949 stipulates that no electric power station may be constructed without the simultaneous installation of equipment for dust and ash absorbtion. Factories processing non-ferrous metals must likewise be equipped with devices for the absorbtion of dusts and gases containing compounds of sulphur, arsenic and fluorine. Coal-tar distilleries are required to install the apparatus necessary for the absorbtion of H_2S and other sulphurous gases. In iron and steel works, provision must be made for the filtration of gases from blast furnaces and for the utilisation of these gases as fuel; and factories using solvents must be equipped with a recovery plant.

The organisational structure of air pollution control in the USSR was approved in 1950. Among the authorities responsible for supervision, are the services of the State Sanitary Inspectorate in case of industrial undertakings subordinate to a Republic and the local stations of the Sanitary and Epidemiological Department, in case of undertakings subordinate to a city.

Newly built industrial undertakings, or factories which have undergone transformation, cannot begin operations, until the necessary permit is issued by the authorities regarding the efficiency of air pollution control devices installed.

Also, the order of 1950, makes the approval of building plans subject to the submission of particulars of the production methods envisaged, raw materials to be used, quantity and composition of liquid, solid and gaseous emissions, height of chimney stacks, etc. While examining the projects, the supervisory authorities consider protected areas, meteorological conditions and topography, the efficacy of control apparatus and possible changes in the quantity and quality of pollution. The sanitary authorities give their final sanction only after having ascertained that all their observations have been taken into account in the drawing up of plans. Regular sanitary inspection is carried out by the local stations of the Sanitary and Epidemiological Department.

22.11.4 France

Included in regulations governing industrial firms in France, is a clause prohibiting the discharge into the air of dense smoke, vapour, dust and toxic and corrosive gases likely to be detrimental to public health, agricultural production, public monuments and scenic beauty. Chimneys must be of sufficient height to ensure satisfactory dilution and dispersion of gas in the air.

22.11.5 Netherlands

Under the Air Pollution Act of the Netherlands, a licence shall be issued subject to the conditions necessary for the prevention and limitation of air pollution. The conditions attaching to a licence may include:

(a) the obligation to apply methods for the prevention and limitation of air pollution as specified therein
(b) upper permissible limits to the emission of pollutants from the establishment or to the concentration of pollutants emitted by the establishment into the environment
(c) the obligation to measure the emission of pollutants and to measure the concentration of pollutants in the surrounding area
(d) the obligation to make the results of the measurements referred to under (c) available to the government bodies designated by the condition

Regulations may be made by a General Administrative Order prescribing the way in which the obligations referred to in sub-section (c) and (d) must be discharged, and regarding the reimbursement of those costs incurred by the holder of the licence in discharging the obligations which he cannot reasonably be expected to bear, whether wholly or in part.

The conditions attaching to a licence may contain the obligation to satisfy further requirements set by government bodies designated in the condition. When such a requirement is set, the time at which the obligation is to commence with regard to that requirement shall be stated.

A condition may also be attached to a licence under which the holder of the licence is obliged to furnish all the persons employed in his establishment with instructions in writing intended to preclude any actions on their part which might result in the establishment being set up or run in a manner that is inconsistent with the licence, or in a provision attaching to the licence being contravened.

22.12 The Environment (Protection) Act, 1986

The salient features of the Act are that it extends to the whole of India and the Central Government shall have the power to take all such measures as it deems necessary or expedient for the purpose of protecting and improving the quality of the environment and preventing, controlling and abating environmental pollution.

Under the Act, the Central Government may frame rules with respect to the following:

1. The standards of quality of air, water or soil for various areas and purposes
2. The maximum allowable limits of concentration of various environmental pollutants (including noise) for different areas
3. The procedures and safeguards for the handling of hazardous substances
4. The prohibition and restrictions on the handling of hazardous substances in different areas
5. The prohibition and restrictions on the location of industries and the carrying on of processes and operations in different areas
6. The procedures and safeguards for the prevention of accidents which may cause environmental pollution and for providing for remedial measures for such accidents

Under section 9 of the Act, where the discharge of any environmental pollutant in excess of the prescribed standards occurs or is apprehended to occur due to any accident or other unforeseen act or event, the person responsible for such discharge and the person-in-charge of the place at which such discharge occurs or is apprehended to occur shall be bound to prevent or mitigate the environmental pollution caused as a result of such discharge and shall also forthwith intimate the fact of such occurrence or apprehension of such occurrence, and be bound, if called upon, to render all assistance to such authorities or agencies, as may be prescribed.

Under Section 10 of the Act, any person empowered by the Central Government in this behalf shall have a right to enter, at all reasonable times with such assistance as he considers necessary, any place for the purpose of performing any of the functions of the Central Government entrusted to him.

Every person carrying on any industry, operation or process of handling any hazardous substance shall be bound to render all assistance to the person empowered by the Central Government under sub-section (i) for carrying out the functions under that sub-section and if he fails to do so without any reasonable cause or excuse, he shall be guilty of an offence under this Act.

Section 11 of the Act deals with the power to take samples and the procedure to be followed in connection therewith.

Under Section 15 of the Act, whoever fails to comply with the Act, is punishable with imprisonment for a term which may extend to five years or with a fine which may extend to one lakh rupees, or with both, and in case the failure continues, with additional fine which may extend to five thousand rupees for every day during which such failure continues after the conviction for the first such failure.

If the failure continues beyond a period of one year after the date of conviction, the offender shall be punishable with imprisonment for a term which may extend to seven years.

Under Section 19, no court shall take cognizance of any offence under this Act except on a complaint made by:

(a) The Central Government or any authority or officer authorised in this behalf by the Government, or
(b) Any person who has given notice of not less than 60 days, in the manner prescribed, of the alleged offence and of his intention to make a complaint, to the Central Government or the authority or officer authorised as aforesaid

Under Section 22, no Civil Court shall have jurisdiction to entertain any suit or proceeding in respect of anything done, action taken or order or direction issued by the Central Government or any other authority or officer in pursuance of any power conferred by or in relation to its or his functions under this Act.

22.13 Conclusions

Though there have been considerable advances in dealing with problems of air pollution, with many legislative measures adopted in some countries, particularly in the United Kingdom and the USA, excessive and often unnecessary pollution is prevalent in many industrial areas. The legislative measures have not always been effective, partly because they lacked precision and legal action has been beset with difficulties, and partly because there has been insufficient public insistence of the mitigation of pollution. In the present state of knowledge, complete elimination of pollution is not practicable, but there could be much greater improvement than has so far been achieved. Such improvement will only be achieved by better legislation more effectively administered, increasing knowledge based on intensified research and development work, and a more determined public opinion. As new industrial processes are developed and conditions change, modification in legislation and methods of enforcement will be required from time to time; but it must be recognised, that no one set of preventive measures will meet all needs everywhere owing to differences in the conditions in various countries and localities.

QUESTIONS

1. What are the main features of the Air (Prevention and Control of Pollution) Act, 1981 of the Government of India?
2. Compare the functions of Central Pollution Control Board with the functions of State Pollution Control Boards in the area of Air Pollution Control.
3. What are the industries specified in the schedule under Air Act, 1981?
4. What are the main approaches in the UK and USA in air pollution control through legislation?
5. Explain briefly the salient features of the Air Pollution Control Acts in the following countries.
   (a) USSR
   (b) The Netherlands
6. Write short notes on:
   (a) Air Act, 1981 (India)
   (b) Clean Air Act, 1956 (UK)
   (c) Air Pollution Control Areas
   (d) The Environment (Protection) Act, 1986.

23

Air Pollution in Indian Cities

23.1 Introduction

An attempt has been made in this Chapter to depict the status of air pollution in some cities of India.

A national air monitoring programme has been undertaken by the National Environmental Engineering Research Institute (NEERI), in accordance with the World Health Organisation's suggestion to assess air quality within the framework of selected parameters. Madras, Bombay, Delhi, Hyderabad, Calcutta, Kanpur, Ahmedabad, Jaipur and Cochin are the centres chosen for the purpose.

WHO chose India in 1978 as the nodal point in South-East Asia for air monitoring. The main purpose of the programme is to alert the public and authorities about any abnormal increase in pollution levels and to ultimately enthuse local agencies to take the programme into their hands.

Four parameters are chosen for monitoring pollution levels. The programme covers the determination of the quantum of suspended particles and that of sulphur dioxide, an assessment of the sulphation rate and of the dust fall. The programme includes collection of meteorological information for data interpretation.

Increasing evidence the world over seems to prove that certain concentrations of particulate matter, sulphur dioxide, carbon monoxide, photochemicals, oxidants and hydrocarbons produce serious carcinogenic, mutagenic and other effects.

In order to limit the deterioration in air quality and improve it further, scientists and technologists have been formulating control measures and ascertaining the magnitude of air pollution from time to time.

This necessitates the establishment of a continuous air quality monitoring network to obtain concurrent and comparable data on both gas and dust levels in the atmosphere.

23.2 National Air Quality Monitoring

The air quality monitoring programme has been initiated by NEERI on a regular basis since 1978. Table 23.1 gives a summary of the air quality with respect to certain common air pollutants in nine cities in India.

Table 23.1 *Annual Summary of Air Quality for 1981*

City	SPM	SO_2	SR (sulphation rate) mg SO_3 ... / 100 cm^2/day	DF (dust fall) MT/km^2/month
Ahmedabad	217*	31*	0.10**	91**
Bombay	184	38	0.45	11
Calcutta	522	90a	0.45*	34*
Delhi	433	31*	0.20*	31*
Hyderabad	143**	30**	—	—
Jaipur	308	6	0.18	19
Kanpur	432	13	0.19	34
Madras	155	8	0.33	—
Nagpur	199	12	0.22	34

*6-7 months average
**4-5 months average
a: scanty data

It can be seen from Table 23.1, that air pollution in many metropolitan cities in India, far exceeds the 150 mg per metre cube standard (for SPM) laid down by the WHO.

Wind roses, geographical features, and meteorology indicate the following observations in respect of these cities.

Bombay: The presence of active pollutants and discharges in the area can result in high atmospheric corrosivity.

Calcutta: The city can be classified as dusty.

Delhi: The commercial area is found to be more polluted as compared with other areas. High SPM values are recorded during the years.

Hyderabad: Air borne dust problems are indicated.

Jaipur: Natural dust due to frequent desert storms is the major source of high concentration of suspended particulate matter.

Kanpur: Air pollution problems can arise due to different wind patterns.

Madras: Air pollution problems can arise at active centres.

Nagpur: Ambient air quality is satisfactory.

23.3 National Scene

A conference on Air Pollution held at Bombay in 1982 revealed some interesting facts:

1. It was found that the lung capacity of children in the age group of 8-10 years was reduced by 20% in Nagpur due to pollution from manganese and power plants.

2. Effective pollution control measures can bring about metamorphic change. Chembur (in Bombay), which was once considered as 'gas chamber'

now has clean air. Emission of sulphur dioxide has been drastically reduced from 210 tons at Chembur due to measures taken such as (1) use of low sulphur fuel (2) dust control and (3) use of new technology for nitric acid plants.

Several studies have shown that the once smooth surface of the priceless Taj Mahal is already suffering from 'stone cancer' caused by corrosive gases in the polluted atmosphere of Agra. A dense green belt of suitable tree species should be created around the Taj Mahal to absorb atmospheric pollution as the value of trees as air scrubbers has been demonstrated by various scientific studies.

3. In a palletisation plant at Mandovi in Goa 99% of air pollution has been reduced through pollution control measures. This plant was shifted from the harbour area to the backwaters at a cost of Rs. 40 lakh. Scrubbers worth Rs. 3 crore and three plantations at a cost of Rs. 30 lakh have resulted in a unique blending of nature with the plant. Only less than half kg of particulates per ton are processed now.

4. Two of the lacunae in the Act were (1) that it did not provide for coordination between air quality and water pollution and urban planning and (2) incentives like tax rebate and financial credit were not provided.

5. Forestation is very valuable in polluted areas. Interaction between air and plant surface results in the removal of pollutants. During the day when photosynthesis exceeds respiration the oxygen supply in the air is replenished and odours are also removed.

23.4 Delhi Scene

According to a recent survey, suspended particles of waste matter in the urban environment in this country are two to five times more than in the advanced West. India with far fewer automobiles than the US has the same level of urban pollution because cars here (which are mostly old) produce four times more carbon monoxide than those in the USA. It is estimated that in Delhi, the incidence of chronic bronchitis is six to fourteen times more than the average elsewhere in India because of the thermal plants.

23.5 Hyderabad Scene

NEERI has briefed the Hyderabad Urban Development Authority on an air quality survey in Hyderabad to determine the level of pollution to define the areas of pollution according to the levels, and finally to develop mathematical models to predict pollution levels, identify the pollution sources, and suggest control measures.

NEERI proposes to prepare air quality maps which will help decide the proper location of industries. NEERI also proposes to take up an air quality monitoring programme in Vishakhapatnam where industrialisation is apace.

Environment scientists suggest that air filters should be used to reduce the vehicular emission of carbon monoxide and carbon dioxide. They also suggest that autocatalytic convertors should be used, as in the US to effectively reduce carbon dioxide and carbon monoxide emission.

At coal based thermal plants and cement factories, electrostatic precipitators should be installed to absorb dust particles. Similarly, for chemical industries, bag filters and wet scrubbers should be used to dilute the gases.

NEERI has fabricated sampling gadgets adaptable to local conditions. Sulphur dioxide and samples of suspended particles are collected every tenth day—a sampling frequency of three times a month.

About ten years ago, the air quality in Hyderabad was good in the light of US standards. But industrialisation in some pockets of the city and the increase in vehicular traffic have contributed to the deterioration of air quality over the city.

23.6 Bangalore City: A Case Study

The influence of climatological factors like wind speed, wind direction, temperature and precipitation on pollution control is significantly large.

The wind speed plays an important role in dispersion of pollutants. During the last few years, many multistoreyed buildings have been constructed in the city of Bangalore. These tall buildings obstruct wind movement resulting in winds with low speed. This decreases the dispersion of pollutants. Moreover, there may be dangerous consequences due to the formation of pockets of high pollutants in the valleys (which, are many in Bangalore). If proper control measures are not taken in future, Bangalore city may turn out to be another Los Angeles.

Another interesting factor is the wind direction. A fixed pattern of wind direction has been observed in Bangalore which shows that the wind is in the north-east direction during January-April and October-December, whereas it blows in the reverse direction, that is south-west, during May-August. This gives an indication that future industrial growth of the city should not be permitted in the above two directional zones. Otherwise it is possible that the pollution emitting industries established in these zones would cause major atmospheric pollution over the city. Hence, the city authorities should be careful in forming industrial layouts and controlling construction of high-rise buildings.

Taking into consideration the prevailing low temperature and weaker winds during the early hours of morning and night in winter, the possibility of frequency of formation of the inversion layer is likely to be high. This would result in the accumulation of pollution during nights or early hours of the morning. This observation will have an important bearing on the location of pollution creating industries that may come up in the future.

Precipitation also plays a role in cleaning the atmosphere by rainout and washout processes. Rainout occurs when substances participate in the formation of rain droplets and subsequently fall to the ground during precipitation periods. For example, particulate substances can act as condensation nuclei for droplet formation, and then fall from the atmosphere when the droplets grow to a size sufficient to cause them to fall to the ground. Washout occurs when substances are collected by falling raindrops. Gases with relatively high affinities for water (i.e., sulphur dioxide) can be partially removed from the atmosphere by washout during precipitation periods. Precipitation, while cleaning the atmosphere, creates water pollution which is again a problem. As far as suspended particulate matter is concerned precipitation effectively reduces its concentration in air.

In the city, the pollution caused by automobiles has exceeded the limitations specified by the Central Board. It is more harmful because of the lead content. The suspended particulate matter and lead present in the air over commercial areas are solely due to the large number of automobiles.

A control in the number of petrol driven vehicles is of utmost necessity. After minimising the vehicles depending on petrol, the vehicles driven using diesel can be slowly reduced; since they let out a lot of smoke, which is a nuisance. The alternative arrangements could be as follows:

1. Introduce electric trains inside Bangalore city in order to provide economic and convenient transportation to the public.

2. The Government can also give sufficient incentives and encouragement for research and development activities in the field of solar powered engines.

3. Lastly, peak-hour congestion and its consequent high pollution levels might be relieved by changing regular working hours, thereby spreading work trips more evenly. Overtime work staggering involves making small systematic shifts in the work hours of employees so that currently underutilised travel times are more adequately used. Conceivably, this could result in decreased peak hour congestion, greater average speeds and reduced air pollution.

As it is impossible to exercise control over the air contaminants once they are emitted into the atmosphere, it would be necessary to identify the sources of pollution, their potential contribution to pollution in the atmosphere, and also to study the existing control equipment for their performance efficiency and adopt them to suit the particular situation. This involves the study of assessing the emissions from different industries with a view to ascertain the suitability of pollution control equipment for industries.

A regular programme of air quality monitoring of Bangalore is done by the Karnataka State Pollution Control Board. Now the Board has taken four parameters viz., suspended particulate matter, sulphur dioxide, oxides of nitrogen and inorganic lead for regular monitoring. It is also necessary to include other parameters like hydrocarbons, carbon monoxide, etc. This is important because, Bangalore has developed into a large industrial city with different types of industries already established and with plans to establish

Table 23.2 *Suspended Particulate Matter Levels in Bangalore*

Unit: $\mu g/m^3$

1985	Jayanagar			Ananda Rao Circle			AMCO Batteries			KS and DL			GIBU		
Months	TS	Max	AM	TS	Max	AM	TS	Max	AM	TS	Max	AM	TS	Max	AM
April	4	141.23	84.6	14	533	360.36	2	102	86.67	2	498	260.1	3	127	99.50
May	5	74.26	56.78	17	942	451.58	4	351	149	4	857.0	311.5	4	163.5	100.78
June	4	147.3	58.0	19	1017	575.4	4	318.8	229.7	2	371.74	271.0	4	86.0	59.48
July	4	94.5	62.12	18	801.80	468.71	4	483.5	271.42	5	65.88	40.23	3	84.4	34.1
August	5	119.7	70.14	24	855.4	504.29	4	199.7	162.10	4	209.80	102.07	3	22.9	106.7
September	4	187.0	90.42	21	804.10	343.38	4	595.80	258.27	3	89.50	49.98	4	67.3	37.52

TS = Total Samples; max = maximum; AM = Arithmetic Mean

Table 23.3 *Sulphur Dioxide (SO_2) Levels in Bangalore*

Unit: $\mu g/m^3$

1985	Jayanagar			Ananda Rao Circle			AMCO Batteries			KS and DL			GIBU		
Months	TS	Max	AM	TS	Max	AM	TS	Max	AM	TS	Max	AM	TS	Max	AM
April	4	11.40	5.58	14	19.40	6.97	2	33.60	21.23	2	14.43	8.80	3	18.75	7.92
May	5	5.91	2.46	17	13.33	5.87	4	78.33	36.38	4	14.81	4.93	4	22.20	9.63
June	4	10.60	5.06	19	20	6.75	4	33.50	13.95	2	4.40	3.0	4	10.10	4.21
July	4	26.3	8.77	18	14.40	4.81	4	91.1	64.8	5	4.4	2.7	2	8.8	5
August	5	3.5	2.77	24	13.3	5.44	4	51.10	28.84	3	4.3	3.8	3	3.5	2.7
September	4	1.4	1.24	21	10.0	4.35	4	14.4	9.98	3	6.30	3.2	4	4.4	1.95

TS = Total Samples; max = maximum; AM = Arithmetic Mean

Table 23.4 *Levels of Oxides of Nitrogen (NO_x) in Bangalore*

1985	Jayanagar			Ananda Rao Circle			AMCO Batteries			KS and DL			GIBU		
Months	TS	Max	AM	TS	Max	AM	TS	Max	AM	TS	Max	AM	TS	Max	AM
April	4	14.2	5.34	14	9.79	3.77	2	8.95	6.97	2	11.2	7.50	3	7.61	3.22
May	5	7.11	6.07	17	17.0	8.20	4	26.70	12.18	4	12.10	7.70	4	3.80	2.67
June	4	5.0	3.10	19	79.16	11.92	4	17	7.89	2	5.0	4.31	4	3.70	2.52
July	4	22.14	6.5	18	5.8	3.29	4	29.1	18.4	5	3.3	1.98	2	5.70	4.27
August	5	5.8	2.34	24	8.3	3.44	4	7.5	4.55	3	5.0	4.6	3	3	1.96
September	4	5.5	3.59	21	15.8	6.74	4	36.6	15.14	3	4.16	2.85	4	3.3	2.33

TS = Total Samples
max = maximum
AM = Arithmetic Mean

Note: **(1) All values are expressed in micrograms per cubic metre (μg/m³)**
(2) Monitoring was done for 8 hours period, once in a week.

Table 23.5 *Statistical Analysis of Data Collected at a Sampling Station*
Place: Jayanagar, Bangalore   Sampling Frequency: Every Eighth Day   Unit: Micrograms per Metre Cube

Month	Sample size	SPM			Sulphur dioxide			Nitrogen oxides			Lead		
		AM	GM	SGD	AM	GM	SGD	AM	GM	SGD	AM	GM	SGD
July'84	3	46.20	44.45	1.42	0.0	1.00	1.00	3.19	2.60	2.32	0.0	1.00	1.00
Aug.'84	4	66.47	66.27	1.36	0.0	1.00	1.00	13.79	7.41	4.71	0.02	0.02	5.64
Sep.'84	5	85.58	82.38	1.37	1.48	1.62	1.79	3.37	2.05	3.18	0.01	0.08	9.84

Where, SPM = suspended particulate matter
AM = Arithmetic mean
GM = Geometric mean
SGD = Standard geometric deviation

Note: When the values of AM, GM and SGD are printed as 0.0 it indicates that only traces of the parameter were observed

more industries. It has a population of nearly four million with a large number of buses, trucks, motor and other vehicles to cater to the needs of the large population.

It would be appropriate at this stage to suggest studies on the air quality during the winter which is likely to be the most sensitive season for pollution accumulation. This is evident from the study of meteorological factors such as wind pattern, temperature, and rainfall over a period of time for Bangalore. The statement is also true of other cities in India where the effects of pollution are determined to be the worst during the winter season. The evaluation of smoke from heavy vehicles, would be another significant contribution in checking the future smoke nuisance.

Tables 23.2, 23.3, and 23.4, respectively show typical levels of suspended particulate matter, sulphur dioxide and oxides of nitrogen at five chosen sites in Bangalore during the period April-September 1985, as monitored by Karnataka State Pollution Control Board in Bangalore. Table 23.5 shows a typical statistical analysis of the data collected at one of the sampling stations in Bangalore.

QUESTIONS

1. Write brief notes on the status of air pollution in Delhi, Bombay, Calcutta, Hyderabad and Bangalore.
2. Discuss the status of air pollution due to automobile exhausts in the following cities: (a) Bangalore (b) Calcutta.
3. Discuss the status of air pollution due to industries in Hyderabad and Bombay.

24

Miscellaneous Current Topics

Miscellaneous current topics of interest like 'acid rain', 'green house effect' 'indoor pollution' and 'occupational diseases' have been discussed in this chapter.

24.1 Acid Rain

Rain is the most important source of fresh water from time immemorial. But in the last few decades, rain water has become progressively more acidic. It is the result of massive air pollution—the price of industrialisation. On account of acid rainfall, today more than 200 lakes in New York State are 'dead'. Nearly 5000 lakes of Sweden are fishless. About 4000 lakes in Canada are unable to support trout due to a high degree of acidification, and if the environmental pollution continues unabated, about 48,000 of Canada's lakes may 'perish' by the end of the century. It is reported that about 30% of West Germany's forests are dying a slow death, apparently caused by acid rain. In our country, acid rainfall has been reported from Bombay and Pune. Thus, acid rain is becoming an international nightmare.

24.1.1 What is Acid Rain

The term 'acid rain' is used to describe all precipitation—rain, snow, and dew—which is more acidic than normal. The destruction by acid rain is not confined only to lakes and forests. It corrodes buildings, monuments, statues, and metals. Further, it poses a serious threat to human health since it contaminates drinking water.

An acid gives a characteristic sour test to water. Acidity is measured on a pH scale, ranging from 0 to 14. Neutral solutions will have a pH of 7. For example, distilled water has a pH of 7. From pH 7 up to 14, solutions are alkaline (basic). The higher the pH, the more alkaline the liquid. From pH 7 down to 0, indicates acidity. Thus lower the pH, higher will be the acidity. Lime juice and vinegar have a pH of about 3. As you all know, the pH scale is logarithmic. That means each unit on the scale is a factor of 10 higher or lower than the next unit. For example, an acid of pH 3 is 10 times stronger than one of pH 4, and 100 times stronger than one of pH 5.

Generally, 'clean' rain is slightly acidic as it dissolves varying amounts of carbon dioxide, which is a common constituent of the earth's atmosphere. The lowest pH value of rain is 5.6 when it is 'clean'. This is the pH level produced by carbonic acid if distilled water were in equilibrium with atmospheric carbon dioxide. Acid rain on precipitation is defined as one which has a pH less than 5.6.

In many parts of Scandinavia, Western Europe, Canada, and the USA, rain has become increasingly acidic. Latest reports indicate that even countries in the southern hemisphere like Australia, Thailand, Brazil, etc., are being affected by acid rainfall. Measurements made before the forties' did not show any acidification or precipitation. But now, rain having a pH of 4.6 is common in Eastern North America. This means, it is 10 times more acidic than the 'clean' rain. In some places, pH as low as 3.7 or even less has been reported.

24.1.2 Causes of Acid Rain

Acidification of our environment is manmade. Large quantities of sulphur and nitrogen oxides are emitted into the atmosphere from the chimneys of industrial plants. They react with the moisture in the air to form sulphuric and nitric acids which fall as acid precipitation far away. Acid rain, therefore is a cocktail of sulphuric and nitric acids. Currently, about 60–70% of the acidity is thought to be due to sulphuric acid and 30–40% to nitric acid.

Burning of fossil fuels for power generation is a major contributor (60–70%) of sulphur dioxide to the atmosphere. Smelting of sulphide ores, particularly of lead, zinc, and copper is also responsible for an increase in the sulphur dioxide content of the atmosphere. The major contributors of oxides of nitrogen are automobile exhaust, power houses, and smelters.

Further, about three decades back most of the air pollutants were close to the ground and transported only through short distances. Consequently, there was heavy air pollution in urban and industrial zones, while rural areas were generally not affected. But now air pollutants are carried long distances in the atmosphere, which has resulted in transfrontier air pollution. Moreover, during the long journey, significant chemical and physical changes occur to produce more hazardous products. Acid rain is the result of a combination of air pollutants over a long distance and their transformation.

The ecological impact of acid rain is quite serious. It is likely to produce irreversible changes. The acidification of streams and lakes affects aquatic animals and plants. High acidity results in the killing of fish, reduced growth, and perhaps reproductive failure. Change in pH prevents hatching of fish eggs. Large scale destruction of trout and salmon have been reported, due to high acidity.

Green algae and many forms of bacteria which are essential to aquatic life/system are killed due to acidity. Also, at low pH the rate of decomposition of organic matter is reduced. This results in the accumulation of organic matter in the water bodies like lakes and streams, which increases the degree of water pollution. It has been reported that prolonged acidic conditions result in the total collapse of aquatic systems, and makes them sterile. Also, acidity increases beyond the tolerable limits the concentration of heavy metals such as lead, cadmium, copper, and zinc in water.

Acid rain affects vegetation, and soil in many ways. The growth of trees is adversely affected by acid rain. It affects forests, and results in consequent vanishing of greenery. These affects are now being studied in detail. In general, acid rain causes damage to leaves, and accelerates leaf surface erosion. It may also affect germination of seeds.

Similarly, acidity affects soil. A plant nutrient such as potassium is gradually leached out of the soil. At the same time, a toxic element like zinc accumulates due to acid rain. Beneficial micro-organisms in the soil are reduced. The population of earthworms which are popularly called 'farmer friends', is reduced as they cannot tolerate an acidic environment.

24.1.3 Remedy

One of the simple solutions to the problem is to neutralise the acid with lime. It is a short-term measure, and it is required to be repeated periodically. As such, it is quite expensive, especially when large areas of water bodies have to be limed. Further, we are not sure whether large-scale liming will create its own ecological problems, and affect the ecological system.

Probably, the best way to overcome this problem is reduce the emission of sulphur dioxide and oxides of nitrogen from industries. In other words, it requires effective air pollution prevention, and control, for both stationary sources (like industries) and mobile sources (like automobiles) of pollution.

24.1.4 Recent Reports

Here are some examples of the effects of acid rain in different parts of the world. Acid rain from the atmospheric pollution caused by industries in USA is reported to be killing the flora and fauna in rivers and lakes in Canada. Thousands of lakes and rivers in Canada from the province of Ontario to Quebec are biologically 'dead' due to the high level of acid in the water. It is estimated that about 48,000 lakes will be incapable of maintaining life within 20 years, if acid precipitation continues. According to one estimate, about 50% of the acid rain is caused by sulphur dioxide transported over the US border from coal-fired plants as far away as the Ohio Valley. According to another report, sulphur dioxide also flows the other way, killing fish life in about 300 lakes in New York's Adriondack mountains.

Acid rain is a big environmental problem in Scandinavian countries. Thousands of their lakes are reported to have been sterilised due to acid precipitation. It has also affected their forests.

It is estimated that nearly 30% of West Germany's forests are dying a slow death. It is causing timber losses and agricultural damage.

In our own country, systematic work on measuring rain water acidity has been carried out only during the last year or so. These studies are being done by the Bhabha Atomic Research Centre (BARC), Bombay, and also by the network of background pollution monitoring stations established in India by the World Meteorological Organisation. Studies show that the rain water acidity at Bangalore, Delhi, Hyderabad and Nagpur has remained more or less constant (the rains are not acidic). Pune, however shows a trend towards higher acidity in rain water. According to another report, observations of rain of pH 4.5 and 4.8 were made in 1974 at Trombay and Chembur respectively in Bombay. The marble of the Taj Mahal has been reported to be affected due to the air pollutant sulphur dioxide from the nearby Mathura Oil Refinery. Some environmentalists call it 'stone cancer'!

24.2 Greenhouse Effect

24.2.1 Effect of Carbon Dioxide

Man's inadvertent activities are increasing the carbon dioxide content of the atmosphere which may probably bring about drastic changes in the world climate in the next 50 years. Increasing global consumption of fossil fuels is steadily increasing the carbon dioxide level in the atmosphere since the beginning of the twentieth century. It has been estimated, that the world's consumption of fossil fuels results in the release of nearly 9×10^9 tonnes of carbon dioxide per year.

The primary source of carbon dioxide is the consumption of fossil fuels. The secondary source is the oxidation of carbon compounds in marshes and forests by natural degradation. Manufacture of cement is another factor contributing to increase of carbon dioxide in the atmosphere.

In the natural 'carbon cycle', atmospheric carbon dioxide is normally removed by two major 'sinks'—plants which absorb CO_2 during the process of 'photosynthesis' and the oceans. But the problem has arisen because we are producing CO_2 faster than the sinks can absorb it. Therefore, the level of CO_2 in the atmosphere is increasing.

This has an influence on climate through what is called the 'greenhouse effect'. The layer of carbon dioxide acts like the glass in a greenhouse because, while the sun's radiation can pass through it and heat up the earth, it stops that same heat being re-emitted by the earth. In other words, CO_2 is transparent to shortwave radiation from the sun but absorbs the longer wave radiation from the earth. The net result of all this is, that there is a danger of the earth gradually being heated up.

If the CO_2 content of the atmosphere is doubled, which can occur by the middle of the next century, then it will result in increase in the present world temperature by about 3.6°C. It is estimated that if the earth continues to warm up, all the glaciers will recede and the ice caps in the Antarctic and the Arctic will begin to melt. Consequently, the sea level will rise by a few metres, and most of the cities on the sea shore may be submerged.

Ecologists are of the opinion that the situation will be worse because the additional inputs of CO_2 to the atmosphere due to the destruction of forests and of soil organic matter will reduce the volume of the biomass sink.

The situation in India is particularly alarming. Historically, the Gangetic plains, the Deccan, and the Central Plateau and Rajasthan desert were covered with dense vegetation. Estimates in 1912 showed that over 36% of the Indian continent was covered with forest. But, today it is just about 20%. The combination of deforestation and increasing fossil fuel discharge has been responsible for the increasing deviation in the climate with decreasing rains.

A similar situation prevails in the climatology of large urban areas. Here the decline of trees combined with the heat-sink effect of the buildings is enhanced many times by the accumulation of CO_2 in the local atmosphere. Studies in cities like Bombay, Delhi, and Calcutta have revealed that the heat-island effect raises night temperatures by 4–6°C above open areas.

However, there are some factors which may counteract the CO_2 effect. The warming of the atmosphere will increase cloud cover and this will reduce the amount of incoming solar radiation. Agricultural and industrial operations as well as deforestation may generate a large quantity of dust which in turn can reduce the effective radiation and thus maintain the atmospheric temperature more or less at the same level.

To overcome the ecological crisis due to CO_2, community based forestry schemes should be undertaken. A good example of this is China, where land under forests has been increased from 5% in 1949 to 12.7% in 1978.

24.2.2 Greenhouse Threat from Trace Gases

According to a recent study by a team of atmospheric scientists in the USA, tiny quantities of over 30 rare gases threaten to warm the earth's atmosphere even more rapidly than carbon dioxide in the next 50 years. Many of these trace gases are industrial by-products. According to the investigators, the trace gases will more than double the warming effect of carbon dioxide in the years to come. The trace gases have been found to be surprisingly very effective at trapping heat that would otherwise radiate into space. It has been reported that a single molecule of some chlorofluoro carbons absorbs as much heat as 10,000 molecules of carbon dioxide. The significant trace gases that cause the greenhouse effect are methane, nitrous oxide, and a variety of members of the chlorofluorocarbon family.

24.3 Indoor Air Pollution

The inside environment of houses often has a higher level of air pollution than the surroundings in cities and towns. Unfortunately, indoor pollution has not been given much importance, although most people spend as much as 80–90% of their time indoors. The various causes of indoor pollution are use of traditional fuel for cooking, tobacco smoke, temperature, humidity, micro-organisms and allergens. The use of formaldehyde in construction materials and poor ventillation due to energy conservation measures in new homes and offices are also responsible for indoor pollution.

24.3.1 Indoor Pollution due to Burning of Traditional Fuels

About 50% of the world's households are using traditional fuels such as, firewood, animal dung, coke, etc., for cooking.

The common pollutants which cause indoor air pollution problems due to combustion of fuels are particulate matter, oxides of sulphur, oxides of nitrogen, carbon monoxide, hydrocarbons, and organic and odour causing chemicals. The emission quantity of these pollutants depends upon the type of fuel used, type of stove or furnace used, feed rate, amount of additional air and operating conditions.

In our country, over 80% of the time of housewives is spent in an indoor environment of which 4–6 hours is spent in kitchens for the purpose of cooking.

In order to determine the emission of sulphur dioxide and total suspended particulate matter during cooking, a short duration atmospheric survey was undertaken at the cooking place of low income houses, chosen randomly in different localities in the city of Lucknow. The survey was conducted by the Industrial Toxicology Research Centre, Lucknow. Coal, firewood and animal dung were being used in these houses for cooking purposes. It was found that emissions of sulphur dioxide and total suspended particulate matter in all the three fuels were much higher than the average annual standard for ambient air quality.

24.3.2 Indoor Pollution due to Pollens

Studies have revealed that pollens, fungal spores, and various types of dusts pollute indoor air, and cause allergic diseases, especially asthma. Further, studies have indicated that an environment around an asthmatic has an important role in triggering an attack of asthma. Pollens from trees, shrubs, herbs, grasses, and weeds causes allergy.

24.3.3 Indoor Pollution due to Artificial Building Materials and Poor Ventilation

Indoor air pollution, especially in new energy-efficient homes and offices, is making many people around the world sick. The hazard is becoming more serious with the growing use of man-made building materials, some of which emit harmful vapours and with energy conservation measures that reduce ventilation. In some advanced countries, the problem has become something of concern for citizens who have shockingly found that no official agency is in charge and no standards exist for control of indoor air pollution.

The most serious problems so far have been reported in new and remodelled office buildings and homes with energy-saving features, and in mobile homes. Most involve formaldehyde in construction materials, such as particle board and indoor plywood, and in urea formaldehyde foam insulation.

Formaldehyde vapours leak into the air when the temperature rises and they combine with other contaminants into a mix that can cause headaches, respiratory irritations, watery eyes, nausea, diarrhoea, skin irritations, and heart problems.

Similar is the case with office workers, who suffer from eye irritation, headaches etc., due to indoor pollution.

The following precautionary measures should be taken to overcome this special problem of indoor pollution.

1. Cautionary labels should be attached to construction materials that contain formaldehyde resins
2. Municipal health authorities should be authorised to test air within homes when a physician suspects formaldehyde or other vapours might be damaging a resident's health
3. The air pollution control regulations, should have provision for compulsory testing of products at the point of manufacture to assure that they will not pollute indoor air
4. Architects and construction engineers must make sure that air flow is not reduced to a danger point in the quest for energy efficiency

As most of us spend a large part of our life indoors, indoor air pollution is one of the major problems we have to solve, and as such, all efforts should be made to eliminate or minimise the deleterious effects of indoor air pollution.

24.4 Occupational Diseases

Industrial workers may be exposed to various types of occupational hazards. From the point of view of air pollution, chemical hazards are significant because they result in various types of occupational diseases.

Chemical agents act in three ways: local action (e.g., dermatitis), inhalation (dusts and gases), and ingestion of chemical substances such as lead, mercury, arsenic, zinc, chromium, cadmium, etc.

The main chemical hazard is because of inhalation of dust particles. The size of dust particles varies between 0.1 and 150 μ. The various sources of dust particles are discussed in Chapter 2. Particles smaller than 3 μ are directly inhaled into the lungs and are mainly responsible for pneumoconiosis. Pneumoconiosis is a lung disease, which may gradually cripple a man by reducing his working capacity due to lung fibrosis and other complications.

Following are the occupational diseases associated with the inhalation of dust particles.

1. Silica	Silicosis
2. Asbestos	Asbestosis
3. Cotton dust	Byssinosis
4. Coal dust	Anthracosis
5. Sugar cane dust	Bagassosis

6. Tobacco	Tobaccosis
7. Iron	Siderosis
8. Hay or grain dust	Farmer's lung

24.4.1 Silicosis

Silicosis is caused by inhalation of dust containing free silica or silicon dioxide (SiO_2). The workers in various industries like the mining industry (coal, mica, gold, silver, lead, zinc, manganese, and other metals), pottery and ceramic industry, sand blasting, building and construction work, rock mining etc., are exposed to this hazard.

The incidence of silicosis depends upon the chemical composition of the dust, size of particles, duration of exposure, and individual susceptibility. The higher the concentration of free silica in the dust, the greater the hazard. Particles between 0.5 and 0.3 microns are the most dangerous because they easily reach the interior of the lungs. The longer the duration of exposure, the greater the risk of developing silicosis. It is found that the incubation period may vary from a few months up to six years of exposure, depending upon the above factors.

The main symptoms of silicosis are cough, and pain in the chest. According to one opinion, in some cases silicosis may lead to 'silico-tuberculosis'.

There is no effective treatment for silicosis. The only way silicosis can be controlled is by

1. Rigorous dust control measures (like isolation, hydroblasting, good housekeeping, personal protective measures, etc.)
2. Regular physical examination of workers

Silicosis was made a notifiable disease under the Factories Act, 1948, and the Mines Act, 1952.

24.4.2 Asbestosis

Asbestos, once acclaimed as a magic mineral for its fire extinguishing properties, is now under fire as a health hazard. The inhalation of asbestos dust causes a disease known as 'Asbestosis', which harms the lungs. Asbestos has also been reported to cause lung cancer, and mesothelioma—cancer of the lining of the lungs, which are usually fatal maladies. Asbestos also causes ailments like chest pain, cough, and perhaps asthma.

The risk of lung cancer has been observed to be particularly high in cigarette smokers among workers exposed to asbestos.

Asbestos fibres are usually from 20–500 μ in length and 0.5–50 μ in diameter. Asbestos is used in the manufacture of asbestos cement, roof tiles, textiles such as fire-proof clothing, gaskets, insulation, brakes in automobiles, etc.

It is the insidious nature of asbestos that makes the hazard particularly dreaded. A few month's exposure may be sufficient for the invisible dust particles to penetrate the human body. Then they work like a time bomb, taking as long as 25–30 years, and in some cases even 40 years to manifest their presence by causing an illness that may, by then, be incurable.

An effective method of combating the asbestos hazard is to reduce the concentration of its airborne fibre particles at the work place and to limit duration of work with asbestos. For this purpose, many countries have prescribed occupational exposure limits. But the threshold limits vary from country to country as do the methods of monitoring. For example, in Britain, the limit prescribed is 0.5 fibre of white asbestos per millilitre (f/ml) of air. Similarly, the limit for blue and brown asbsetos is 0.2 f/ml. In many other countries, the limits are 1.0 f/ml for white asbestos, and 0.5 f/ml for blue asbestos.

In our country, at present there is no legislation to control the asbestos hazard and there is an urgent need to bring forward suitable legislation to counter it.

In the USA, there is a drive to remove asbestos from school buildings, as the fire proof insulation in many school buildings is considered to be a cancer threat and the cause of other diseases in the children and school employees.

Preventive measures to overcome the asbestos menace involve rigorous dust control measures, use of protective equipment by workers exposed to the hazard, and regular medical examination of workers exposed to asbestos.

24.4.3 Byssinosis

It is due to inhalation of cotton fibre dust over long periods of time. It is a disabling lung disease. The symptom is chronic cough ending in chronic bronchitis (respiratory disorder).

India has a large number of textile industries employing nearly 35% of factory workers. About 20% of textile mill workers are victims of byssinosis.

QUESTIONS

1. What is acid rain? What are the causes for acid rain?
2. What are the effects of acid rain?
3. How is the problem of acid rain overcome?
4. Explain the effects of acid rain in
   (a) Canada
   (b) Scandinavian countries
   (c) West Germany
5. "The ecological impact of acid rain is quite serious". Discuss.
6. Describe the phenomenon of 'greenhouse effect' due to carbon dioxide.
7. Explain the greenhouse threat from trace gases.
8. Discuss the problem of 'indoor pollution'. Suggest a suitable remedy for it.
9. Write notes on:
   (a) Silicosis
   (b) Asbestosis
   (c) Byssinosis
   (d) Anthracosis
10. What is silicosis? What are its symptoms? How can it be controlled?

REFERENCES

1. "Acidification", Ministry of Housing, Physical Planning and Environment, Ministry of Agriculture and Fisheries, The Hague, Netherlands, 1973.
2. "Action for Environmental Quality", US Environmental Protection Agency, Washington, 1973.
3. Agarwal, P.D., "Electricity not such a Clean Fuel", *SAE Journal*, vol. 79, Sec. 2, 1971, p. 38.
4. Agate, J.N., "Industrial Fluorosis", *Medical Research Council (Brit) Mem.*, No. 22, 1949.
5. "Air Monitoring Programme Design for Urban and Industrial Areas', *GEMS*, WHO, Offset Pubn. No. 33, Geneva, 1977.
6. "Air Quality Criteria and Guides for Urban Air Pollutants", *Tech. Report Series No. 506*, WHO Pubn., Geneva, 1972.
7. Air (Prevention and Control of Pollution) Act 1981, Government of India, New Delhi.
8. "Air Pollution", *Monograph Series*, No. 46, WHO Pubn., 1961.
9. "Air Pollution Meteorology", U.S Dept. of Health, Education and Welfare, Washington DC, 1971.
10. "Air Pollution Control", *CPHERI*–Short Course Manual, Nagpur, 1973.
11. "Air Pollution", World Health Organisation, Geneva, 1961, p. 32.
12. "Air Pollution Act", Published by the Ministry of Housing, Physical Planning and Environment, The Netherlands, 1983.
13. "Air Quality Criteria for Particulate Matter", US Dept. of Health, Education and Welfare, National Air Pollution Control Admn., Pubn. No. AP-50, Washington DC, 1969.
14. Air Quality Status in Bangalore City, *Namma Parisara*, vol. 1, 2 and 3, Aug., Sept., and Oct., 1985.
15. "Air Quality Monitoring", A Course Manual, *NEERI*, Nagpur, 1978.
16. Alexander, T., "Some Burning Questions About Combustion", *Fortune*, Feb., 1970.
17. "Ambient Air Quality Standards", Operating Manual, *Envirotech*, New Delhi.
18. "Analysing and Interpreting Air Monitoring Data" *GEMS*, WHO, Geneva, 1980.
19. Analytical Methods Manual Ch. 7, Manufacturing Chemists Association Inc., Washington DC.
20. Andonyev, S. and O. Filipyev, *Dust and Fume Generation in the Iron and Steel Industry*, Mir Publishers, Moscow, 1977.
21. Anon, "Nitrous Fume", Dept. Sci. Ind. Research (UK), *Leaflet 5*, 1939.
22. Anon, "Air Pollution in Donora Pa", *Pub. Hlth. Bull.*, 1949, p. 306.
23. Anon, "Mortality and Morbidity During the London Fog of December 1952", Ministry of Health Dept., *Report 95*, *HMSO*, London, 1954.
24. Anon, "Industrial Plants and Stations Show Progress in Pollution Control", *Power*, 114, 27, 1970.
25. Ansdale, R.F., "The Wankel RC Engine—Design and Performance", Butterworth and Co. Ltd., London, 1968, p. 119.
26. Arceivala, S.J., C.V. Sabnis and P.K. Yennawar (Eds), "How to Conduct Air Pollution Surveys?", *NEERI*, Nagpur.
27. Ashar N.G., "Pollution Control with Profit", *Proc. Symp on Env. Pollution*, Nagpur, 1973, p. 335.
28. "A Study of Air Pollution in Bangalore", Regional Occupational Health Centre, *Annual Report, 1980*, p. 126.
29. "Atmospheric Pollutants", *Tech. Report Series*, No. 271, WHO, Geneva, 1964.
30. Baker, R.A. and R.C. Doerr, "Catalytic Reduction of Nitrogen Oxides in Automobile Exhaust", *J. Air Pollution Control Board*, Oct. 1964.

31. Beckman, E.W. et al., "Exhaust Emission Control by Chryster the Cleaner Air Package", Trans. SAE, vol. 75, *Paper No. 660107, 1967.*
32. Bennett, P.A. et al., "Reduction of Air Pollution by Control of Emissions for Automotive Crankcases", Trans. SAE, vol. 68, 1960.
33. Bethea, R.M., *Air Pollution Control Technology*, Van Nostrand Reinhold Co., New York, 1978.
34. Bhave, V.R., "When Thermistor is Used for Temperature Inversion Studies", Proceedings, *Symposium on Reliability of Electronics through Environmental Engineering*, Bangalore, 1971, pp. 29-48.
35. Bhave V.R., "Motor Aspirated–Radiation Shielded Temperature Sensors for Temperature Inversion Studies", *Indian J. Environ. Hlth.*, vol. 17, No. 4, 1975, pp. 355-360.
36. Bibben, R.J. and I.G. Young, *Systems Approach to Air Pollution Control*, Wiley Eastern Publication Pvt. Ltd., New Delhi, 1973.
37. Bishop, C.A., "EJC Policy Statement on Air Pollution and its Control", *Chem. Eng. Prog.*, vol. 53, 1957, p. 146.
38. Blokker, P.C., "Control of Atmospheric Pollution in the Petroleum Industry", Stitchting Concawe, The Hague.
39. Blokker, P.C., "Major Aspects in Air Pollution Monitoring in Urban and Industrial Areas", Stitchting Concawe, The Hague, 1973.
40. Briggs, G.A., "Plume Rise—A Critical Review", *Journal Nuclear Safety*, vol. 12, No. 1, 1971.
41. Brower, M., "Basic Data on Air Pollution Control in Los Angeles Country Air Pollution Control District", USPHS, 1956.
42. Bryan, R.J. and J.C. Romanovsky, "Instrumentation for Air Pollution, Instr. and Automation, vol. 29, 1956, p. 2432.
43. Byers, H.R., *General Meteorology*, McGraw-Hill Book Co., Inc., New York, 1944.
44. Cadle, R.D. and H.S. Johnston, "Chemical reactions in Los Angeles Smog", *Proc. 2nd National Air Pollution Symp.*, Stanford Research Inst., Los Angeles, California, 1952.
45. Cantwell, E.N. et al., "A Progress Report on the Development of Exhaust Manifold Reactors", *Paper presented at SAE,* International Auto. Engng. Congress, Detroit, Jan. 1969.
46. Caplan, J.D., "Causes and Control of Automotive Emissions", *Auto Div. Proceedings, Inst. Mech. Engrs.*, London, 1962-63, p. 242.
47. Caplan J.D., "Smog Chemistry Points the Way to Rational Vehicle Emission Control", *Trans* SAE, 74, 1966.
48. "Carburettors or Injection?", *Automobile Engineer*, Oct. 1971, pp. 30-31.
49. Central Board for Prevention and Control of Water Pollution, "Emission Regulations Part I and II", Comprehensive Industry Document Series, New Delhi, 1985.
50. Central Board for Prevention and Control of Water Pollution, "MINAS for Various Industries", Comprehensive Industry Document Series, New Delhi, 1985.
51. Cholak, J., "The Nature of Pollution in Number of Industrial Communities", *Proc. 2nd National Air Pollution Symp.*, Stanford Research Inst., Los Angeles, California, 1952.
52. *Cleaning our Environment, The Chemical Basis for Action*, American Chemical Society, Washington DC, 1969.
53. Climatological Tables of Observations in India, IMD, 1880–1940.
54. Climatological Tables of Observations in India, IMD, 1931–1960.
55. Cornelius W. et al., "Ammonia as an Engineer Fuel", Energy Depot Concept, *SP-263*, SAE, New York, 1964.
56. Cross, F.L. and H.E. Hesketh, "Handbook for the operation and Maintenance of Air Pollution Control Equipment", *Technomic Pubn.* 1975.

57. Daniel, W.A., "Flame Quenching at the Walls of an Internal Combustion Engine", Sixth International Symposium on Combustion, New York, 1957, p. 886.
58. Davies, C.N., "Particle in Atmosphere—Natural and Man-made", *Jr. Atm. Env.* vol 8, No. 11, Nov. 1974, p. 1078.
59. Denient and Wentworth, "Exhaust Gas Hydrocarbons—Genesis and Exodus", SAE *Progress in Tech.* Series 6. (1964), p. 204.
60. Derycke, D., "Catalytic Purification of Car Exhaust", *Auto Div. Proceedings,* Inst. of Mechanical Engineers, London, 1962–63, p. 220.
61. Desai, H.B., "Air Pollution Control Technology in Petroleum Refineries". Proc. *Symp. Air Pollution Control Techniques,* CLI, CPHERI and SOCLEN, Bombay Sep. 1973.
62. "Dispersion for a complex of Stacks", Stichting Concawe, The Hague, 1974.
63. Dixit, D.K., "Air Pollution due to Automobiles", *Indian J. Environ Hlth,* vol. 17, No. 3, 1975, pp. 250–264.
64. Drinker, P. and T. Hatch, *Industrial Dust,* 2nd edn., McGraw-Hill Book Co., Inc., New York, 1954.
65. Ebersole, G.D. and I.A. McReynolds, "An Evaluation of Automobile Total Hydrocarbon Emissions", *Trans.* SAE 75, 1967.
66. Ehlers, V.M. and E.W. Steel, *Municipal and Rural Sanitation,* 6th edn., McGraw-Hill Pubn., New York, 1965.
67. Elmer Robinson and Robert C. Robinson, "Gaseous Sulphur Pollutions from Urban and Natural Sources", *J. of Air Pollution Control Association* vol. 20, No. 4, April 1970, p. 234.
68. Elmer Robinson et al., "Gaseous Nitrogen Compounds Pollutants from Urban and Natural Sources", *J. of Air Pollution Control Association,* vol. 20, No. 5, May 1970, p. 304.
69. Eltinge, L. et al., "Potentialities of Further Emission Reduction by Engine Modifications", Trans. SAE, 77, 1968.
70. Engineer, N.B. and V.C. Doshi, Air Pollution in Cement industry, *Proc. Symp. Air Pollution Control Techniques,* CLI, CPHERI and SOCLEEN, Bombay, Sep. 1973.
71. "Environmental Analysis of the Varnium Fuel Cycle", Pt. II *Nuclear Power Reactors,* Washington DC, (US Environmental Protection Agency, 1973).
72. "Environment survey of the uranium fuel cycle", *Report Wash*—1248 (Washington, DC (US Atomic Energy Commission, 1974).
73. Fagley, W.S. et al., "Maintenance and Automobile Exhaust Second Report", SAE National Meeting, *Paper No. 486 A,* Detroit, March 1962.
74. Faith, W.L., *Air Pollution Control,* John Wiley and Sons, Inc., New York, 1959.
75. Faith, W.L. and Arthur A. Atkinson, "Air Pollution" 2nd edn.. Wiley-Interscience, New York, 1972.
76. Felt, A.E. and R.V. Kerley, "Engines and Effects of Leadfree Gasoline", SAE Paper No. 710 367, 1970.
77. Giles, W.S. and W.H. Updike, "Influence of Low Lead Fuels on Exhaust Valve Performance", *SAE Paper No. 710 674,* Aug., 1971.
78. "Glossary of Air Pollution", *WHO Regional Pubn.* European Series No. 9, 1980.
79. Gopalakrishnan, K.V. et al., "Air Pollution due to Automobiles", *Proceedings of Seminar,* "A Study of Environmental Pollution in Madras City", June, 1971.
80. Gray, J.T. et al., "Ammonia Fuel Engine Compatibility and Combustion" *Trans.* SAE 75, 1967, p. 785.
81. "Guide to the Appraisal and Control of Air Pollution", *APHS,* 1969.
82. Gupta, C.P. and P.K. Goel, "Automotive Vehicle Emission", *J. Inst. Engrs.* (I) Part ME 4, 1973, p. 141.
83. Gupta, D.R. and C.P. Gupta, "Air Pollution by Exhaust Hydrocarbon Emission from Automobiles", *J. Inst. Engrs.* (I), vol. 51, 1971, p. 9.
84. Gupta, S.D. and N.V. Iyer, "Evaporative Emissions from Gasoline Powered Vehicles—Petroleum and Hydrocarbons", July 6, 1971.

85. Hagen–Smit, A.J., "Chemistry and Physiology of Los Angeles Smog", *Ind. Eng. Chem.*, vol. 44, 1952.
86. Hanumanth Rao, S., "Air and Water Pollution Control",—*Paper presented at the Institution of Engineers (I)*, Karnataka Centre, Bangalore, May 1984.
87. Harmon, R.A., "Gas Turbines—An Industry with World-wide Impact", *Mechanical Engineering*, March 1973.
88. "Health Aspects Related to Indoor Air Quality", *EURO Reports and Studies No.* 21, 1979.
89. Hesketh, H.E., "Understanding and Controlling Air Pollution", Ann Arbor Science Publishers, Inc., Michigan, 1972.
90. Hesketh, Howard E. "Understanding and Controlling of Air Pollution", Ann Arbor Science Publishers, Inc. Michigan, 1972, p. 36.
91. Hopper W.C., "The Problem of Gasoline Engine Exhaust Control", *Report No. 12/72*, Stichting Concawe, The Hague, 1972.
92. Hulis, T.A. et al., "Spark-ignition Engine Operation and Design for Minimum Exhaust Emission", *Paper No. 660 405, SAE Mid-Year Meeting*, Detroit, 1966.
93. Hulis, T.A. and H.A. Nickol, "Influence of Engine Variables on the Carburation System ", *Paper No. 660 114, SAE Automotive Engineering Congress*, Detroit, Jan., 1966.
94. Hulis, T.A. and H.A. Nickol, "Influence of Engine Variables on Exhaust, Oxides of Nitrogen Concentrations from a Multicylinder Engine", *Paper No. 670 482, SAE Mid-Year Meeting*, Chicago, May 1967.
95. Hunter, D.C. and H.C. Wohlens "Air Pollution experiments for Jr. and Sr. High School Science Classes", 2nd edn, Air Pollution Control Association, Pittsburgh, USA.
96. IS: 4167, 5182 (Part I, II, IV, V, VI, VIII, IX, XIV, XV), 8829, 9078, 8118, 9057, 9260, 10195.
97. Izmerov, N.F., "Control of Air Pollution in the USSR", Publ. Hlth, *Paper, No. 54, WHO*, Geneva, 1973.
98. Jackson, M.W. et al., "The Influence of Air-Fuel Ratio, Spark Timing and Combustion Chamber Deposits on Exhaust Hydrocarbon Emission", *Paper Presented at SAE National Meeting*, Detroit, March, 1962.
99. Jackson, M.W., "Effects of Some Engine Variables and Control Systems on Composition and Reactivity of Exhaust Hydrocarbons", *Trans. SAE*, 75, 1967.
100. Jackson, M.W. and R.I. Everett, "Effect of Fuel Composition on Amount and Reactivity of Evaporative Emissions", *SAE International Auto. Eng. Congress*, Detroit, Jan. 1969.
101. Jacobs, M.B., *Chemical Analysis of Air Pollutants*, Interscience Publishers, Inc., New York, 1964.
102. Joe. O Led Batter, "Air Pollution, Part (A) Analysis", Marcel Deckker, Inc., New York, 1972, p. 7.
103. Johnstone, H.F., "Gas Cleaning and Purification", 'Encyclopedia of Chemical Technology', R.E. Kirk, and D.F. Othmer, (Eds), 7, Inter Science Publishers, Inc., New York, 1951, p. 83.
104. Johnstone, H.F., "Properties and Behaviour of Air Contaminants in Air Pollution", L.C. McCabe (Ed.), McGraw-Hill Book Co., Inc., New York.
105. Junge, C.E., *Air Chemistry and Radioactivity* Academic Press, New York, 1963.
106. Kataz, M., "Measurement of Air Pollutants", *Guide to the Selection of Methods*, WHO. Geneva, 1969.
107. Kataz, M., "Atmospheric Pollution Studies with Particular Reference to the Detroit-Windsor Area", *Air Repair*, vol. 4, 1955, p. 176.
108. Keise, M., "Chronic Hydrogen Arsenide Poisoning", *Arch Exptl. Pathol. Pharmakol*, vol. 186, 1937, pp. 337–376.
109. Kenneth, W. and Cecil F. Warner, "Air Pollution, Its Origin and Control", IEPA Dun-Donhally Publishers, New York, 1976.

110. Lal, M.M. and Akhlaq Husain, "Short Term Study of SO_2 and TSPM in Indian Kitchens", *Paper Presented at the National Seminar on Assessment and Management of Pollution*, New Delhi, 1984.

111. Larson; G.P., C.I. Fischer, and W.J. Hamming, "Evaluating Sources of Air Pollution", *Ind. Eng. Chem.*, vol. 45, 1953, p. 1070.

112. Lawther, P.J., A.E. Martin and E.T. Wilkins, "Epidemiology of Air Pollution", *Publ. Hlth. Pap. No. 15*, WHO, Geneva, 1962.

113. Littman, F.E. and R.W. Benoliel, "Continuous Oxidant Recorder", *Anal Chem.*, vol. 25, 1953, p. 1480.

114. Louis Theodore, "Air Pollution Control for Hospitals and Other Medical Facilities", Garland STPM Press, New York, 1981, p. 5.

115. Lucas, D.H., D.J. Moore and G. Spurr, "The Rise of Hot Plume from Chimneys', *Intern. J. Air and Water Poll.*, vol. 7, 1963.

116. Lund, H.F., "Industrial Pollution Control Hand Book", McGraw-Hill Book Co., New York, 1971.

117. Luthra, S.P. et al., "Experimental Investigation into the Suitability of Ammonia as SI Engine Fuel", *J. Inst. Engrs.*, vol. 51, 11, Pt. ME 6, 1971, p. 235.

118. Mage, J. and G. Batta, "Results of the Investigation into the Cause of Deaths which Occurred in Meuse Valley during Fog of December 1930", Chim. and Ind., Paris, vol. 27, 1932, pp. 961–975.

119. Magill, P.L. and R.W. Benoliel, "Air Pollution in Los Angeles Country', *Ind. Eng. Chem.*, 44, 1347, 1952.

120. Magill, P.L., F.R. Holden and C. Acklay, "Air Pollution Handbook", McGraw-Hill Book Co., Inc., New York, 1956.

121. Magill, P.L., D.H., Hutchinson, and J.M. Stormes "Hydrocarbon Constituents of Exhaust Gases", *Proc. 2nd National Pollution Symp.*, Stanford Research Inst., Los Angeles, California, 1952.

122. Mahadevan, R. and N. Venkataramani, "Petrol Injection for Automotive Emission by Modifications of the Carburation System", *Paper No. 660 114, SAE Automotive Engineering Congress*, Detroit, Jan. 1966.

123. Mahadevan, K., "Air Pollution by Combustion Engines", *Proceedings of the Symposium*, "A Study of Environmental Pollution in Madras City", June, 1971.

124. Maira, S.V., G. Ramkumar, N.S. Manthapurwar and V. Raman, "Assessment of Emission from Stack Gases", *Proc. Symp. Air Pollution Control Techniques*, CLI, CPHERI and SOCLEEN, Bombay, Sep., 1973.

125. Mallette, F.S. (ed), "Problems and Control of Air Pollution', Reinhold Publishing Co., New York, 1955.

126. Mancy. K.H., "Instrumental Analysis for Water Pollution Control", Ann Arbor Science Publishers Inc., Michigan, 1972.

127. Manthapurwar, N.S. and V. Raman, "Field Studies on Particulate Emission from Oil Fired Boilers", *J. of Inst. of Engineers' (India)*, Environment Engg. Division, vol. 65, Feb. 1985, p. 84.

128. Manual Bhagal, "Man is a Chemical Note, Aswins in a Chemical Soup", *Industrial Times*, April 30–May 13th 1984, p. 10.

129. Mathur, H.B., G.S. Jha, and R.K. Bakshi, "Control and Particulate Emissions from Iron and Steel Industry", *Proc. Symp. Air Pollution Control Techniques*, CLI, CPHERI and SOCLEEN, Bombay, Sept. 1973.

130. Mathur H.B. and J.P. Subramanyam, "Control of Engines Through Exhaust Gas Recirculation", *Ind. Jour. of Air Pollution Control*, vol. 5, No. 1, 1984.

131. Mathur, M.L. and R.P. Sharma, "A Case for Unleaded Gasoline: Reduced Air Pollution and Better Engine Life", *J. Inst. Engrs.*, vol. 53, 1972, pp. 57 63.

132. McCabe, L.C. (Ed), *Air Pollution*, McGraw-Hill Book Co., Inc., New York, 1952.

133. McJones, R.W. and R.J. Corbeil, "Natural Gas Fuelled Engines have Lower Exhaust Emissions", *SAE Journal*, vol. 78, Sec. 6, 1970, pp. 31–34.

134. Meetham, A.R., *Atmospheric Pollution*, 2nd edn., Pergamon Press, London, 1956, p. 215.
135. Mehrotra, S.K. and S.P. Mene, "Air Pollution Problems Pertaining to Aluminium Industry", *Proc. Symp. on Env. Pollution*, Nagpur, 1973, p. 279.
136. Menon, M.P.S., "Asthma—A Medical Enigma", National Book Trust, India, New Delhi, 1982.
137. "Meteorology and Atomic Energy", US AEC Division of Technical Information, 1968, pp. 227–283.
138. Meteorological and Environmental System, *Catalogue*, CLIMATE Instruments Company, USA.
139. "Methods of Air Sampling and Analysis", 2nd edn., American Public Health Association, 1977.
140. Middleton, J.T., J.B. Kendrick, Jr. and E.F. Darley, "Damage to Vegetation from Polluted Atmospheres", *Paper presented in 22nd mid-year Meeting, Div. of Refining Am. Petroleum Inst.*, Philadelphia, May, 1957.
141. Moses H. and J.E. Carson, "Stack Design Parameters Influencing Plume Rise", *Jour. of Air Pol. Control Association*, USA, vol. 18, 1968.
142. Neporozhny P.S. (ed.), "Thermal Power Plants and Envl., Control", Mir Publishers, Moscow, 1984.
143. News items from newspapers, *The Hindu*, *Indian Express* and *Deccan Herald*, Bangalore.
144. Nicholis, J.E. et al., "Inlet Manifold Water Injection for Control of Nitrogen Oxides—Theory and Experiment", *Paper No. 690 018, SAE International Auto. Engg. Congress*, Detroit, Jan, 1969.
145. "Air Quality in Selected Cities in India", NEERI, 1978, 1979 and 1980.
146. Nelson, E.E., "Hydrocarbon Control for Los Angeles by Reducing Gasoline Volatility", *Paper No. 690 087, SAE International Auto Engg. Congress*, Detroit, Jan. 1969.
147. Newhall, H.K., "Control of Nitrogen Oxides by Exhaust Recirculation—A Preliminary Theoretical Study", *Paper No. 670 495, SAE Mid-Year Meeting*, Chicago, May 1967.
148. Nou, K.E. and T. Joseph, *Industrial Air Pollution Control*, Ann Arbor Science Publishers Inc., Michigan, 1973.
149. Obert, E.F., "Internal Combustion Engines", International Text Book Company, Scranton, Pennsylvania, 1968.
150. O'neill D., "Switch to Unleaded Fuel Offers Benefits, Poses Problems", *SAE Journal*, vol. 78, Sec. 9, 1970, p. 20.
151. "Our Polluted Planet", Ambassador College Press, Pasadena, California, 1968.
152. Padmanabha Murthy, B. "Meteorological Condition in Air Pollution Techniques", *Proc. Symp. Air Pollution Control Techniques*, CLI, CPHERI and SOCLEEN, Bombay, Sept., 1973.
153. Pahnke, A.J. and J.F. Conte, "Effect of Combustion Chamber Deposits and Driving Conditions on Vehicle Exhaust Emissions", *Trans. SAE Paper No. 690 017.*
154. Park, J.E., "Preventive and Social Medicine", 3rd edn., 1972.
155. Parker, A, "Coal in Relation to Atmospheric Pollution", Department. Sci. Ind. Research (UK), Invest Atmospheric Pollution, *26th Report*, 1949, p. 116.
156. Pasquill, F., "Atmosphèric Diffusion", Van Nostrand and Co. Inc. USA, 1962.
157. Patel, C.B. et al., "Fly Ash Fallout from a Thermal Power Station in Post-rain Period", *Journal of Institution of Engineers (India)*, Environmental Engg., Division, vol. 65, Feb., 1985, p. 27.
158. Paul F. Pennelly, "Primary and Secondary Particulates as Pollutants", *J. of Air Pollution Control*, vol. 20, No. 1, p. 697.
159 Perkins, H.C., "Air Pollution", McGraw-Hill Kogakusha Ltd., Tokyo, 1974.

160. Phair, J.J., "The Epidemiology of Air Pollution", *Air Pollution Handbook*, Magill, P.L. et al., (ed.), McGraw-Hill Book Co., Inc., New York, 1956.
161. Pratt, D.C.F. and A. Rutherford, "Removal of Hydrogen Sulphide from Exhaust Air from a Viscose Staple Fibre Factory", Chem. and Ind. (London), vol. 41, 1955, p. 1281.
162. "Pollution of Indoor Air", article on environment, *The Hindu*, Madras, Nov. 9, 1980
163. Power Generation Systems—Energy System Engineering, by Editors of '*Power*', McGraw-Hill Book Co., USA, 1967, p. 156.
164. Prabu Prasad, N., "Long-term Air Quality Monitoring of Atmosphere over Bangalore City", *Dissertation*, UVCE, Bangalore, 1982.
165. Pratap-Mowli et al., "Ambient Air Quality of a Developing Industrial Area", *Proc. Int. Cong. Env. Pollution Control*, Tirupati, Jan. 1986, p. 149.
166. Raja Rao Capt. S., "Laws and Regulations for Environmental Protection", *Paper Presented at Indian Institute of Science*, Bangalore, 1985.
167. Raj Chengappa, "The Boom City", *India Today*, April 15, 1983, p. 50.
168. Ramprasad, G. et al., "Emission of Lead from Automobiles in Ahmedabad", *Ind. Journal of Industrial Medicine*, vol. 19, June 1973.
169. Ramprasad, G. et al., "Qualitative and Quantitative Aspects of Dust Pollution in Small Scale Flour Mills", Food Industries and the Environment, *Int. Symp.*, Budapest, Hungary, 1982.
170. Rao, H.V.N., "Air Pollution Control Legislation—A Critical Review", *Paper Presented at the National Seminar*, "Assessment and Management of Pollution", New Delhi, Feb. 1984.
171. Rao, H.V.N., "Design of Control Equipments for Air Pollution", paper presented at special symposium on "Pollution Control and Abatement", *The Electrochemical Society of India*, Bangalore, July, 1987.
172. Rao, M.N., "Analytical Instrumentation—a Powerful Tool in Air Pollution Studies", *Air Pollution Seminar*, Durgapur, 1969.
173. Rao, M.N., "Role of Vehicle Emissions in Air Pollution", *Paper Presented at the Air Pollution Seminar*, Institution of Engineers (India) and International Centre, Durgapur, 1969.
174. Rao, M.N., "Current Trends in the Control of Air Pollution and Industrial Dust Problems", *Paper Presented at the Seminar*, Air Pollution in City Area—VJTI, Bombay, 1972.
175. Rao, M.N., "Air Pollution as a Fluid Mechanics Problem", *Paper Presented at the Symposium on Environmental Pollution*, Nagpur, Jan., 1973.
176. Rao, M.N., "Recent Trends in Sampling and Assessment of Stack Effluents", *Paper Presented at the Symposium on Environmental Pollution*, Nagpur, Jan. 1973.
177. Rao, M.N., "Environmental Pollution and Trace Quantity Engineering", *Paper Presented at Symposium*, "Curative Measures for Pollution Abatement", Durgapur, June 1973.
178. Rao, M.N. and N. Kurmaiah, "Significance of Trace Quantity Engineering in Pollution Studies", *Paper Presented at Symposium*, "Recent Developments in the Field of Public Health Engineering", Hyderabad, October, 1973.
179. Rao, M.N., "Recent Trends in the Methods of Air Pollution Control in the Steel Industry", *Paper Presented at the All India Seminar*, "Environmental Impact and Developmental Activities", Hyderabad, December, 1977.
180. Rao, M.N., "Industrial Pollutants in India—A Case Study", *Paper Presented at the Indian Institute of Management*, Bangalore, May, 1981.
181. Rao, M.N., "Effects of Air Pollution on Human Beings, Animals and Plants", *Paper Presented at the Two-tier Air Pollution Programme*, Organised by the National Productivity Council, Bangalore, March, 1982.
182. Rao, M.N., "Status of Air Pollution in India", *Paper Presented at the Two-tier*

Air Pollution Programme, Organised by the National Productivity Council, Bangalore, March, 1982.

183. Rao, M.N., "Sources of Air Pollution", *Paper Presented at Two-tier Air Pollution Programme*, Organised by National Productivity Council, Bangalore, March, 1982.
184. Rao, M.N., "Management of Industrial Wastes", *Paper Presented at Workshop*, "Dimensions and Policy Implications of Environmental Management", Organised by IIM, Bangalore, June, 1982.
185. Rao, M.N., "Air Pollution Source Monitoring", *Paper Presented in the QIP Course on Modern Instrumental Techniques for Pollution Monitoring*, IISc., Bangalore, Oct., 1983.
186. Rao, M.N., "Air Pollution Monitoring", *Paper Presented at the Institution of Engineers (I)*, Bangalore, May, 1984.
187. Rao, M.N., "Lead Pollution Control—A Case Study", *Proc. Int. Cong. Env. Pollution Control*, Tirupati, Jan., 1986, p. 116.
188. Raza, S.H. and Jayanath Chapala, "Relationship between Leaf Area Damage and Secondary Producer Activity at VAN Deer Sanctuary and BBN Savanna Scrub Grass Lands", *Ind. Jour. of Air Pollution Control*, vol. 5, No. 1, 1984.
189. Razanov, V.A., "Current Problems of Establishing Hygienic Standards", *WHO Inter-regional Seminar*, Moscow, 1967.
190. "Removing Sulphur Dioxide from Stacks", *Chem. Eng. Desk Book*, Issue, April 27, 1970, p. 173.
191. *Report of the Sub-committee Appointed by the Panel of the Cement Industry*, Cement Industry Assn., Bombay, 1973.
192. *Report of ILO on Asbestos*, Geneva, 1985.
193. *Report on Release of Radioactivity in Effluents and Solid Waste from Nuclear Power Plants for 1972*, Washington, DC, US Atomic Energy Commission, 1973.
194. Rounds, F.G. et al., "Some Effects of Engine Fuel Variables on Exhaust Gas Hydrocarbon Content", *Trans SAE* 65, 1957, p. 608.
195. Salariya, K.S. and S.L. Jindal, "Gravity Settling Chamber—A Working Design for Particulate Emission Control", *Proc. Int. Cong. Env. Pollution Control*, Tirupati, Jan., 1986, p. 141.
196. Sastry, C.A., "Air Pollution", Chemical Engg. Education Development Centre, IIT, Madras, 1975.
197. Sastry, C.A. and T.K. Srinivasan, "Air Pollution in Madras City", CPHERI Report, 1973.
198. Sastry, C.A., T.V. Ramakrishna and N. Balasubramanian, "A Course Manual on Sampling and Analysis of Particulates for Metallic Constituents", IIT Madras, Nov., 1983.
199. Satishbabu, K.R., "Imparting Pleasant Smell to Leather Using Perfumes", *B. Tech. Project*, Madras University, 1972.
200. Scheffler, C.E., "Combustion Chamber Surface Area—A Key to Exhaust Hydrocarbons", *Trans. SAE, 75, Paper No. 660 111, 1967.*
201. Schmidt, W.A. and C.R. Flodin, "Fundamental Principles, Design, Application Performance and Limitations of Electrical Precipitation Equipment", "Air Pollution", L.C. McCabe (Ed.), McGraw-Hill, New York, 1952.
202. "Selected Methods of Measuring Air Pollutants", WHO Offset Pub. No. 24, 1976.
203. Senson, J.D., "Reduction of Nitrogen Oxides in Automobile Exhaust", *Paper No. 690 019, SAE, International Auto. Engg. Congress*, Detroit, Jan., 1969.
204. Sen, G.J., J.A. Borgos and J.G. Olin, "Monitoring Particulate Emissions" *Chemical Engineering Progress*, vol. 67, No. 10, 1971, p. 83.
205. Shepherd, M., "Rapid Determination of Small Amounts of Carbon Monoxide in Air", *Anal Chem.* vol. 19, 1947, p. 77.

206. Shepherd, M., S.M. Rock, R. Howard and J.M. Stormes, "Isolation, Identification and Estimation of Gaseous Pollutants of Air", *Anal. Chem.*, vol. 23, 1951, p. 1431.
207. Shrivatsava, J.B., "Key Note Address", *Proc. Symp. Air Pollution Control Techniques*, CLI, CPHERI and SOCLEEN, Bombay, Sept., 1973.
208. Silverman, L., J.L. Whittenberger and J. Muller, "Physiological Response of Man to Ammonia in Low Concentrations", *J. Ind. Hyg. Toxicol*, vol. 31, 1949, p. 74.
209. Shivaji Rao, T., *Report on Taj Mahal*, Andhra University, Waltair, 1981.
210. Singh, P.P. et al., "Formation and Control of Oxides of Nitrogen Emissions from Gas Turbine Combustion System", *J. Eng. Power*, Trans. ASME, 94, vol. No. 4, 1982, p. 271.
211. Sinha, J.K. "Dust Hazards at Cement Works", *J. Institution of Engineers*, 1978.
212. Sinha, J.K., Pollution from Cement Industry, *Proc. 3rd Cement Industry Operation Seminar*, New Delhi, 1973.
213. Smith M.E. (Ed.), "Recommended Guide for the Prediction of the Dispersion of Air-borne Effluents", ASME United Engineering Centre, New York, 1968.
214. Smith, W.J. and K.W. Stinson, *Fuels and Combustion*, McGraw-Hill Book Co., Inc., New York, 1952.
215. South-West Research Institute, "Air Pollution Survey of Houston Area", *Tech. Rept. No. 4 for Houston Chamber of Commerce*, Texas, 1957.
216. Stanford Research Institute, "The Smog Problem in Los Angeles Country", *Third Interim Report*, Waste Oil and Gas Association, Los Angeles, California, 1950.
217. "Standard Methods for the Sampling and Analysing of Atmospheres", Am. Soc. for Testing and Materials, 1971.
218. "State of the Environment", *Report UNEP*, 1980.
219. Stern, A.C. (Ed.), "Air Pollution", vol. 1, 2nd edn., Academic Press, New York, 1968.
220. Stern, A.C. (Ed.), "Air Pollution", vol. 2, 2nd edn., Academic Press, New York, 1968.
221. Stern, A.C. (Ed.), "Air Pollution", vol. 3, 2nd edn., Academic Press, New York, 1968.
222. Stern, A.C. (Ed.), "Air Pollution", vol. 4, 2nd edn., Academic Press, New York, 1968.
223. Stern, A.C. (Ed.), "Air Pollution", vol. 5, 2nd edn., Academic Press, New York, 1968.
224. Strauss, W. (Ed.), "Air Pollution Control" Part-I, Wiley-Interscience: New York, 1971.
225. Strauss, W. (Ed.), "Air Pollution" Part-II, Wiley-Interscience, New York, 1972.
226. Strauss, W. (Ed.), "Air Pollution" Part-III, Wiley-Interscience, New York, 1978.
227. Suess, M.J. and S.R. Craxford (Eds), "Manual on Urban Air Quality Management", WHO, Regional Pub. European Series No. 1, 1976.
228. Sukla, S.C., "Problem of Air Pollution in India", *All India Conference on Air Pollution*, Bombay, September 1980, p. B-1.
229. "Sulphur in the Atmosphere", Pergamon, Press, *Proceedings of the International Symposium held in Dubrovnic*, Yugoslavia, Sept., 1977.
230. Sundaresan, B.B., "Control of Air Pollution from Nitric Acid Manufacturing Plants", *Proc. Symp. Air Pollution*, College of Engineering, Guindy, Madras, 1972.
231. Sutton, O.G., *Micrometeorology*, McGraw-Hill Book Co. Inc. New York, 1953.
232. "Technology and the Control of Automotive Air Pollution", Part 2, *Report*, Panel of Electrically Powered Vehicles to the Commerce Technology Advisory Board, *SAE Journal*, 1968, p. 42.
233. Teller, A.J., "Air Pollution Control", *Chemical Engineering Desk Book*, May 1972.

234. Temple, R.G., "Control of Internal Combustion Engine, Air Pollution Control Part I, Wiley-Interscience New York, 1971, p. 179.

235. The Sources of Air Pollution and Their Control, Pub. Hlth. Serv. Publ. No. 1548, Government Printing Press, Washington DC, 1966.

236. Thrinavakkarasu, E., "Ambient Air Quality Monitoring of Atmosphere over Bangalore City", *ME Thesis*, submitted to Bangalore University, 1985.

237. Thom G.C. and W.R. Ott "Air Pollution Indices—A Compendium and Assessment of Indices Used in the United States and Canada", Council of Envl. Quality and EPA, USA, 1975.

238. Thomas, M.D., R.H. Hendricks and G.R. Hill, "Some Chemical Reactions of Sulphur Dioxide After Absorption by Alfalfa and Sugar Beet', *Plant Physiol*, vol. 19, 1944, p. 212.

239. Training Manual—"Meteorological Instrumentation in Air Pollution", Institute for Air Pollution Training, EPA. Research Triangle Park, USA, paras 6.1 to 6.9.

240. "Urban Air Pollution with Particular Reference to Motor Vehicles", *Tech. Report Series*, No. 410, WHO, Geneva, 1969.

241. Vouk, V.B., "WHO and Criteria for Air Quality", *WHO Inter-regional Seminar*, Moscow, 1967.

242. Wade, D.T., "Factors Influencing Vehicle Evaporative Emissions", *Paper No. 670 126, Presented at SAE International Auto Engg. Congress*, Detroit, January 1967.

243. WHO, Air Quality Criteria and Guide for Urban Air Pollutants, *Report of a WHO Export Committee*, WHO Technical Dept. Series No. 506, 1972.

244. Willet, H.C., "Descriptive Meteorology", Academic Press, Inc., New York, 1944.

245. Wintringahn, J.S. et al., "Car Maintenance Expense in Owner Service with Leaded and Nonleaded Gasoline", *SAE Paper No. 720 499*, 1972.

246. Wohlers, H.C., "Air Pollution—The Problem, the Source and the Effects", Drexel Institute of Technology, Philadelphia, 1969.

247. Yennamwar, P.K., S.N. Dixit, V.L. Pampattiwar, J.M. Dave and S.J. Arceivala, "Short Term Air Quality Surveys in four Major Cities of India", *Env. Hlth.*, vol. 12, 1970, p. 355.

248. Yennamwar, P.K. and C.A. Sastry, "Air Pollution Problems in Fertiliser Industries in Tamil Nadu", *CPHERI Report*, 1972.

249. Yennamwar, P.K. et al., "A Course Manual on Air Pollution Control", NEERI, Nagpur, 1979.

APPENDICES

APPENDIX I

Glossary of Words and Terms

Abatement: Reduction and control of air pollution.

Absorption: The transfer of one or more constituents from a gaseous mixture to a liquid, in which the absorbed material may dissolve physically or react chemically.

Acid rain: Refers to all precipitation-rain, snow, and dew which is more acidic than normal.

Adsorption: Removal of impurities from a gas stream by concentration on the surface of a solid or liquid.

Aerosol: A colloidal system in which the dispersion medium is a gas and the dispersed phase is a solid or liquid (usually they are less than 50 μ).

Air monitoring: The process of detection and measurement of pollutants in air.

Air pollution: The presence in ambient atmosphere of substances, generally resulting from the activity of man, in sufficient concentration, present for a sufficient time and under circumstances to interfere significantly with comfort, health or welfare of persons or with full use or enjoyment of property.

Air pollution index: An arbitrarily derived mathematical combination of air pollutants which gives a single number trying to describe the ambient air quality. (The index usually holds good for the area in which it is derived.)

Air quality criteria: Scientific information about the levels of air pollution and the duration of exposure which result in adverse effects on health and welfare.

Air quality standards: The concentration of a particular pollutant that is permitted in the atmosphere to minimise adverse effects.

Ambient air quality: A physical and chemical measure of the concentration of contaminants in the ambient atmosphere. The quality is usually monitored over a specific period.

Bag house: An air pollution control device that traps gas borne particulates by forcing the gas through filter bags.

Carcinogen: A substance which can cause living tissue to become cancerous.

Chimney: A structure with an opening or outlet from or through which any air pollutant may be emitted.

Collector: A device for removing and retaining pollutants (particularly particulates) from air or other gases.

Coning: A type of plume which is shaped like a cone. This takes place in a near neutral atmosphere (adiabatic condition) when the wind velocity is greater than 32 km/h.

Contaminant: Unwanted material usually harmful or of a nuisance value or both.

Control equipment: Any apparatus, device, equipment or system to control the quality and manner of emission of any air pollutant and includes any device used for securing the efficient operation of any industrial plant.

Cost/benefit ratio: An economic evaluation index relating control costs to benefits.

Cyclone: A type of particulate collector which depends upon centrifugal force for its action.

Dew point: For air containing water vapour, it is the temperature at which liquid water begins to condense for a given state of humidity and pressure, as the temperature is reduced.

Dispersion: The mixing of gases containing a high concentration of pollutants with the ambient air.

Dry bulb temperature: The actual temperature of the gas. Measured with a conventional thermometer.

Dust: Solid particles predominantly larger than colloidal ones and capable of temporary suspension in air or other gases. They do not tend to flocculate except under electrostatic forces; they also do not diffuse but settle under the influence of gravity.

Dust fall: The weight of particulate matter deposited due to gravitational action per unit time per unit area of the surface.

Effective stack height: It is the actual stack height plus the plume rise.

Emission: Discharge of air pollutants.

Emission inventory: A list of pollutants emitted from a source by type and amount (e.g., tonnes per day).

Emission standard: Rule or law to regulate or control the amount of a particular pollutant that may be discharged to the outdoor atmosphere from its source.

Fall out: A radioactive pollutant in the air caused after the explosion of a nuclear device, its degree of contamination depending on several factors, such as distance, wind, and power of the device.

Fanning: A type of plume emitted under extreme inversion conditions where mostly lateral diffusion takes place with very little vertical mixing.

Flue gas: Gaseous emissions discharged through a flue or stack.

Fly ash: The unburned particulate matter discharged from a stack.

Fog: Visible aerosols in which the dispersed phase is liquid.

Fume: The solid particles generated by condensation from the gaseous state, generally after volatilisation from melted substances, and often accompanied by a chemical reaction, such as oxidation. Fumes flocculate and sometimes coalesce. Popularly, the term is used in reference to any or all types of contaminants and, in many laws or regulations with the added qualification that the contaminants have some unwanted action.

Fumigation: The phenomenon in which pollutants that are aloft in the air are brought rapidly to ground level when the air destabilises.

Grab sample: Sampling instantaneously.

Grain: A unit of weight equivalent to 64.8 milligrams or 1/7000 of a pound.

Impinger: A sampling instrument adopting the principle of impingement for the collection of particulate matter. Usually, this term is used for specific instruments, the 'midget' impinger and the 'standard' impinger.

Impinger, midget: A specific instrument employing wet impingement, using a liquid volume of 10 ml and gas flow of 2 litres per minute.

Impinger, standard: A specific instrument employing wet impingement, using a liquid volume of 75 ml and a gas flow of 28 litres per minute.

Incineration: Combustion of solid, liquid, or gaseous wastes under controlled conditions.

Indicators of air pollution: Physical conditions, such as, smoke, mist, fog, smog, etc., which usually indicate polluted atmospheric conditions.

Inertial separators: Air pollution control equipment that utilises the principle of inertia to remove particulate matter from a stream of air or gas.

Inversion: Condition in the atmosphere in which air temperature increases with elevation. Under this condition, the atmosphere is said to be in stable equilibrium.

Isokinetic: A condition of sampling in which the flow of gas into the sampling device (at the opening or face of the inlet) has the same flow rate and direction as the ambient atmosphere being sampled.

Lapse rate: The decrease of atmospheric temperature with height.

Lofting: A type of plume which occurs when an inversion exists only below the plume and the plume is inhibited from mixing downward.

Looping: A type of plume which has a wavy character. It occurs in a highly unstable atmosphere because of rapid mixing.

Mass concentration: Concentration expressed in terms of mass of a substance per unit volume of gas or liquid.

Micrometeorology: Study of minute variations in atmospheric conditions confined to an area of a few square kilometres and up to an elevation of 500–1000 metres in the atmosphere—governs the transmission and diffusion of air pollutants.

Mixing height: Height above the earth's surface to which released pollutants will extend, primarily through the action of atmospheric turbulence.

Odour threshold: The lowest concentration of a substance in air at which its odour is perceptible.

Oxidants and ozone: In the field of air pollution, ozone and oxidants are the products of photochemical reactions between nitrogen oxides and some organic pollutants. The concentrations of ozone and oxidants in the atmosphere are an index of photochemical smog.

Particulate matter: Solid or liquid particles in a finely divided state capable of temporary suspension in air or other gases.

Particle size: The size of liquid or solid particles expressed as the average or equivalent diameter.

Photochemical reaction: Any chemical reaction that is initiated as a result of absorption of light.

Photochemical smog: A type of air pollution resulting from photochemical reactions.

Plume: The path and extent in the atmosphere of the gaseous effluent released from a source, usually a stack.

Pollutant: Same as 'contaminant'.

ppb: A unit of measure of the concentration of gases in air expressed as parts of the gas per billion (10^9) parts of the air-gas mixture, normally both by volume.

ppm: A unit of measure of the concentration of gases in air expressed as parts of the gas per million (10^6) parts of the air-gas mixture, normally both by volume.

Precipitator, electrostatic: An air pollution control equipment used to remove particulates under the influence of an electrostatic field.

Pressure drop: The differential pressure between two points in a system. The resistance to flow between the two points.

Probe: A tube used for sampling or for measuring pressures at a distance from the actual collection or measuring apparatus. It is commonly used for reaching inside stacks and ducts.

Reentrainment: It is a phenomenon wherein dust is collected from the air stream and then is returned back to the air stream.

Relative humidity: The ratio of the actual vapour pressure of the air to the saturation vapour pressure.

Ringelmann chart: A series of charts, numbered from 0 to 5, that simulate various smoke densities, by indicating different percentages of black. A Ringelmann No. 1 is equivalent to 20% black and Ringelmann No. 5 is 100% black. It is used for measuring the density of smoke coming from stacks, by comparison.

Sampler, high volume: A device capable of collecting the particulates at a high rate.

Sampling: A process consisting of the withdrawal or isolation of a fractional part of a whole. In air or gas analysis, it means the separation of a portion of an ambient atmosphere with or without the simultaneous isolation of selected components.

Scrubber: A type of apparatus used in sampling and in gas cleaning in which the gas is passed through a space containing wetted packing or spray.

Settling chamber: Simplest type of particulate collector where particles in an air or gas stream settle due to gravity.

Smog: A mixture of smoke and fog.

Smoke: Small gas-borne particles resulting from incomplete combustion, consisting predominantly of carbon and other combustible material and present in sufficient quantity to be observable independently of the presence of other solids.

Smust: A mixture of smoke and dust. It means a combined effect of both.

Soot: Fine carbon particles containing small, varying amounts of complex hydrocarbons and mineral matter from the combustion of fossil fuels.

Stack: A vertical duct for the discharge of products of combustion or process emission to the atmosphere.

Synergistic effect: This term is used to characterise the combined effect of air pollutants when that effect is greater than or different from what would be expected if the known effects of each of the pollutants were simply added together.

Survey, air pollution: Air monitoring of a given geographical area for the purpose of determining the nature, sources, extent, and effects of air pollution.

Temperature lapse rate: The rate at which temperature changes with increasing height of the atmosphere.

Venturi: A tube with a constricted middle, at which point the velocity of a stream of air or gas increases, with decrease in pressure.

Visibility: It is a measure of the degree of clearness of the atmosphere, which is usually expressed as the distance of the farthest object towards the horizon which may be clearly seen by the unaided normal eye.

Wash out: A process of removal of particulate and soluble matter from air by water drops during rain precipitation.

Wind rose: Diagramatic representation to show the distribution of wind direction experienced at a given location over a considerable period.

APPENDIX II

List of Indian Standards Relating to Air Pollution

Glossary of terms relating to air pollution		4167–1966
Methods for measuring air pollution:		
Part I	Dustfall	5182 (Part I)–1969
Part II	Sulphur dioxide	5182 (Part II)–1969
Part III	Radioactivity (particulate) in air	5182 (Part III)–1970
Part IV	Suspended matter	5182 (Part IV)–1973
Part V	Sampling of gaseous pollutants	5182 (Part V)–1975
Part VI	Nitrogen oxides	5182 (Part VI)–1975
Part VII	Hydrogen sulphide	5182 (Part VII)–1973
Part VIII	Sulphation rate	5182 (Part VIII)–1976
Part IX	Oxidants	5182 (Part IX)–1974
Part X	Carbon monoxide	5182 (Part X)–1976
Part XI	Benzene	5182 (Part XI)–1982
Part XII	Polynuclear aromatic hydrocarbons in air particulate matter	5182 (Part XII)–1974
Part XIV	Guidelines for planning the sampling of atmosphere	5182 (Part XIV)–1979
Part XV	Mass concentration of particulate matter in the atmosphere	5182 (Part XV)–1974

Part XVI	Recommended practice for collection by filtration and determination of mass, number, and optical sizing of atmosphere particulates	5182 (Part XVI)-1980
Part XVII	C_1 to C_5 hydrocarbons in air by gas chromatography	5182 (Part XVII)-1979
Part XVIII	Continuous analysis and automatic recording of the oxidant content of the atmosphere	5182 (Part XVIII)-1974
Part XIX	Chlorine	5182 (Part XIX)-1982
Part XX	Carbon disulphide	5182 (Part XX)-1982

Micrometeorological techniques in air pollution studies, guidelines for	8829–1978
Nitric acid and nitrogenous fertiliser industries, limits for gaseous emissions from	9005–1978
Petroleum refineries, limits for gaseous emissions from	8636–1977
Ringelmann and miniature smoke charts, code for use of	9078–1979
Smoke emission levels for diesel vehicles	8118–1976
Sulphuric acid and phosphatic fertiliser industries, limits for gaseous emissions from	8635–1977
Vehicles powered by spark ignition engines, emission limits for carbon monoxide	9057–1979
Gaseous emissions from man-made fibre plants, limits for:	
Part I Cellulosic fibres	9233 (Part I)-1979
Guide for units used in Air Quality Measurements	9260–1980
Code for practice for Work Environment Monitoring (air-borne contaminants)	9679–1980
Control of air pollution in petroleum refineries	10,179–1982
Limits for ambient air quality	
Part I: Sulphur dioxide and suspended particulate matter	10,193–1982

APPENDIX III

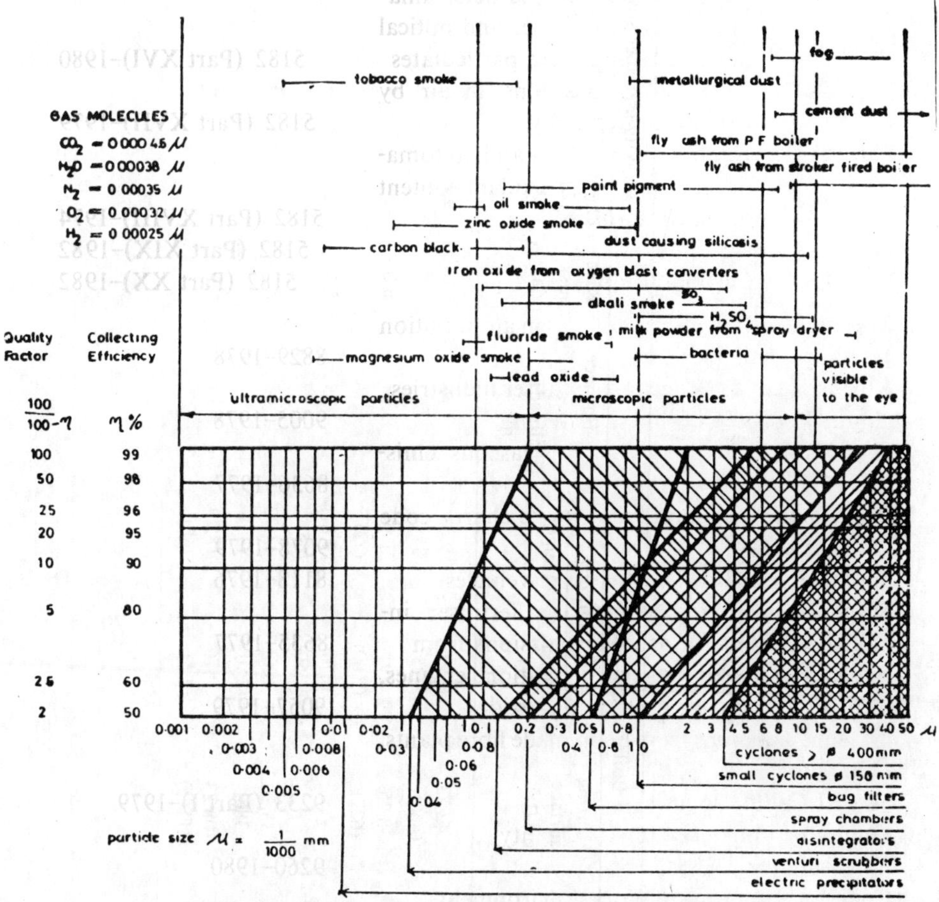

Particle size of some industrial dusts and the working ranges of most common types of dust collectors.

(Courtesy: Flakt India Ltd.)

APPENDIX IV

TABLE 1: *Design of Gravity Settling Chamber*

No.	Type	Boiler specifications							Gravity chamber size		
		Size *m* (ft)	Working pressure g/m² (psi)	Heating surface m² (ft²)	Chimney Diameter m (ft)	Chimney Height	Furnace area (inclined) m²	Rice husk t/day	Length (*L*) m	Width (*W*) m	Height (*H*) m
1	Cornish	1.83 × 6.18 (6 × 20)	8.44 (120)	42.73 (460)	0.99 (3.25)	19.81 (65)	3.96	7.0	9.1	4	5
2	Lancashire	1.83 × 7.31 (6 × 24)	7.03 (100)	58.25 (627)	0.91 (3.00)	21.94 (72)	5.54	6.0	9.1	4	5
3	Coachran	2.44 × 4.88 (8 × 16)	7.03 (100)		0.76 (2.50)	19.81 (65)	3.75	5.5	7.1	4	4
4	Vertical tubular spin wood	1.68 × 3.35 (5.5 × 11)	7.03 (100)	35.30 (380)	0.61 (2.0)	13.72 (45)	3.48	5.0	9.1	3	3
5	Multitubular marine	1.83 × 3.66 (6 × 12)	5.62 (80)	108.23 (1165)	0.76 (2.5)	18.29 (60)	3.11	6.0	9.1	4	5

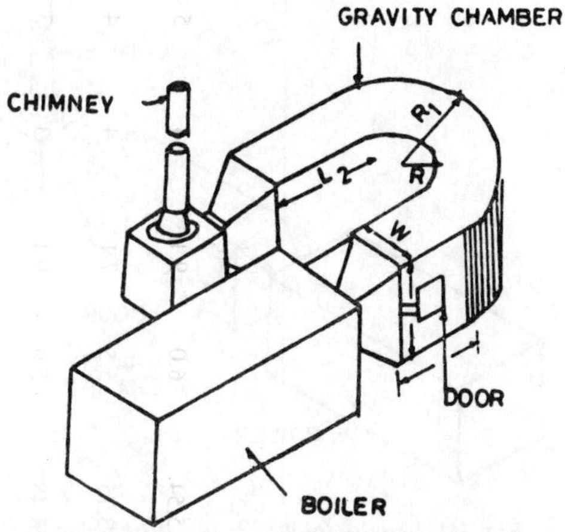

Fig. I *U-shaped gravity settling chamber*

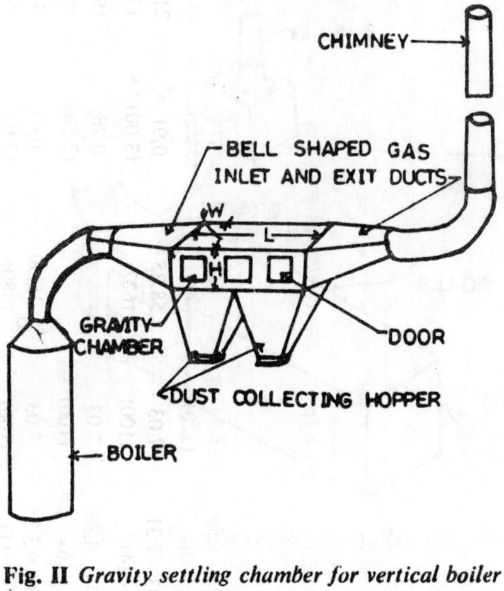

Fig. II *Gravity settling chamber for vertical boiler*

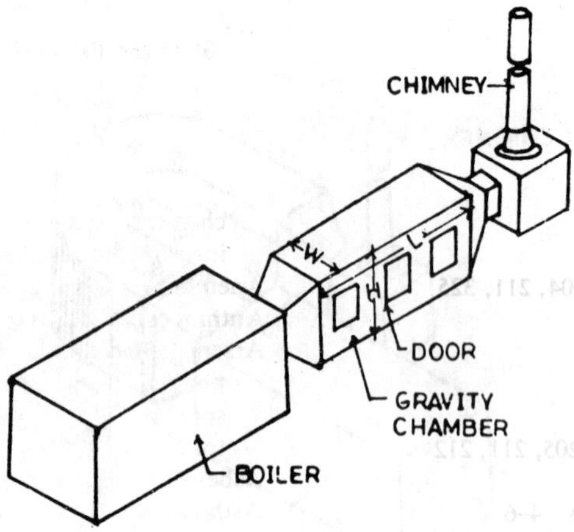

Fig. III *Longitudinal gravity settling chamber*

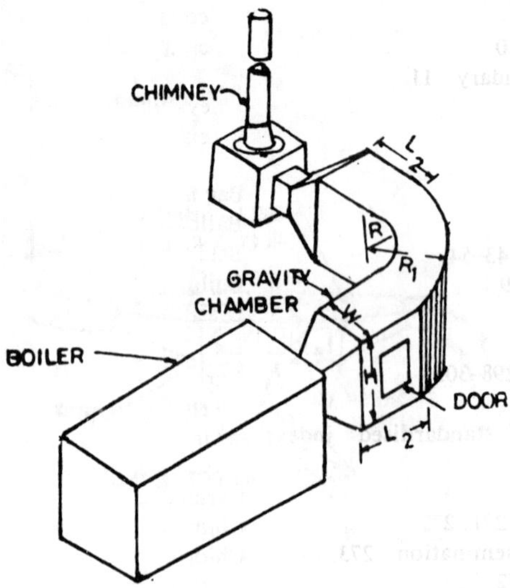

Fig. IV *L-shaped gravity settling chamber*

Index